AF786896

Engineered 2D Materials for Electrocatalysis Applications

Online at: https://doi.org/10.1088/978-0-7503-5719-7

Engineered 2D Materials for Electrocatalysis Applications

Edited by
Chandra Sekhar Rout
Centre for Nano and Material Sciences, Jain University (Deemed-to-be University), Jain Global Campus, Kanakapura, Bangalore 562112, Karnataka, India

IOP Publishing, Bristol, UK

© IOP Publishing Ltd 2024

All rights reserved. No part of this publication may be reproduced, stored in a retrieval system or transmitted in any form or by any means, electronic, mechanical, photocopying, recording or otherwise, without the prior permission of the publisher, or as expressly permitted by law or under terms agreed with the appropriate rights organization. Multiple copying is permitted in accordance with the terms of licences issued by the Copyright Licensing Agency, the Copyright Clearance Centre and other reproduction rights organizations.

Permission to make use of IOP Publishing content other than as set out above may be sought at permissions@ioppublishing.org.

Chandra Sekhar Rout has asserted his right to be identified as the editor of this work in accordance with sections 77 and 78 of the Copyright, Designs and Patents Act 1988.

ISBN 978-0-7503-5719-7 (ebook)
ISBN 978-0-7503-5717-3 (print)
ISBN 978-0-7503-5720-3 (myPrint)
ISBN 978-0-7503-5718-0 (mobi)

DOI 10.1088/978-0-7503-5719-7

Version: 20240201

IOP ebooks

British Library Cataloguing-in-Publication Data: A catalogue record for this book is available from the British Library.

Published by IOP Publishing, wholly owned by The Institute of Physics, London

IOP Publishing, No.2 The Distillery, Glassfields, Avon Street, Bristol, BS2 0GR, UK

US Office: IOP Publishing, Inc., 190 North Independence Mall West, Suite 601, Philadelphia, PA 19106, USA

To everyone who pursues research in materials science, catalysis and electrocatalysis.

Contents

Preface

After the technological revolution, work efficiency increased substantially, and it continues to develop rapidly today. The ground-breaking 'There is plenty of room at the bottom' speech given in 1959 by physicist Richard Feynman opened the way to nanotechnology. Research on nanotechnology has gained increasing attention and investment, leading to advancements in various areas. The investigation and development of materials in various dimensions remain an important topic of research. From a material chemistry perspective, the term 'dimensionality' refers to the number of spatial dimensions in which the material's properties differ or behave in a particular way. Typically, materials are classified into four dimensions such as zero-dimensional (0D), one-dimensional (1D), two-dimensional (2D), and three-dimensional (3D) materials. In the nano dimension, the optoelectronic, magnetic, and physicochemical properties are drastically different from bulk materials. Researchers are attempting to comprehend and control the special characteristics that each dimension offers for a variety of functions.

The discovery of graphene not only launched an exciting era of 2D material research but also paved the way for advances in nanoscience and nanotechnology. Graphene, the first 2D wonder material, was isolated and characterized in 2004 by A Geim and K Novoselov at the University of Manchester in the United Kingdom. 2D materials are substances that have reduced dimensionality, with a lateral thickness on the order of nanometers or even atomic scale. They have a structure like a sheet with finite thickness in one dimension and limitless extension in the other two. 2D materials have unique features that differ from their bulkier counterparts. Typically, they exhibit strong covalent bonding within the sheet, with van der Waals forces between sheets, resulting in unique mechanical, electrical, and optical properties. Quantum confinement effects in 2D materials are especially significant because of their reduced aspect dimensions. In the early stage of 2D materials, researchers successfully utilized the potential of graphene in a variety of applications, including optoelectronic, spintronics, energy storage, energy conversion, biomedical devices, and more. The pursuit of atomically thin sheets of other layered materials has been sparked by the success of graphene. The family of 2D materials is constantly developing, greatly broadening the range of potential phenomena that can be investigated in two dimensions. In addition to graphene, the family of 2D materials covers materials like graphitic carbon nitride (g-C_3N_4), hexagonal boron nitride (h-BN), black phosphorus (BP), and different classes like transition metal dichalcogenides (TMDs), MXenes, perovskites, metal–organic framework (MOFs), layered double hydroxides (LDHs), metal oxides, and so on. The structure, composition, and functions of each material in these several classes of 2D materials are unique from others. These materials cover a wide spectrum of electronic structures from zero bandgap graphene to direct bandgap BP. The advanced optoelectronic characteristics of 2D materials, together with their good mechanical features like robustness and flexibility, make them suitable components for the development of a new generation of high-performance devices.

Over the past several decades, there has been significant progress in designing 2D material-based electrocatalysts for different reactions. These reactions are hydrogen evolution reaction (HER), oxygen evolution reaction (OER), oxygen reduction reaction (ORR), CO_2 reduction reaction (CO_2 RR), N_2 reduction reaction (N_2 RR), and methanol oxidation reaction (MOR). Researchers are interested in 2D materials for electrocatalytic applications because of their unique and superior physicochemical characteristics. A modified or engineered electrocatalyst frequently exhibits some distinctive properties compared to bare or pure 2D materials catalyst. This book provides a concise overview of recent advances in the engineering and structural modification of 2D materials at the nanoscale, as well as their applications in electrocatalytic applications. In this book, engineered 2D material-based electrocatalysts for electrocatalytic applications will be discussed. The subject matter will cover everything from the introduction of 2D materials, numerous engineering approaches to modifying these materials, principles of electrocatalytic reactions, and the role of engineered 2D materials in reactions, and also addressing challenges and opportunities.

In chapter 1, we provide an introduction and discuss the importance of 2D materials. Also, we thoroughly discuss their unique optoelectronic, magnetic, mechanical, and catalytic properties. In chapter 2, we discuss the numerous engineering approaches such as defects, alloying, doping, strain and stress engineering, morphology tuning, edge engineering, and heterostructuring used to modify the properties of 2D materials. Then in upcoming chapters (chapters 3–8), we explore the recent advances in engineered 2D materials for OHER, OER, ORR, CO_2 RR, N_2 RR, and MOR. These chapters discuss the mechanism of each reaction, recent advancements, and opportunities in electrocatalysis with engineered 2D materials. Finally, in chapter 9, we discuss the prospects for engineered 2D materials as electrocatalysts.

Acknowledgments

This treatise *Engineered 2D Materials for Electrocatalysis Applications* is nothing but the tough grind of the many experts who compiled the chapters and have gone through a lot of backbreaking assignments. The authors are obligated and thankful to the anonymous reviewers who not only appraised the work thoroughly but also enhanced its quality. The immense support from the IOP Publishing team, especially Caroline Mitchell and Isabelle Defillion was instrumental and significant. The authors are most grateful for the love of their families, friends and, above all, the grace from above.

Editor biography

Chandra Sekhar Rout

Professor Chandra Sekhar Rout is a full Professor at Centre for Nano and Material Sciences (CNMS), Jain University. Before joining CNMS, He was a DST-Ramanujan Fellow at IIT Bhubaneswar, India (2013–17). He received his BSc (2001) and MSc (2003) degrees from Utkal University and his PhD from JNCASR, Bangalore (2008) under the supervision of Professor C N R Rao, Bharat Ratna. He did his postdoctoral research at National University of Singapore (2008–09), Purdue University, USA (2010–12) and UNIST, South Korea (2012–13). He was awarded the prestigious Ramanujan Fellowship and Young Scientist award of Department of Science and Technology, Government of India in 2013, emerging investigator award 2017 from Elsevier, IAAM medal 2017 from International Association of Advanced Materials 2017, Young researcher award from Venus International Foundation in 2015. His research interests include preparation and characterization of 2D layered materials and its hybrids for chemical sensors and biosensors, supercapacitors and energy storage devices, field emitters and electronic devices. He has authored more than 200 research papers in international journals and edited six books. His current h-index is 54 with total citations >12 000. He was ranked among the top 2% of scientists in India by the Stanford study in 2020–22. He serves as a board member of various reputed journals and is associate editor of '*RSC Advances*' a journal of Royal Society of Chemistry and '*American Journal of Engineering and Applied Sciences*' of Science Publications. He has completed several sponsored projects funded by the government of India and has supervised PhD, postdoc and MSc students. He has delivered more than 50 invited talks in various national and international conferences and travelled to several countries such as USA, UK, Israel, Singapore, South Korea, Brazil and Russia etc for collaboration purposes.

List of contributors

K Namsheer
Centre for Nano and Material Sciences, Jain University (Deemed-to-be University), Jain Global Campus, Kanakapura, Bangalore 562112, Karnataka, India

Mansi Pathak
Centre for Nano and Material Sciences, Jain University (Deemed-to-be University), Jain Global Campus, Kanakapura, Bangalore 562112, Karnataka, India

Komal N Patil
Material Chemistry and Catalysis, Debye Institute for Nanomaterials Science, Utrecht University, 3584 CG Utrecht, The Netherlands

Abhinandan Patra
Centre for Nano and Material Sciences, Jain University (Deemed-to-be University), Jain Global Campus, Kanakapura, Bangalore 562112, Karnataka, India

Sithara Radhakrsihnan
Centre for Nano and Material Sciences, Jain University (Deemed-to-be University), Jain Global Campus, Kanakapura, Bangalore 562112, Karnataka, India

Chandra Sekhar Rout
Centre for Nano and Material Sciences, Jain Univesity (Deemed-to-be University), Jain Global Campus, Kanakapura, Bangalore 562112, Karnataka, India

K A Sree Raj
Centre for Nano and Material Sciences, Jain Univesity (Deemed-to-be University), Jain Global Campus, Kanakapura, Bangalore 562112, Karnataka, India

Pratik V Shinde
Department of Molecular Sciences and Nanosystems, Ca' Foscari University of Venice, Via Torino 155, Mestre 30172, Italy

Vitthal M Shinde
Department of Chemistry, Annasaheb Waghire College, Otur Post, Junnar Taluka, Pune District, Pune 412409, India

Chapter 1

Introduction of 2D materials

Pratik V Shinde and Chandra Sekhar Rout

Due to their distinct physicochemical characteristics, two-dimensional (2D) nanomaterials have attracted significant interest in a wide range of applications. The catalog of 2D materials has grown since the discovery of graphene and now includes black phosphorus, hexagonal boron nitride, transition metal dichalcogenides, metal oxides, perovskites, MXenes, graphitic carbon nitride, etc. Therefore, this chapter is a comprehensive discussion of several classes and structural variations of 2D materials. In addition, the variation of 2D material properties with respect to their structure is illustrated. The properties such as electrical, optical, magnetic, mechanical, and catalytic are briefly discussed. Finally, a summary and future outlook for 2D materials is given.

1.1 Introduction

Since the technological revolution, work efficiency has increased substantially, and it continues to develop rapidly today. The groundbreaking 'There's plenty of room at the bottom' speech given in 1959 by physicist Richard Feynman opened the way to nanotechnology. Research on nanotechnology has gained increasing attention and investment, leading to advancements in various areas. The investigation and development of materials in various dimensions remain an important topic of research. From a material chemistry perspective, the term 'dimensionality' refers to the number of spatial dimensions in which the material's properties differ or behave in a particular way. Typically, materials are classified into four dimensions, namely zero-dimensional (0D), one-dimensional (1D), two-dimensional (2D), and three-dimensional (3D) materials. In the nano dimension, the optoelectronic, magnetic, and physicochemical properties are drastically different from bulk materials. Researchers are attempting to comprehend and control the special characteristics that each dimension offers for a variety of functions.

The discovery of graphene not only launched an exciting era of 2D material research but also paved a path for advances in nanoscience and nanotechnology.

doi:10.1088/978-0-7503-5719-7ch1

© IOP Publishing Ltd 2024

Graphene, the first 2D wonder material, was isolated and characterized in 2004 by A Geim and K Novoselov at the University of Manchester in the United Kingdom [1]. 2D materials are substances that have reduced dimensionality, with a lateral thickness on the order of nanometers or even atomic scale. They have a structure like a sheet with finite thickness in one dimension and limitless extension in the other two. 2D materials have unique features that differ from their bulkier counterparts. Typically, they exhibit strong covalent bonding within the sheet, with van der Waals forces between sheets, resulting in unique mechanical, electrical, and optical properties [2, 3]. Quantum confinement effects in 2D materials are especially significant because of their reduced aspect dimensions. In the early stage of 2D materials, researchers successfully utilized the potential of graphene in a variety of applications, including optoelectronic, spintronics, energy storage, energy conversion, biomedical devices, and more [4–8]. The pursuit of atomically thin sheets of other layered materials has been sparked by the success of graphene. The family of 2D materials is constantly developing, greatly broadening the range of potential phenomena that can be investigated in two dimensions. In addition to graphene, the family of 2D materials covers materials like graphitic carbon nitride (g-C_3N_4), hexagonal boron nitride (h-BN), black phosphorus (BP), and different classes like transition metal dichalcogenides (TMDs), MXenes, perovskites, metal–organic framework (MOFs), layered double hydroxides (LDHs), metal oxides, and so on [3, 9–13]. The structure, composition, and functions of each material in these several classes of 2D materials are unique from others. These materials cover a wide spectrum of electronic structures from zero bandgap graphene to direct bandgap BP [14]. The advanced optoelectronic characteristics of 2D materials, together with their good mechanical features like robustness and flexibility, make them suitable components for the development of a new generation of high-performance devices.

In this chapter, we present a complete account of the structure and properties of 2D materials. Initially, we provide a brief introduction to the various dimensions of materials along with features of 2D materials. In the next section, we briefly introduce the potential properties of 2D materials, including electrical, optical, magnetic, mechanical, and catalytic properties. We hope the study and exploration of these diverse classes of 2D materials continue to increase our understanding of their distinct characteristics and open up new prospects for varied applications.

1.2 Different classes of 2D materials

There are various classes of 2D materials, each with its own set of characteristics and possible uses. Here we have provided the structural introduction and significant properties of major classes of 2D materials (figure 1.1).

1.2.1 Graphene

A single sheet of carbon atoms organized in a 2D honeycomb lattice makes up graphene. The unique structure of graphene possesses an extraordinary combination of properties. It is exceedingly lightweight and flexible while being extremely strong

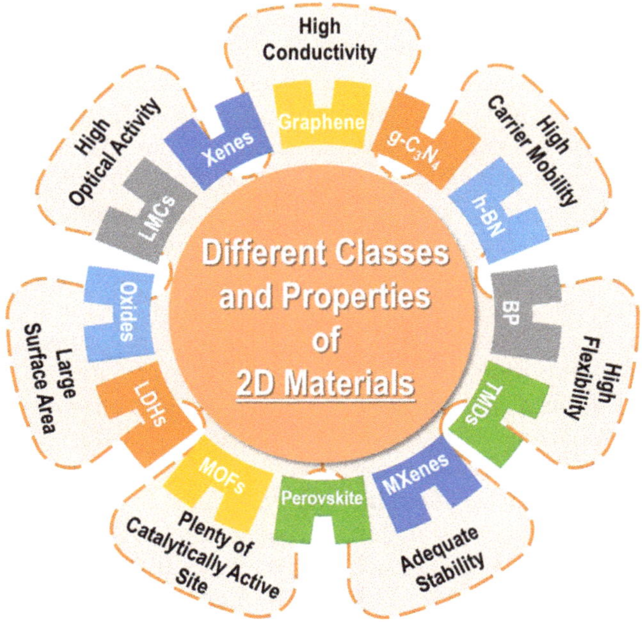

Figure 1.1. Schematic representing the diverse classes of 2D materials and their attractive properties.

—about 200 times stronger than steel [15]. Graphene exhibits a large theoretical specific surface area of 2630 $m^2\ g^{-1}$ and strong mechanical strength of approximately 1 TPa [16]. It shows high electric conductivity, outstanding heat conductivity, and transparency over a broad spectrum of wavelengths [17]. Moreover, it possesses high electron mobility at a room temperature of 15 000 $cm^2\ V^{-1}\ s^{-1}$, optical transparency of $\approx 97.7\%$, and intriguing thermal conductivity of 4.84×10^3–$5.30 \times 10^3\ W\ m^{-1}\ K^{-1}$ [18–20]. Because of these promising characteristics, graphene is used for a wide range of multiple applications like transistors, batteries, supercapacitors, solar cells, water splitting, etc [21–25]. However, there are obstacles to be addressed before graphene can be effortlessly incorporated into a variety of commercial applications. A lot of research is still being done on areas including cost-effectiveness, precision control over the tuning of its properties, and mass production.

1.2.2 Graphitic carbon nitride (g-C_3N_4)

The structure of polymeric g-C_3N_4 is formed from earth-abundant non-metals like carbon, nitrogen, and hydrogen. The only assembly of C–N bonds in the ideal g-C_3N_4 lacks any electron localization in the π state [26]. The g-C_3N_4 is thought to be the most stable allotrope of carbon nitrides (C_xN_y) because of its resemblance to graphene in terms of the layered structure. Tri-s-triazine (C_6N_7) rings joined by carbon–nitrogen (C–N) bonds form a repeating unit of g-C_3N_4 [27]. It is a chemically robust, economically viable semiconductor with a good visible light response up to 460 nm [28]. A greater amount of light can be absorbed due to its bandgap, which is about 2.7 eV [29]. Moreover, the structure of g-C_3N_4 allows for a substantial surface area and exhibits great chemical stability and thermal stability with modifiable electronic properties [30]. Its unique structure, tunable characteristics, and environmental friendliness make it a potential material for a wide range of technological advances.

1.2.3 Hexagonal boron nitride (h-BN)

Boron nitride (BN) is one of the most promising inorganic materials, which is found in crystal forms such as sp^3-bonded cubic and wurtzite forms; and sp^2-bonded hexagonal and rhombohedral forms [31]. Among these, h-BN with a structure analogous to that of graphene formed of sp^2 hybridized, strong covalent, and strongly polarized B-N bonds along the plane [32]. The layers of alternating hexagonal B and N atoms make up its atomically flat structure, which is held together by van der Waals interactions. It is frequently referred to as 'white graphene' due to its structural similarities to graphene and its white color. The h-BN is an electrical insulator with a wide bandgap of approximately 5–6 eV [33]. Also, it shows high in-plane thermal conductivity of 600 W m^{-1} K, hardness of 1.3–1.5 GPa, Young's modulus of 36.5 GPa, and chemical stability [34].

1.2.4 Black phosphorus (BP)

In comparison to white, red, and violet phosphorus, BP is the most stable phosphorus allotrope [35]. BP atoms are tightly bonded in the plane by covalent bonds, producing a puckered honeycomb structure, and such layers are stacked together by weak van der Waals interactions. Through sp^3-hybridized orbitals, each phosphorus atom is connected to three surrounding phosphorus atoms [10]. BP has a sizable bandgap and is a semiconductor material. With decreasing thickness from bulk to single layer, BP reveals a layer-dependent bandgap that can be tuned from 0.3 to 2 eV [36]. Unlike TMDs, which have a crossover from the indirect to direct bandgap, BP always has a direct form of energy band structure across a wide range of thicknesses [35].

1.2.5 Transition metal dichalcogenides (TMDs)

TMDs are a subclass of 2D materials that compose themselves by sandwiching transition metals between layers of chalcogen atoms. The general formula of TMDs is MX_2, where M represents a transition metal such as Mo, W, Nb, V, etc and X is a chalcogen element like S, Se, or Te. They specifically show strong in-plane covalent connections and weak out-of-plane van der Waals forces. The facile exfoliation into atomically thin flakes is made possible by the weak interactions between layers. Each layer normally has a thickness of 6–7 Å [37]. The most commonly found polymorphs of TMDs are 1T (trigonal), 2H (hexagonal), and 3R (rhombohedral). TMDs exhibit diverse electronic properties such as insulating (HfS_2), semiconducting (MoS_2 and WS_2), semimetallic (WTe_2 and $TiSe_2$), and metallic (NbS_2 and VSe_2). The main reason behind this is transition metals with d-electrons in variable numbers from group 4 to group 10 fill the non-bonding d bands to different levels [37]. TMDs have a tunable bandgap, which is essential for adjusting their optical and electrical characteristics to meet the needs of different applications. They are desirable for use in flexible electronics and thermal applications because of their great mechanical flexibility and thermal conductivity. TMDs have fascinating possibilities for technology advancement and enable novel products with improved performance in electronics, optoelectronics, energy, sensors, and several other fields.

1.2.6 MXenes

MXenes are a swiftly enlarging class of 2D materials with a broad spectrum of potential applications due to their unique combination of characteristics. MXene is produced through selective etching of the parent material known as the MAX phase. The typical formula for the metal carbide/nitride or carbonitride structure 'MAX' is $M_{X+1}AX_n$, where $n = 1$, 2, or 3. Here, A stands for an element from group 13 or group 14 (Al or Si), M is a transition metal, and X is either carbon and/or nitrogen. 2D structures with a metal carbide/nitride core and useful surface terminations are created as a result of this procedure. The word 'MXene' refers to the chemical formula $M_{n+1}X_nT_x$, where M signifies the transition metal, X for carbon or nitrogen, T for surface terminations (like -OH, -F, or -O), and n for the number of layers of the transition metal. Remarkable features unique to MXenes include their large surface area, hydrophilic surface groups, catalytically active basal planes with exposed metal sites, and amazing electrical conductivity (10 000 S cm^{-1}) [38]. Moreover, their excellent mechanical strength, good thermal stability, and tunable surface chemistry make them more special. The most studied and often used MXene is $Ti_3C_2T_x$, which is renowned for its 2D properties, metallic conductivity, and hydrophilicity.

1.2.7 Perovskites

Perovskites have captured the interest of researchers working in a variety of sectors including optoelectronics, medicine, sensors, and so on [12, 39, 40]. The mineral $CaTiO_3$ was discovered by geologist Gustav Rose in the Ural Mountains in 1839, and it was further characterized by Russian mineralogist Count Lev Alekseyevich von Perovski [41]. Perovskite materials are compounds having the general chemical formula ABX_3, where A and B are cations and X is an anion (oxygen or halogen). In ABX_3, the octahedron of X ions surrounds the B ion, while A lies in the center of the cube. Oxide and halide perovskite, two of the many varieties of perovskites, are frequently utilized in electrical gadgets. The general formula of oxide perovskites is ABO_3, and most of the metallic elements are found stable in this form [42]. In metal halide perovskites A is a monovalent cation (either organic (e.g. $CH_3NH_3^+$ (MA), $CH(NH_2)_2^+$ (FA)) or inorganic (e.g. Cs^+)), B is a divalent metal cation (typically Pb^{2+}), and X is a halogen anion (Cl^-, Br^-, I^-) [43]. Different crystallographic arrangements emerge from the substitution or partial replacement of cations or anions in perovskites, which distorts the cubic structure. An amazing diversity of electrical, optical, and chemical properties can be produced by a variety of phase transitions caused by changes in those atomic arrangements [44–46]. Their enormous magnetoresistance, superconductivity, piezoelectricity, pyroelectricity, and multiferroic property, present excellent chances to combine the requirements for sensors, catalyst electrodes, fuel cells, etc [12, 47–52].

1.2.8 Metal–organic framework (MOFs)

MOFs are crystalline compounds composed of metal ions or clusters coupled with organic ligands. It is simple to exfoliate 2D MOFs into thin, stacked layers thanks to their layered structure, which has strong covalent bonds in the in-plane direction and weak interlayer interactions. These materials exhibit unique properties by fusing the characteristics of MOFs with the structural benefits of 2D materials. The properties like ultrahigh porosity, a large surface area, mechanical flexibility, significant conductivity, and plenty of active sites make them promising candidates for applications like sensors, catalysis, and energy storage [53–55]. In order to increase the activity and endurance of catalysts, 2D MOF nanostructures offer long-term stability and strong electrical conductivity [53]. Through the careful selection of metal ions and organic ligands, MOFs' chemical and structural features can be tailored to meet the requirements of a particular application.

1.2.9 Layered double hydroxides (LDHs)

LDHs have attracted research attention because of their diverse but well-defined structural characteristics. The unique structure of LDHs is made up of positively charged metal hydroxide layers interlaced with charge-balancing anions. The general formula of LDH is $[M^{II}{}_{1-x}M^{III}{}_x(OH)_2](A^{n-})_{x/n}{\cdot}mH_2O$, where, M^{II} represents divalent metals cations (e.g. Mg^{2+}), M^{III} represents trivalent metal cations (e.g. Al^{3+}), A^{n-} represents interlayered anions (e.g. Cl^-), and X denotes the molar ratio of trivalent metal ions to the sum of trivalent and divalent metal ions [56]. A negatively charged anion located between the layers of metal hydroxide maintains the overall charge neutrality of LDH. The charge density and charge distribution may be directly impacted by the molar ratio and spatial arrangement of M^{III} and M^{II} on the major lamellae [56]. The spacing between layers can be influenced by the volume, valence, and amount of anions. LDH compounds with a wide range of chemical characteristics are made possible by combining various divalent and trivalent cations into the brucite-like layers and intercalating different anions between the interlayers. Due to their excellent physicochemical properties, including large surface area, appropriate mesopore distributions, and customizable structure, attracted considerable interest as potential materials for a variety of applications [57–59]. Their large surface area offers a lot of accessible sites for adsorption, surface modifications, and chemical reactions.

1.2.10 Metal oxides

Metal oxides are a class of material in which metal cations bond with oxygen anions. In many technological devices, metal oxides play a key role due to structural anisotropy, complex surface chemistry, and distinctive electronic structures. Metal oxide exhibits properties like high surface area, significant electrical conductivity, tunable bandgaps, fast ion diffusion, and favorable redox chemistry [60, 61]. Although layered metal oxides have minimal interlayer van der Waals interactions, they have significant intralayer covalent connections. The versatility of their prospective uses is provided by the physicochemical qualities of these materials, which depend on the element compositions and structural arrangements associated. MoO_2 comes in three polymorphs—the hexagonal, tetragonal, and monoclinic phases—and typically has a stable monoclinic structure [62]. The MoO_6 octahedra in the MoO_2 crystal are connected by Mo–Mo bonds, which cause the Mo electrons to delocalize in the conduction band and exhibit metallic conductivity. The mostly studied MoO_3 oxide is found in different polymorphs such as α-MoO_3 (orthorhombic phase), β-MoO_3 (monoclinic phase), and h-MoO_3 (hexagonal phase) [63]. α-MoO_3 exists in a layered structure and is the thermodynamically stable phase, whereas the other two phases are metastable. There are several prominent crystal structures MnO_2 shows, such as α-MnO_2, β-MnO_2, γ-MnO_2, δ-MnO_2, and λ-MnO_2 [64]. In single-layered MnO_2, one Mn layer is sandwiched between two O layers. Each Mn coordinates to six O atoms, resulting in edge-sharing MnO_6 octahedra [65]. Aside from that, metal oxides such as WO_3, Nb_2O_5, V_2O_5, and others have been extensively researched in a variety of fields [66–68].

1.2.11 Layered metal chalcogenides (LMCs)

The metals from the p-block of the periodic table such as Sn, Ga, and Ge metals bond to chalcogen atoms (Se, S, or Te) in a layered structure. The layers are held together by relatively weak van der Waals forces. SnS_2 shows rich polytypism including 2H, 4H, and 18R due to various S–Sn–S layer periodicity sequences [69]. SnS found in two phases such as α-SnS and β-SnS and differentiated from each other based on the Sn–S bond length. At room temperature, α-SnS is the stable form. The two stable phases of 2D gallium sulfide are the hexagonal GaS (layered) phase and the monoclinic Ga_2S_3 phase [70]. The tunable bandgap, high carrier mobility, and good thermal stability of metal chalcogenides make them attractive for applications [69–72].

1.2.12 Xenes

The monoelemental and atomically thin 2D materials are referred to as 'Xenes'. For example, the materials researched under a class of Xenes are germanene, borophene, arsenene, stanene, antimonene, gallenene, silicene, tellurene, etc. The letter 'X' stands for the name of a specific chemical element ranging from group III to group VI, while the word 'ene' is derived from the sp^2-hybrid alkene bond, which stands for alkene. Because sp^3 bonding is preferred over sp^2 bonding, Xenes form a buckled structure rather than a flat honeycomb (sp^2) lattice [73, 74]. The upper and lower atoms buckled because of the large distance between the atoms [75]. Xenes have many fascinating optoelectronic properties such as strong nonlinearity, wide optical response, fast recovery time, outstanding photothermal effect, greater photoelectric effect, and high carrier mobility [76]. Depending on the substrate, chemical functionalization, and strain, their electrical structure can range from simple insulators to semiconductors with tunable gaps to semimetallic [73]. Xenes are advantageous for upcoming electrochemical applications due to their enormous surface-active site density, quick reaction kinetics, high specific capacities, and flexibility [77]. The experimental exploration of Xenes properties as well as applications is still in its early stages, and additional efforts are needed to fully utilize their potential.

1.3 Properties of 2D materials

For the creation of next-generation technology, it is essential to understand and take advantage of the properties of 2D materials. Here, in this section, we have provided a brief discussion of the properties of 2D materials.

1.3.1 Electrical

Based on their unique composition, structure, and dimensions, 2D materials can display a wide range of electrical properties. They have a variety of electronic structures, including semimetals, metals, topological insulators, narrow-gap semiconductors, and large-band insulators. Graphene is a zero bandgap semimetal or semiconductor [78]. All carbon atoms' p orbitals are aligned perpendicular to the sp^2 hybridization plane and create a delocalized π bond that moves through the entire sheet of graphene [79–82]. At room temperature, since π electrons freely flow in the plane, graphene exhibits high conductivity. Moreover, it shows superconductivity, high carrier rate, semi-integer quantum Hall effect, and bipolar electric field effect [83]. Figure 1.2(a) shows the electronic band structure of graphene, where at the Fermi level, the valence band meets the conduction band in a Dirac-core-like gapless energy structure [84]. The coordination environment of the transition metal and its number of d-electrons have a significant impact on the electrical structure of TMDs. They have configurable bandgaps that switch from indirect to direct when changing from a bulky structure to a single layer (figure 1.2(b)) [85]. The shift in band structure with layer number is caused by quantum confinement. For example, the bandgap of MoS_2 modifies from ~1.2 to ~1.8 eV for a single layer [86]. The electrical characteristics of TMDs range from insulators (HfS_2), semiconductors (MoS_2 and WS_2), semimetals (WTe_2 and $TiSe_2$), to true metals (NbS_2 and VSe_2) [37]. The gap between gapless graphene and semiconducting TMDs is filled by BP. Its direct bandgap varies from 0.3 eV in bulk structure to 2 eV in monolayer structure [36]. Moreover, BP also exhibits high charge carrier mobility of ~1000 $cm^2\ V^{-1}\ s^{-1}$ and ambipolar transport characteristics [87]. The pristine h-BN has a wide bandgap of around 5–6 eV and behaves as an insulator [33]. Also, its electrical resistivity is 3.0×10^7 Ωcm [34]. The $Ti_3C_2T_x$ MXene is the most conductive of the several MXene structures, however, the Mo- and V-based MXenes behave like semiconductors [88]. It has been predicted that surface terminations, which are added during MXene synthesis, regulate metal-to-insulator transitions [88–90]. As shown in figures 1.2(c)–(e), the loss of adsorbed species, intercalants, and terminating species during annealing has an impact on the MXene sample resistance [88]. Theoretical investigations have indicated that surface functionalization decreases the

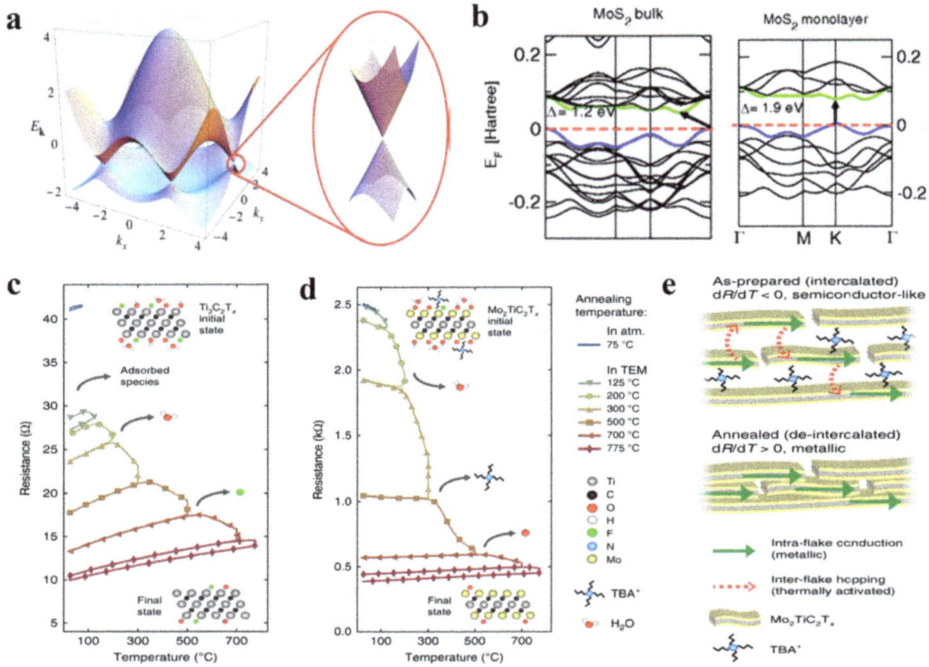

Figure 1.2. (a) The electronic band structure of graphene. Reprinted (figure 3) with permission from [84], Copyright (2009) by the American Physical Society. (b) Band structures of bulk and monolayer MoS_2. Reprinted (figure 3) with permission [85] Copyright (2011) by the American Physical Society. (c–d) Evolution of electronic properties of MXene with *in situ* vacuum annealing for $Ti_3C_2T_x$ and $Mo_2TiC_2T_x$. (e) Schematic showing the effect of intercalants on conduction through multi-flake $Mo_2TiC_2T_x$. Reproduced from [88] CC BY 4.0.

$Ti_3C_2T_x$ density of states at the Fermi level, implying a reduction in the charge carrier density and thus a reduction in conductivity. As shown in figures 1.3(a)–(c), Chen *et al* studied the strain-controlled electronic properties of CrS_2 TMDs [91]. The 1T′ phase can become a spin-up or spin-down half-metal in response to a tensile or compressive strain. The 1T′ phase is an indirect bandgap semiconductor, having spin-up and spin-down electron energies of 0.26 and 1.92 eV, respectively.

Due to the majority of the linking organic groups being insulators with modest π-orbital conjugation, MOFs are typically thought to be poor conductors [92]. Theoretical studies predicted that, by tailoring Zn^{2+} ions in MOFs with Co^{2+} ions, the bandgap can be tuned from semiconducting to metallic states [93]. Also, by altering the size of the secondary building unit cluster and switching up the conjugation of the organic linker, Zn-based MOFs' bandgaps can be tailored [94]. Depending on the type of ligands, metal ions, and their arrangement, 2D MOFs can display various charge transport methods, such as hopping or band-like transport [95, 96]. The single-layer metal oxides exhibit modest to wide bandgaps of around 1.22–6.48 eV and high carrier mobility up to 8540 $cm^2\ V^{-1}\ s^{-1}$ (for InO), as well as

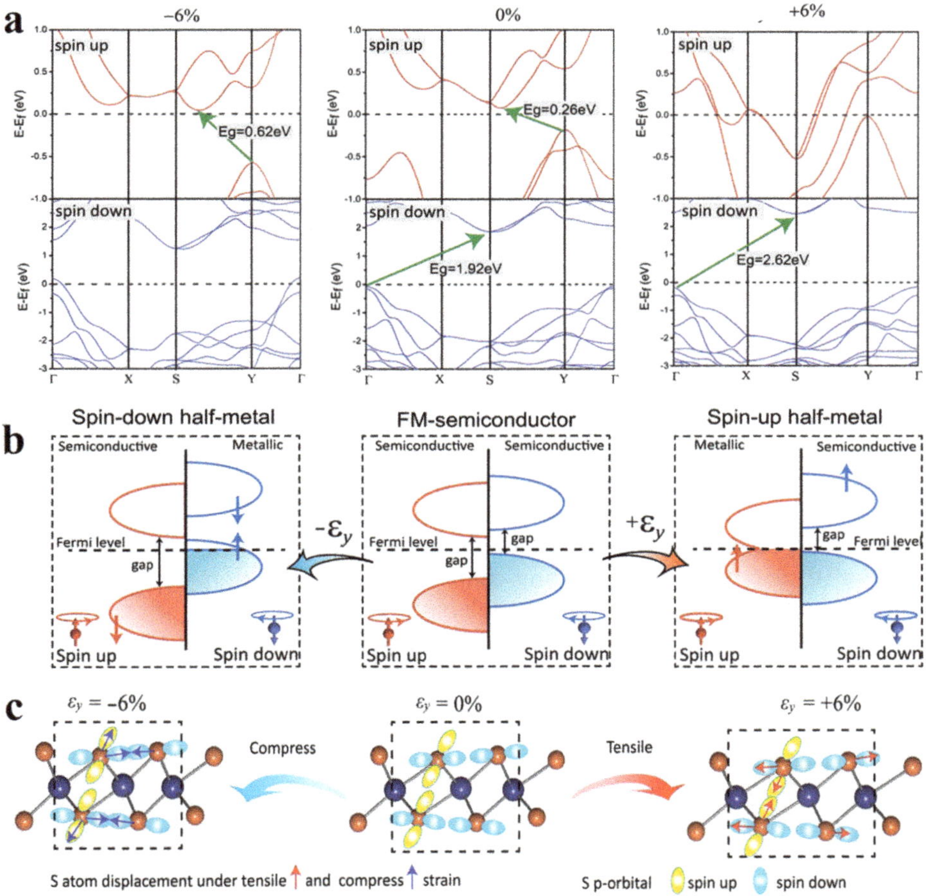

Figure 1.3. (a) Under −6%, 0%, and +6% strain in the *y*-direction, the spin-polarized band structure of 1T′ phase of CrS_2. (b) A representation of the changes in electron structure caused by tensile ($+\varepsilon_y$) and compressive ($-\varepsilon_y$) strain. (c) A diagram showing how spin-polarized valence band bottom (VBM) orbitals alter when subjected to compressive or tensile strain. Reproduced from [91] CC BY 4.0.

improved oxidation resistance [97]. Perovskite bandgaps can be tuned by modifying the thickness of the layer, which further changes the effective electron–hole confinement [98]. The indirect bandgap of GaS is 2.5 eV in the bulk, whereas it is 3.43 eV in the monolayer [70]. The bandgap of GeS can be efficiently tuned by applying an external strain, enabling the emission wavelength to be modulated, as per theoretical studies [99]. The orthorhombic polymorph of SnS has a bandgap of about 1.3 eV, while the cubic polymorph has a bandgap of 1.5–1.7 eV [100]. With the help of an external strain and electric field, the electric properties of the GaS–SnS_2 heterostructure can be efficiently controlled [101]. The GaS–SnS_2 heterostructure is a semiconductor with an indirect bandgap of 1.82 eV and a type-II band alignment that allows the photo-generated carriers to be easily separated.

1.3.2 Optical

The optical properties of 2D materials depend on their composition, structure, thickness, and surface chemistry. Graphene absorbs approximately 2.3% of the incident red light and 2.6% of the incident green light, according to its linear optical characterization [102, 103]. The optical properties of graphene revealed intriguing properties that are not only dependent on substrate and layer but also temperature-dependent [104]. As shown in figures 1.4(a)–(c), Razzhivina *et al* studied the chiral optical properties of Möbius graphene nanostrips [105]. The dissymmetry factors of tiny twisted graphene nanostrips can reach approximately 0.01. In twisted structure, circular dichroism can only be seen for transitions polarized across the nanostrip, whereas absorption is produced by transitions polarized both across and along the nanoribbon's axis. Due to its poor capacity for light absorption, small specific surface area, and rapid recombination of photo-generated active charges, g-C_3N_4 has a low photocatalytic efficiency [106]. The hBN is optically transparent in the visible region of the electromagnetic spectrum due to the wide bandgap [107]. By varying the twist angle of stacked and twisted hBN multilayers, it is possible to modify the optical properties of hBN thin films [108]. A new moiré sub-bandgap is created with a continually reducing magnitude as a function of the twist angle, resulting in a tunable luminescence wavelength. This is caused by the production of a moiré superlattice between the two interface layers of the twisted films.

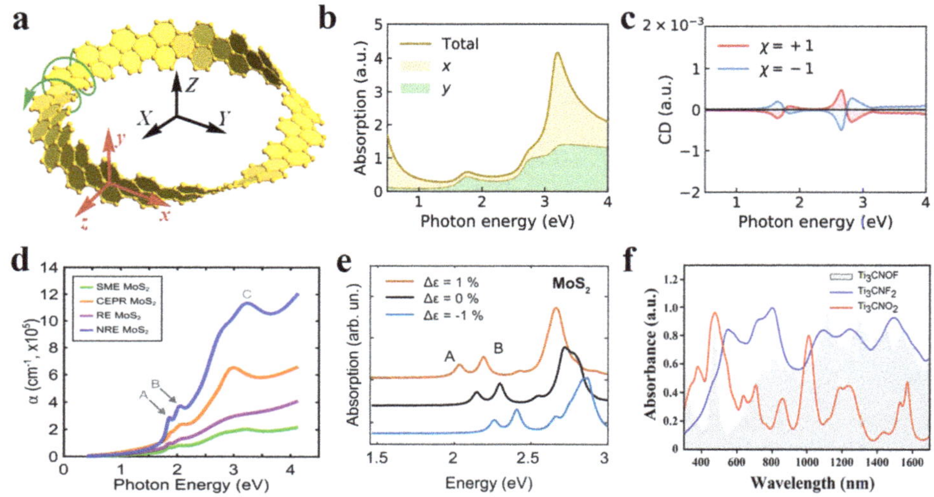

Figure 1.4. (a) Circular graphene nanostrips obtained by twisting the nanoribbon. (b) Absorption spectra of chiral graphene nanostrip. (c) Circular dichroism spectra of chiral graphene nanostrips. Reproduced from [105]. CC BY 4.0. (d) Absorption coefficients of MoS_2 films. Reproduced from [112] CC BY 4.0. (e) BSE absorption spectra under biaxial strain of single-layer MoS_2. Reproduced from [110] CC BY 4.0. (f) Absorption spectra Ti_3CN MXene with various terminations. Reprinted with permission from [115]. Copyright (2022) by the American Chemical Society.

Strain can significantly alter the optical characteristics of single-layer TMDs due to a change in bandgap [109, 110]. In addition to strain, another important aspect in the process of a lowered photoluminescence in TMDs on the substrate is the charge-transfer mechanism linked to the alignment of the energy levels in the substrate and TMD [111]. Busch *et al* studied the exfoliation procedure-based optical properties of solution-deposited MoS_2 films [112]. The exfoliation methods like solvent-mediated exfoliation, chemical exfoliation with phase reconversion, redox exfoliation, and native redox exfoliation have a significant impact on optical properties due to chemical impurities, carrier doping, flake dimensions, and lattice strain (figure 1.4(d)). For tensile strain on MoS_2, $MoSe_2$, WS_2, and WSe_2 materials, a redshift of the bandgap that reaches a 95meV/% for WS_2 [110]. The substrate extension/compression-dependent bandgap shifts are observed for $MoSe_2 < MoS_2 < WSe_2 < WS_2$ in that sequence. Figure 1.4(e) shows the Bethe–Salpeter equation (BSE) absorption spectra under biaxial strain of single-layer MoS_2. The linear dependence of both the A and B excitons was observed. Based on theoretical studies, the optical absorption in BP is strongly dependent on thickness, doping, and light polarization [113]. The MXene can be tuned optically in a nonlinear way, exhibiting absorption peaks from the ultraviolet to the near-infrared spectrum [114]. A negative ground-state absorption due to Pauli blocking has been detected in transition absorption studies for the Ti_3CN MXenes [115]. Additionally, the multiphoton absorption effect caused a substantial nonlinear optical response conversion for Ti_3CN MXenes from saturable absorption to reversed saturable absorption. Figure 1.4(f) displays the optical spectra of O^-, F^-, and mixed terminated Ti_3CN MXene, which reveal a typical metallic appearance with significant infrared absorption. Tin chalcogenides exhibit narrow bandgaps, high absorption coefficients ($\sim 10^5$ cm^{-1}), and also absorb light in the visible to the near-infrared range [116, 117]. The optical characteristics of 2D perovskite materials include tuning of the bandgap emission, narrowband emission, and broadband emission wavelength [118–120]. Similar to perovskites, metal oxides have oxygen ions in a favorable position for changes in optical properties to occur [121]. A large bandgap is a characteristic of metal oxides, and as a result, many of them are transparent in visible light [122, 123]. GeS is a layered semiconductor with a direct bandgap of 1.65 eV in the visible region, a high carrier mobility of 3680 cm^2 V^{-1} s^{-1}, and high photoresponse [124, 125].

1.3.3 Magnetic

2D materials show various types of magnetic behavior, like ferromagnetic, antiferromagnetic, or paramagnetic. 2D materials' magnetic properties can be considerably altered by adding magnetic dopants or defects [126, 127]. Theoretical calculations reveal that the magnetic properties of phosphorus-doped graphene can be controlled with the concentration of phosphorus and its configurations [126]. This leads to ferromagnetic and/or antiferromagnetic properties with the transition temperature up to room temperature (figures 1.5(a) and (b)). N-doped graphene has ferromagnetic properties and a high Curie temperature (>600 K) [128]. With an increase in nitrogen content in the samples, the saturation magnetization and coercive field increase. The saturation magnetizations for the sample with the

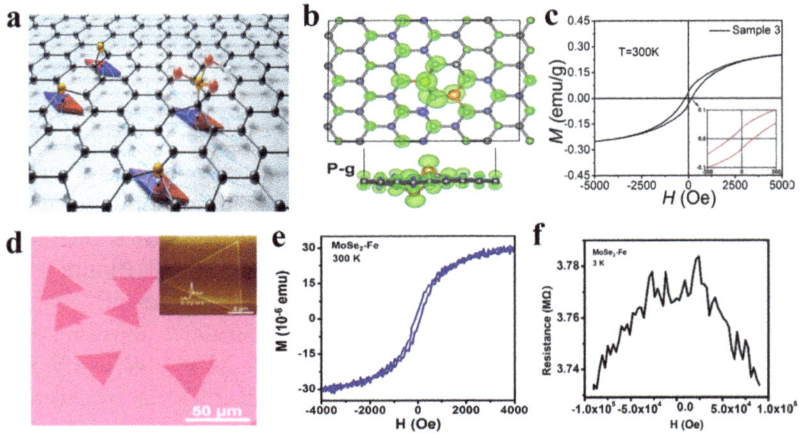

Figure 1.5. (a) Schematic represent the study of magnetic properties of graphene by phosphorus doping. (b) Stable magnetic arrangement (P-g). Carbon and phosphorus atoms are shown in gray and orange, respectively. Reprinted with permission from [126]. Copyright (2020) by the American Chemical Society. (c) Magnetization versus magnetic field (*M–H*) curves of the pristine and N-doped graphene at room temperature. Reprinted from [128], copyright (2016) with permission from Springer Nature. (d) Optical microscope image of the Fe-doped $MoSe_2$ monolayer. Inset: AFM and height profile of monolayer Fe-doped $MoSe_2$ (e) Magnetic hysteresis loops for Fe-doped $MoSe_2$ (f) Magnetoresistance curve under different magnetic fields at 3 K. Reprinted with permission from [129]. Copyright (2022) by the American Chemical Society.

highest nitrogen content are 0.148 emu g^{-1} at 300 K and 0.282 emu g^{-1} at 10 K, while the coercive forces are 544.2 Oe at 10 K and 168.8 Oe at 300 K. At room temperature, the $MoSe_2$ crystals with Fe, Co, and Ni doping exhibit ferromagnetic activity [129]. The Fe-doped $MoSe_2$ field effect transistor exhibits n-type semiconductor properties, demonstrating the formation of a dilute magnetic semiconductor at ambient temperature (figures 1.5(d)–(f)). As shown in figures 1.6(a)–(c), the most varied features are found in CrS_2 TMDs, which have 1T, 1T′, and 2H phases that correspond to antiferromagnetic metal, ferromagnetic semiconductor, and non-magnetic semiconductor, respectively [91]. The magnetic ground states of CrS_2 could change with high strain. The semiconductor Curie temperature (T_C) is from 943.1 K to 1337.6 K, as strain changes from −2% to 3% (figure 1.6(d)). The Heisenberg model's high T_C value (>800 K) indicates that the FM/half-metal state of 1T′ should be stable in the presence of ambient conditions.

Carrasco *et al* investigated the influence of morphologies on the magnetic properties of CoAl and NiFe LDHs [130]. The distortion of the layers leads to a better antiferromagnetic character, lower coercive fields, greater saturation magnetization, ordering temperatures, activation energies, and lesser Mydosh parameters. For samples with an interlayer spacing greater than 25 Å, the magnetic interactions overall in hybrid n-sulfate intercalated NiFe-LDHs changed from ferromagnetism to antiferromagnetism [131]. As the interlayer distance increased, the various magnetic characteristics, including spontaneous magnetization, blocking temperature, and coercive fields gradually decreased. Doping SnO_2 with transition metal

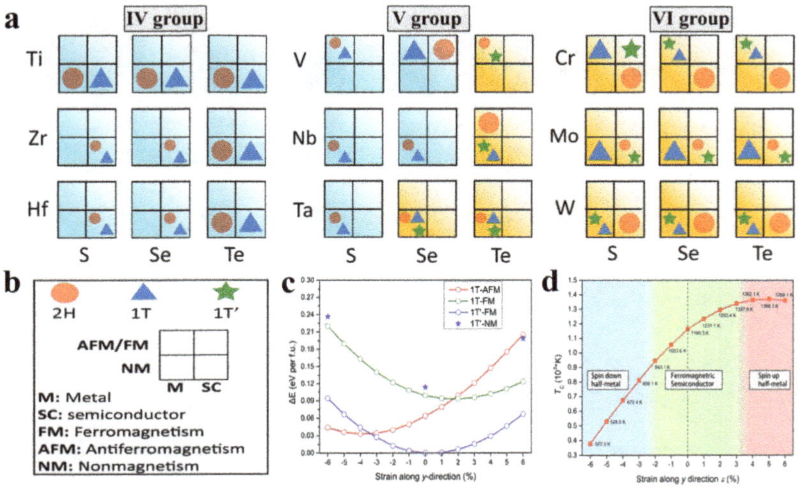

Figure 1.6. (a) Electronic and magnetic properties of IV, V, and VI TMDs. (b) Symbolic indicators of (a). (c) The relative energy of different magnetic state in 1T and 1T′ CrS_2 under strain. (d) T_c of 1T′-CrS_2 under y-direction strain. Reproduced from [91]. CC BY 4.0.

ions, such as Co, Ni, Mn, Fe, etc, can make it an appealing candidate for ferromagnetic semiconductors [132–135]. The Ni-doped SnO_2 powder exhibit ferromagnetic nature and its saturation magnetization is 5×10^{-4} emu g^{-1} and the coercive field is 83–96 sOe [132]. At room temperature, uncapped SnO_2 nanoparticles show a magnetic moment of 0.023 emu g^{-1} [136]. The capping of SnO_2 with cetyl trimethyl ammonium bromide (CTAB) improved its saturation magnetic moment to 0.081 emu g^{-1} by varying the surface electronic configuration. The borate-functionalized 2D amorphous g-C_3N_4 nanosheets (B-C_3N_4) display great ferromagnetism with a Curie temperature around 550K [137]. According to theoretical results, the bridging planar—B(OH)—group is important to the magnetic moment exchange in B-C_3N_4. The paramagnetic to antiferromagnetic transition at ~16.1 K in $Rb_2CuCl_2Br_2$ perovskite as per temperature-dependent magnetization measurement [138]. These transitions depend on the antiferromagnetic exchange interaction present between the two layers. The competition between the interlayer antiferromagnetic interaction and the intralayer ferromagnetic interaction helps to decide the ground state of perovskite.

1.3.4 Mechanical

Due to their distinct atomic structure and low dimensionality, 2D materials display intriguing mechanical specifications. Young's moduli of 2D materials are significantly higher than those of ordinary bulk materials. Graphene has a high Young's modulus of 1 TPa and can withstand strains of more than 25% without damage [139, 140]. The graphene films have greater flexibility and tensile strength that is approximately three times that of pyrolytic graphite sheets [141]. According to the simulated model, the graphene surface exhibits wrinkles when heated, and a PMMA

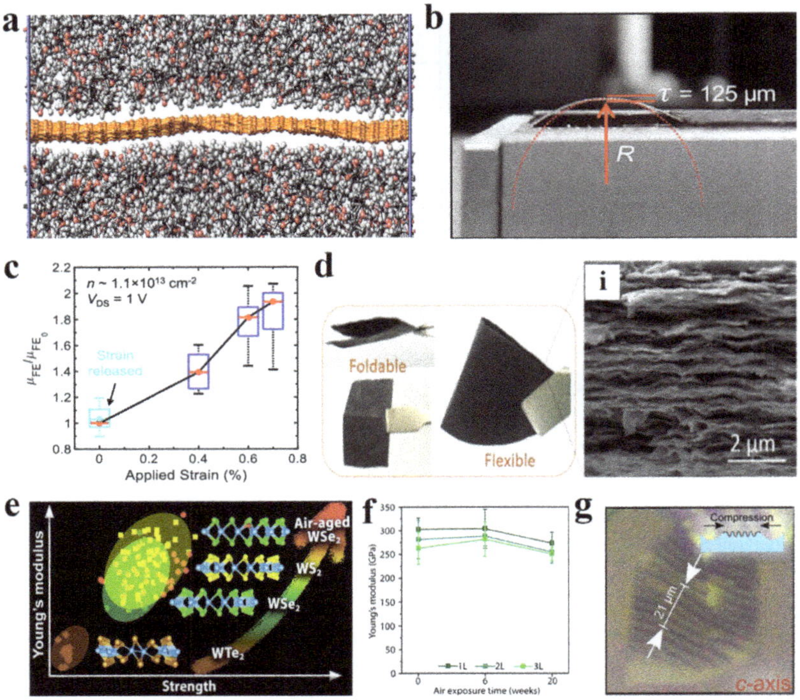

Figure 1.7. (a) Wrinkled graphene surface under tension and thermal conditions. Reprinted from [142], Copyright (2017), with permission from Elsevier. (b) Figure of a bent substrate for calculating the applied strain. (c) Field-effect mobility for eight devices as a function of applied strain, normalized to the beginning (unstrained) values. Reprinted with permission from [145]. Copyright (2022) by the American Chemical Society. (d) Digital images of composites film (70 wt%), shows the foldable and flexible characteristic; inset (i) shows its corresponding SEM image. Reprinted from [146], Copyright (2020), with permission from Elsevier. (e) Mechanical properties of atomically thin tungsten dichalcogenides. (f) Young's moduli of 1–3L WS_2 after air exposure for 6 and 20 weeks. Reprinted with permission from [147]. Copyright (2021) by the American Chemical Society. (g) Optical images of a MoO_3 flake on Gel-Film substrate applying compression along *c*-axes. The inset shows the formation of ripples. Reproduced from [148]. CC BY 4.0.

interphase zone is present close to the graphene surface (figure 1.7(a)) [142]. The nanocomposites' Young's and shear moduli increase with increasing graphene volume proportion and decrease as the temperature rises from 300 K to 500 K. According to first-principles calculations, the monolayer g-C_3N_4 shows highly isotropic mechanical properties and its Young's modulus is 187 N m^{-1} [143]. Monolayer MoS_2 has an in-plane stiffness of 180 ± 60 N m^{-1}, which corresponds to an effective Young's modulus of 270 ± 100 GPa [144]. This is comparable to the strength of steel. The strain between 6% and 11% is where breaking occurs, and the average breaking strength is 15 ± 3 N m^{-1} (23 GPa). Monolayer MoS_2 transistors on flexible substrates with uniaxial tensile strain exhibit strain-enhanced electron mobility [145]. With tensile strain up to 0.7%, the on-state current and mobility are almost doubled, and after the strain is released, the devices recover to their initial state (figures 1.7(b) and (c)). In the addition of 30 wt% poly(3,4-ethylenedioxythiophene)/poly(styrenesulfonate) (PEDOT/PSS) into $Ti_3C_2T_x$ film, the tensile strength is 38.5 ± 2.9 MPa and this increment is as high as 155% compared to $Ti_3C_2T_x$ film (figure 1.7(d)) [146].

As shown in figures 1.7(e) and (f), Falin *et al* studied the intrinsic and air-aged mechanical properties of mono-, bi-, and trilayer (1–3L) WS_2, WSe_2, and WTe_2 materials [147]. The 1L WS_2 has the maximum Young's modulus of 302.4 ± 24.1 GPa and strength of 47.0 ± 8.6 GPa. While Young's modulus and strength of 1L WSe_2 (258.6 ± 38.3 and 38.0 ± 6.0 GPa) and 1L WTe_2 (149.1 ± 9.4 and 6.4 ± 3.3 GPa) are relatively low. In comparison to the other two materials, WS_2's elasticity and strength are most significantly reduced with increasing thickness. The mechanical properties of 1–3L WS_2 and WSe_2 are stable in the air for up to 20 weeks. The modulus and strength of the 1–3L WSe_2 increase with air aging. The MoO_3 forms a rippling pattern under compression along various orientations, with the periodicity depending on the direction [148]. The Young's modulus value along the a-axis is 44±8 GPa and along the *c*-axis is 86±15GPa (figure 1.7(g)). The anisotropy in Young's modulus has a significant impact on optical and mechanical properties. The elastic modulus value of BP nanosheets is 276 32.4 GPa, and it decreases with increasing thickness [149]. Additionally, the effective strain of BP varies between 8 and 17% with a 25 GPa breaking strength. The Young's modulus of few-layer BP along the armchair direction and the zig-zag direction is 35.1 ± 6.3 GPa and 93.3 ± 21.8 GPa, respectively [150]. The printed MOF-hydrogel composites exhibit high strength of 277.6 kPa, modulus of 152.3 kPa, toughness of 744.7 kJ m^{-3}, and can be stretched up to 453.0% of their initial length [150]. By varying the composition and printing parameters, the mechanical properties can be changed.

1.3.5 Catalytic

The unusual electrical and structural features of 2D materials have sparked a surge in catalysis research. There are more active sites available for catalytic reactions in 2D materials because of their large surface area. This makes catalytic sites more accessible and improves the efficiency of reactant adsorption and diffusion [151]. In hybrid or composite structures, graphene plays a crucial role in improving conductivity and enabling faster charge-transfer kinetics for catalytic reactions [151]. Moreover, graphene helps to enhance the stability of the electrocatalyst and synergistically boosts performance with other materials. For different electrocatalytic processes, 2D materials like graphene, gC_3N_4, and MXenes are recognized to have enormous promise as stable supports [7, 11, 152]. The advantages brought about by the synergism of these supports and blend materials include high performance and excellent stability for various catalytic reactions. The bandgap tunable features, availability of different phase structures with semiconducting/metallic nature, and well durability make TMDs class materials attractive for catalysis applications [11, 153, 154]. MXenes are suitable candidates for the formation of hybrid materials due to an unusual combination of metallic conductivity (up to 10 000 S cm^{-1}) and hydrophilicity [155]. The large surface area and adjustable structure of MXenes have shown their tremendous promise as electrocatalysts. The combination of graphene with MXene nanosheets is a fruitful approach as it provides highly active sites for catalytic processes [156–158].

The surface of 2D MOF nanosheets has more exposed metal atoms that serve as active sites for electrocatalysis [159]. These exposed metal sites are coordinatively unsaturated and rich in dangling bonds, which increases the catalytic activity. By selecting suitable organic ligands, metal nodes with optimal oxidation states can be established in 2D MOF nanosheets [160]. Metal species with high oxidation states are advantageous for oxidation reactions like the OER. The 2D MOF promoted mass transfer and fast electron transport, which led to high reaction rates. The attributes of 2D LDHs can be customized to meet the requirements of different catalytic applications because of the tunability of the cations and anions. LDHs also offer flexible layered structures, a large number of exposed edge surface atoms, and unique electronic structures [161].

Increased accessibility of active sites in catalysts is a possible approach for improving the catalytic activity of 2D materials. The engineering 2D materials like doping, defect, alloying, stain, combination with other materials, etc can modify their electronic structure, leading to improved catalytic performance [11, 162–165]. The ideal platforms for 2D materials allow defects, doping, and alloying engineering to modify current catalytic sites or develop new catalytic active sites. The intrinsic differences in electronegativity and atomic size in heteroatom-doped graphene lead to the breakdown of electroneutrality and the introduction of stress and strain. This prompted the formation of additional catalytic sites, which made it easier for intermediates to chemisorb and desorb [162, 166–168]. The designing of 2D/2D heterostructures significantly improves electrical conductivity, electronic states, and chemical properties through interactions between materials, and offers abundant active sites for reactions [169–172]. The strong interfacial interaction of two materials produces multiple high-speed electron transport channels and simultaneously provides an electronic coupling effect, thus increasing activity by accelerating electron transfer. 2D materials played a crucial role in catalysis reactions such as hydrogen evolution reaction (HER), oxygen evolution reaction (OER), oxygen reduction reaction (ORR), CO_2 reduction reaction (CO_2 RR), N_2 reduction reaction (N_2 RR), and methanol oxidation reaction (MOR) (figure 1.8) [169, 173–177]. Future research should concentrate on maximizing the structural superiority of 2D materials and understanding the fundamentals of electrocatalytic processes.

1.4 Conclusion

This chapter sheds light on the various classes of 2D materials, available different structures, and their properties. First, various classes of 2D materials are presented including, graphene, g-C_3N_4, BP, TMDs, LDHs, MOF, hBN, perovskites, MXenes, LMC, metal oxide and Xenes. It briefly detailed how these classes differ structurally from one another and one another's attributes. There is a vast diversity of 2D structures present with thousands of combinations, structures, compositions, and phases. To improve the functionality of 2D materials in applications, it is crucial to

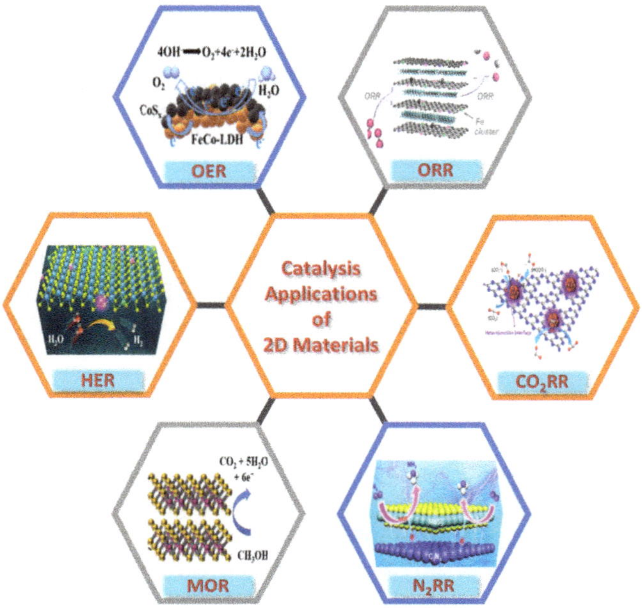

Figure 1.8. Schematic of various electrocatalysis applications of 2D materials. (Top left) reprinted with permission from [174]. Copyright (2021) by the American Chemical Society. (Top right) reprinted with permission from [175]. Copyright (2020) by the American Chemical Society. (Middle left) reprinted with permission from [173]. Copyright (2022) by the American Chemical Society. (Middle right) reprinted with permission from [176]. Copyright (2023) by the American Chemical Society. (Bottom left) reproduced with permission from [177]. Copyright (2021) by the American Chemical Society. (Bottom right) reprinted with permission from [169]. Copyright (2020) by the American Chemical Society.

comprehend the relationship between optimum architecture and their properties. Therefore, the optoelectronic and physicochemical properties are then presented in order to better understand their properties in relation to different changes in structural makeup. The reduced dimensions of 2D materials result in quantum confinement effects, which have an impact on the material's optoelectronic properties. A variety of 2D materials with metallic, semiconducting, and insulating properties are offered under several classes. Due to their atomically thin structure, 2D materials have high optical absorption and emission properties. It is possible to design flexible electronic devices using 2D materials because of their atomic thickness, which sustains a small radius of curvature. A variety of magnetic properties, such as ferromagnetism, diamagnetism, paramagnetism, antiferromagnetism, and spin-dependent moments, can be seen in 2D materials. Furthermore, 2D materials can exhibit superior catalytic performances due to their high surface-to-volume ratio, availability of plenty of active sites, tunable chemical composition, excellent conductivity, and appropriate stability. Therefore, 2D materials opens up amazing opportunities for the creation of next-generation electronics by allowing engineering with different strategies and tuning their properties as per demand.

By modifying the materials' composition, thickness, alloying, doping, defect formation, and strain, it is possible to tailor the properties of 2D materials. Overall, this chapter serves as a resource for comprehending the numerous kinds of 2D materials and the different properties they possess.

Acknowledgments

The authors gratefully acknowledge financial assistance from the SERB Core Research Grant (Grant No. CRG/2022/000897), Department of Science and Technology (DST/NM/NT/2019/205(G)), and Minor Research Project Grant, Jain University (JU/MRP/CNMS/29/2023).

References

[1] Novoselov K S, Geim A K, Morozov S V, Jiang D E, Zhang Y, Dubonos S V, Grigorieva I V and Firsov A A 2004 Electric field effect in atomically thin carbon films *Sci.* **306** 666–9

[2] Shanmugam V *et al* 2022 A review of the synthesis, properties, and applications of 2D materials *Part. Part. Syst. Charact.* **39** 2200031

[3] Shinde P V and Singh M K 2019 Synthesis, characterization, and properties of graphene analogs of 2D material *Fundamentals and Sensing Applications of 2D Materials* (Woodhead Publishing) pp 91–143

[4] Pargoletti E, Hossain U H, Di Bernardo I, Chen H, Tran-Phu T, Chiarello G L, Lipton-Duffin J, Pifferi V, Tricoli A and Cappelletti G 2022 Engineering of SnO_2–graphene oxide nanoheterojunctions for selective room-temperature chemical sensing and optoelectronic devices *ACS Appl. Mater. Interfaces* **12** 39549–60

[5] Khokhriakov D, Karpiak B, Hoque A M and Dash S P 2020 Two-dimensional spintronic circuit architectures on large scale graphene *Carbon* **161** 892–9

[6] Qi F, Li Q, Zhang W, Huang Q, Song B, Chen Y and He. J 2023 Freestanding ReS_2/Graphene heterostructures as binder-free anodes for lithium-ion batteries *ACS Appl. Mater. Interfaces* **15** 21162–70

[7] Shinde P, Rout C S, Late D, Tyagi P K and Singh M K 2021 Optimized performance of nickel in crystal-layered arrangement of $NiFe_2O_4$/rGO hybrid for high-performance oxygen evolution reaction *Int. J. Hydrogen Energy* **46** 2617–29

[8] Güneş F, Aykaç A, Erol M, Erdem C, Hano H, Uzunbayir B, Şen M and Erdem A 2022 Synthesis of hierarchical hetero-composite of graphene Foam/A-Fe_2O_3 nanowires and its application on glucose biosensors *J. Alloys Compd.* **25** 895:162688

[9] Zhang X, Yuan X, Jiang L, Zhang J, Yu H, Wang H and Zeng G 2020 Powerful combination of 2D g-C_3N_4 and 2D nanomaterials for photocatalysis: recent advances *Chem. Eng. J.* **390** 124475

[10] Shinde P V, Kumar A, Late D J and Rout C S 2021 Recent advances in 2D black phosphorus based materials for gas sensing applications *J. Mater. Chem.* C **9** 3773–94

[11] Shinde P V, Mane P, Late D J, Chakraborty B and Rout C S 2021 Promising 2D/2D $MoTe_2/Ti_3C_2T_x$ hybrid materials for boosted hydrogen evolution reaction *ACS Appl. Energy Mater.* **4** 11886–97

[12] Shinde P V, Patra A and Rout C S 2022 A review on the sensing mechanisms and recent developments on metal halide-based perovskite gas sensors *J. Mater. Chem.* C **10** 10196–223

[13] Huo J M, Wang Y, Meng J, Zhao X Y, Zhai Q G, Jiang Y C, Hu M C, Li S N and Chen Y 2022 ··· π interaction directed 2D FeNi-LDH nanosheets from 2D Hofmann-MOFs for the oxygen evolution reaction *J. Mater. Chem.* A **10** 1815–20

[14] Liu S, Huo N, Gan S, Li Y, Wei Z, Huang B, Liu J, Li J and Chen H 2015 Thickness-dependent raman spectra, transport properties and infrared photoresponse of few-layer black phosphorus *J. Mater. Chem.* C **3** 10974–80

[15] Ling S, Wang Q, Zhang D, Zhang Y, Mu X, Kaplan D L and Buehler M J 2018 Integration of stiff graphene and tough silk for the design and fabrication of versatile electronic materials *Adv. Funct. Mater.* **28** 1705291

[16] Shinde P V, Saxena M and Singh M K 2019 Recent developments in graphene-based two-dimensional heterostructures for sensing applications *Fundamentals and Sensing Applications Of 2D Materials* (Woodhead Publishing) pp 407–36

[17] Wang S J, Geng Y, Zheng Q and Kim J K 2010 Fabrication of highly conducting and transparent graphene films *Carbon* **48** 1815–23

[18] Zhang W, Chai C, Fan Q, Song Y and Yang Y 2020 A novel two-dimensional sp-sp^2-sp^3 hybridized carbon nanostructure with a negative in-plane poisson ratio and high electron mobility *Comput. Mater. Sci.* **185** 109904

[19] Ma L, Xu H, Lu Z and Tan J 2022 Optically transparent broadband microwave absorber by graphene and metallic rings *ACS Appl. Mater. Interfaces* **14** 17727–38

[20] Chang L, Wang D, Jiang A and Hu Y 2022 Soft actuators based on carbon nanomaterials *Chem. Plus. Chem.* **87** e202100437

[21] Bai Z, Xiao Y, Luo Q, Li M, Peng G, Zhu Z, Luo F, Zhu M, Qin S and Novoselov K 2022 Highly tunable carrier tunneling in vertical graphene–WS_2–graphene van der Waals heterostructures *ACS Nano.* **16** 7880–9

[22] Gong L, Zhang Y and Li Z 2022 V-MOF@graphene derived two-dimensional hierarchical V_2O_5@graphene as high-performance cathode for aqueous zinc-ion batteries *Mater. Today Chem.* **23** 100731

[23] Mary B C, Vijaya J J, Saravanakumar B, Bououdina M and Kennedy L J 2022 $NiFe_2O_4$ and 2D-rGO Decorated With $NiFe_2O_4$ nanoparticles as highly efficient electrodes for supercapacitors *Synth. Met.* **291** 117201

[24] Lin X *et al* 2022 *In situ* growth of graphene on both sides of a Cu–Ni alloy electrode for perovskite solar cells with improved stability *Nat. Energy* **7** 520–7

[25] Shinde P V, Samal R and Rout C S 2022 Comparative electrocatalytic oxygen evolution reaction studies of spinel $NiFe_2O_4$ and its nanocarbon hybrids *Trans. Tianjin Univ.* **28** 80–8

[26] Zhu J, Xiao P, Li H and Carabineiro S A Graphitic carbon nitride: synthesis, properties, and applications in catalysis *ACS Appl. Mater. Interfaces* **6** 16449–65

[27] Gashi A, Parmentier J, Fioux P and Marsalek. R 2022 Tuning the C/N Ratio of C-rich graphitic carbon nitride (g-C3N4) materials by the melamine/carboxylic acid adduct route *Chem.–A Eur. J.* **28** e202103605

[28] Jiang L, Yuan X, Pan Y, Liang J, Zeng G, Wu Z and Wang H 2017 Doping of graphitic carbon nitride for photocatalysis: a reveiw *Appl. Catal.* B **217** 388–406

[29] Rono N, Kibet J K, Martincigh B S and Nyamori V O 2021 A review of the current status of graphitic carbon nitride *Crit. Rev. Solid State Mater. Sci.* **46** 189–217

[30] Qi K, Liu S Y and Zada A 2020 Graphitic carbon nitride, a polymer photocatalyst *J. Taiwan Inst. Chem. Eng.* **109** 111–23

[31] Izyumskaya N, Demchenko D O, Das S, Özgür U, Avrutin V and Morkoç H 2017 Recent development of boron nitride towards electronic applications *Adv. Electron. Mater.* **3** 1600485

[32] Roy S *et al* 2021 Structure, properties and applications of two-dimensional hexagonal boron nitride *Adv. Mater.* **33** 2101589

[33] Naclerio A E and Kidambi P R 2023 A review of scalable hexagonal boron nitride (h-BN) synthesis for present and future applications *Adv. Mater.* **35** 2207374

[34] Hwang G H, Kwon Y S, Lee J S and Jeong Y G 2021 Enhanced mechanical and anisotropic thermal conductive properties of polyimide nanocomposite films reinforced with hexagonal boron nitride nanosheets *J. Appl. Polym. Sci.* **138** 50324

[35] Zhang Y, Zheng Y, Rui K, Hng H H, Hippalgaonkar K, Xu J, Sun W, Zhu J, Yan Q and Huang W 2017 2D black phosphorus for energy storage and thermoelectric applications *Small* **13** 1700661

[36] Deng B *et al* 2017 Efficient electrical control of thin-film black phosphorus bandgap *Nat. Commun.* **8** 14474

[37] Chhowalla M, Shin H S, Eda G, Li L J, Loh K P and Zhang H 2013 The chemistry of two-dimensional layered transition metal dichalcogenide nanosheets *Nat. Chem.* **5** 263–75

[38] Najam T, Shah S S, Peng L, Javed M S, Imran M, Zhao M Q and Tsiakaras P 2022 Synthesis and nano-engineering of MXenes for energy conversion and storage applications: recent advances and perspectives *Coord. Chem. Rev.* **454** 214339

[39] Abate A 2023 Stable tin-based perovskite solar cells *ACS Energy Lett.* **8** 1896–9

[40] Yang Z, Wang T, Xu X, Yao J, Xu L, Wang S, Xu Y and Song J 2023 Fiber optic plate coupled Pb-free perovskite x-ray camera featuring low-dose-rate imaging toward dental diagnosis *J. Phys. Chem. Lett.* **14** 326–33

[41] Goel P, Sundriyal S, Shrivastav V, Mishra S, Dubal D P, Kim K H and Deep A 2021 Perovskite materials as superior and powerful platforms for energy conversion and storage applications *Nano Energy* **80** 105552

[42] Wang Y, Lv Z, Zhou L, Chen X, Chen J, Zhou Y, Roy V A and Han S T 2018 Emerging perovskite materials for high density data storage and artificial synapses *J. Mater. Chem.* C **6** 1600–17

[43] Lu C H, Biesold-McGee G V, Liu Y, Kang Z and Lin Z 2020 Doping and ion substitution in colloidal metal halide perovskite nanocrystals *Chem. Soc. Rev.* **49** 4953–5007

[44] Whitfield P S, Herron N, Guise W E, Page K, Cheng Y Q, Milas I and Crawford M K 2016 Structures, phase transitions and tricritical behavior of the hybrid perovskite methyl ammonium lead iodide *Sci. Rep.* **6** 35685

[45] Wei W J, Li C, Li L S, Tang Y Z, Jiang X X and Lin Z S 2019 Phase transition, optical and dielectric properties regulated by anion-substitution in a homologous series of 2D hybrid organic–inorganic perovskites *J. Mater. Chem.* C **7** 11964–71

[46] Matsuishi K, Ishihara T, Onari S, Chang Y H and Park C H 2004 Optical properties and structural phase transitions of lead-halide based inorganic–organic 3D and 2D perovskite semiconductors under high pressure *Phys. Status Solidi* B **241** 3328–33

[47] Maeno Y, Hashimoto H, Yoshida K, Nishizaki S, Fujita T, Bednorz J G and Lichtenberg F 1994 Superconductivity in a layered perovskite without copper *Nature* **372** 532–4

[48] Suyal G, Colla E, Gysel R, Cantoni M and Setter N 2004 Piezoelectric response and polarization switching in small anisotropic perovskite particles *Nano Lett.* **4** 1339–42

[49] Liu W, Wang G, Cao S, Mao C and Dong X 2010 Structural, dielectric, and pyroelectric properties of (1− *x*) $PbSc_{0.5}Ta_{0.5}O_3$–(*x*) $PbHfO_3$ ceramics *J. Am. Ceram. Soc.* **93** 3023–6

[50] Saha R, Sundaresan A and Rao C N 2014 Novel features of multiferroic and magnetoelectric ferrites and chromites exhibiting magnetically driven ferroelectricity *Mater. Horizons* **1** 20–31

[51] Hong S, Díez A M, Adeyemi A N, Sousa J P, Salonen L M, Lebedev O I, Kolen'ko Y V, J V and Zaikina 2022 Deep eutectic solvent synthesis of perovskite electrocatalysts for water oxidation *ACS Appl. Mater. Interfaces* **14** 23277–84

[52] Jeerh G, Zou P, Zhang M and Tao S 2022 Perovskite oxide $LaCr_{0.25}Fe_{0.25}Co_{0.5}O_{3-\delta}$ as an efficient non-noble cathode for direct ammonia fuel cells *Appl. Catal.* **B 319** 121919

[53] Khan U, Nairan A, Gao J and Zhang Q 2022 Current progress in 2D metal–organic frameworks for electrocatalysis *Small Struct.* **4** 2200109

[54] Liu Y, Liu J, Pan Q, Pan K and Zhang G 2022 Metal-organic framework (MOF) derived In_2O_3 and g-C_3N_4 composite for superior NO_x gas-sensing performance at room temperature *Sens. Actuators* **B 352** 131001

[55] Wang X, Zhao C, Liu B, Zhao S, Zhang Y, Qian L, Chen Z, Wang J, Wang X and Chen Z 2022 Creating edge sites within the 2D metal-organic framework boosts redox kinetics in lithium–sulfur batteries *Adv. Energy Mater.* **12** 2201960

[56] Yu J, Yu F, Yuen M F and Wang C 2021 Two-dimensional layered double hydroxides as a platform for electrocatalytic oxygen evolution *J. Mater. Chem.* A **9** 9389–430

[57] Gao G, Yang G, Dou M, Kang S, Yin X, Yang H, Yang W, Li D and Dou J 2023 The construction of 3D hierarchical CdS/NiAl-LDH photocatalyst for efficient hydrogen evolution *Int. J. Hydrogen Energy* **48** 2200–10

[58] Liu S, Wan R, Lin Z, Liu Z, Liu Y, Tian Y, Qin D D and Tang Z 2022 Probing the Co role in promoting the OER and Zn–air battery performance of NiFe-LDH: a combined experimental and theoretical study *J. Mater. Chem.* A **10** 5244–54

[59] Zhang R, Dong J, Zhang W, Ma L, Jiang Z, Wang J and Huang Y 2022 Synergistically coupling of 3D FeNi-LDH arrays with $Ti_3C_2T_x$–MXene nanosheets toward superior symmetric supercapacitor *Nano Energy* **91** 106633

[60] Mei J, Liao T, Kou L and Sun Z 2017 Two-dimensional metal oxide nanomaterials for next-generation rechargeable batteries *Adv. Mater.* **29** 1700176

[61] Choi W S, Chisholm M F, Singh D J, Choi T, Jellison Jr G E and Lee H N 2012 Wide bandgap tunability in complex transition metal oxides by site-specific substitution *Nat. Commun.* **3** 689

[62] Ren B, Wang Y and Ou J Z 2020 Engineering two-dimensional metal oxides via surface functionalization for biological applications *J. Mater. Chem.* B **8** 1108–27

[63] Patil K N, Shinde P V, Srinivasappa P M, Nabgan W, Chaudhari N K, Rout C S and Jadhav A H 2022 Rational competent electrocatalytic oxygen evolution reaction on stable tailored ternary MoO_3-NiO-activated carbon hybrid catalyst *Int. J. Energy Res.* **46** 12549–64

[64] Samal R, Chakraborty B and Rout C S 2019 Understanding the phase dependent energy storage performance of MnO2 nanostructures *J. Appl. Phys.* **126** 045112

[65] Chen J, Meng H, Tian Y, Yang R, Du D, Li Z, Qu L and Lin Y 2019 Recent advances in functionalized MnO_2 nanosheets for biosensing and biomedicine applications *Nanoscale Horizons* **4** 321–38

[66] Yang Z, Wang J, Wang J, Li M, Cheng Q, Wang Z, Wang X, Li J, Li Y and Zhang G 2022 2D WO_{3-x} nanosheet with rich oxygen vacancies for efficient visible-light-driven photocatalytic nitrogen fixation *Langmuir* **38** 1178–87

[67] Han D, Xin Y, Yuan Q, Yang Q, Wang Y, Yang Y, Yi S, Zhou D, Feng L and Wang Y 2019 Solution-processed 2D Nb_2O_5 (001) nanosheets for inverted $CsPbI_2Br$ perovskite solar cells: interfacial and diffusion engineering *Sol. RRL* **3** 1900091

[68] Mei Y, Liu Y, Xu X, Zhang M, Dong Y and Qiu J 2023 Suppressing vanadium dissolution in 2D V_2O_5/MXene heterostructures via organic/aqueous hybrid electrolyte for stable zinc ion batteries *Chem. Eng. J.* **452** 139574

[69] Shinde P and Rout C S 2021 Advances in synthesis, properties and emerging applications of tin sulfides and its heterostructures *Mater. Chem. Front.* **5** 516–56

[70] Ahmed S *et al* 2022 Nonlinear optical activities in two-dimensional gallium sulfide: a comprehensive study *ACS Nano.* **16** 12390–402

[71] Ahmed S, Qiao J, Cheng P K, Saleque A M, Ivan M N, Alam T I and Tsang. Y H 2021 Two-dimensional gallium sulfide as a novel saturable absorber for broadband ultrafast photonics applications *ACS Appl. Mater. Interfaces* **13** 61518–27

[72] Shimada T, Ohuchi F S and Parkinson B A 1992 Thermal decomposition of SnS_2 and $SnSe_2$: novel molecular-beam epitaxy sources for sulfur and selenium *J. Vac. Sci. Technol. A: Vac., Surf., Films* **10** 539–42

[73] Khan K, Tareen A K, Wang L, Aslam M, Ma C, Mahmood N, Ouyang Z, Zhang H and Guo Z 2021 Sensing applications of atomically thin group IV carbon siblings xenes: progress, challenges, and prospects *Adv. Funct. Mater.* **31** 2005957

[74] Molle A 2016 Xenes: a new emerging two-dimensional materials platform for nanoelectronics *ECS Trans.* **75** 163

[75] Takeda K and Shiraishi K 1994 Theoretical possibility of stage corrugation in Si and Ge analogs of graphite *Phys. Rev.* B **50** 14916

[76] Liu Z, Liu J, Yin P, Ge Y, Al-Hartomy O A, Al-Ghamdi A, Wageh S, Tang Y and Zhang H 2022 2D Xenes: optical and optoelectronic properties and applications in photonic devices *Adv. Funct. Mater.* **32** 2206507

[77] Wang T, Wang H, Kou Z, Liang W, Luo X, Verpoort F, Zeng Y J and Zhang H 2020 Xenes as an emerging 2D monoelemental family: fundamental electrochemistry and energy applications *Adv. Funct. Mater.* **30** 2002885

[78] Nigar S, Zhou Z, Wang H and Imtiaz M 2017 Modulating the electronic and magnetic properties of graphene *RSC Adv.* **7** 51546–80

[79] Dresselhaus M S, Jorio A and Saito R 2010 Characterizing graphene, graphite, and carbon nanotubes by Raman spectroscopy *Annu. Rev. Condens. Matter Phys.* **1** 89–108

[80] Wang L *et al* 2013 One-dimensional electrical contact to a two-dimensional material *Sci.* **342** 614–7

[81] Reich S and Thomsen C 2004 Raman spectroscopy of graphite *Philos. Trans. R. Soc. Lond.* A **362** 2271–88

[82] Yan J A, Ruan W Y and Chou M Y 2009 Electron-phonon interactions for optical-phonon modes in few-layer graphene: first-principles calculations *Phys. Rev.* B 30**79** 115443

[83] Wang J, Ma F, Liang W and Sun M 2017 Electrical properties and applications of graphene, hexagonal boron nitride (h-BN), and graphene/h-BN heterostructures *Mater. Today Phys.* **2** 6–34

[84] Neto A C, Guinea F, Peres N M, Novoselov K S and Geim A K 2009 The electronic properties of graphene *Rev. Mod. Phys.* **81** 109

[85] Kuc A, Zibouche N and Heine T 2011 Influence of quantum confinement on the electronic structure of the transition metal sulfide TS_2 *Phys. Rev.* B **83** 245213

[86] Yang K, Liu T and Zhang X D 2021 Bandgap engineering and near-infrared-II optical properties of monolayer MoS_2: a first-principle study *Front. Chem.* **9** 700250

[87] Liu Y, Chen M and Yang S 2021 Chemical functionalization of 2D black phosphorus *InfoMat.* **3** 231–51

[88] Hart J L, Hantanasirisakul K, Lang A C, Anasori B, Pinto D, Pivak Y, van Omme J T, May S J, Gogotsi Y and Taheri M L 2019 Control of MXenes' electronic properties through termination and intercalation *Nat. Commun.* **10** 522

[89] Dong L, Kumar H, Anasori B, Gogotsi Y and Shenoy V B 2017 Rational design of two-dimensional metallic and semiconducting spintronic materials based on ordered double-transition-metal MXenes *J. Phys. Chem. Lett.* **8** 422–8

[90] Anasori B, Shi C, Moon E J, Xie Y, Voigt C A, Kent P R, May S J, Billinge S J, Barsoum M W and Gogotsi Y 2016 Control of electronic properties of 2D carbides (MXenes) by manipulating their transition metal layers *Nanoscale Horizons* **1** 227–34

[91] Chen K, Deng J, Yan Y, Shi Q, Chang T, Ding X, Sun J, Yang S and Liu J Z 2021 Diverse electronic and magnetic properties of CrS_2 enabling strain-controlled 2D lateral heterostructure spintronic devices *NPJ Comput. Mater.* **7** 79

[92] Usman M, Mendiratta S and Lu K L 2017 Semiconductor metal–organic frameworks: future low-bandgap materials *Adv. Mater.* **29** 1605071

[93] Choi J H, Choi Y J, Lee J W, Shin W H and Kang J K 2009 Tunability of electronic band gaps from semiconducting to metallic states via tailoring Zn ions in MOFs with Co ions *Phys. Chem. Chem. Phys.* **11** 628–31

[94] Lin C K, Zhao D, Gao W Y, Yang Z, Ye J, Xu T, Ge Q, Ma S and Liu D J 2012 Tunability of band gaps in metal–organic frameworks *Inorg. Chem.* **51** 9039–44

[95] Xie L S, Skorupskii G and Dincă M 2020 Electrically conductive metal–organic frameworks *Chem. Rev.* **120** 8536–80

[96] Sun L, Campbell M G and Dincă M 2016 Electrically conductive porous metal–organic frameworks *Angew. Chem. Int. Ed.* **55** 3566–79

[97] Guo Y, Ma L, Mao K, Ju M, Bai Y, Zhao J and Zeng X C 2019 Eighteen functional monolayer metal oxides: wide bandgap semiconductors with superior oxidation resistance and ultrahigh carrier mobility *Nanoscale Horizons* **4** 592–600

[98] Blancon J C *et al* 2017 Extremely efficient internal exciton dissociation through edge states in layered 2D perovskites *Sci.* **355** 1288–92

[99] Guo G and Bi G 2018 Effect of tensile strain on the band structure and carrier transport of germanium monosulphide monolayer: a first-principles study *Micro Nano Lett.* **13** 600–5

[100] Norton K J, Alam F and Lewis D J 2021 A review of the synthesis, properties, and applications of bulk and two-dimensional tin (II) sulfide (SnS) *Appl. Sci.* **11** 2062

[101] Ren D H, Li Q, Qian K and Tan X Y 2022 Tunable electronic properties of GaS–SnS_2 heterostructure by strain and electric field *Chin. Phys.* B **31** 047102

[102] Nair R R, Blake P, Grigorenko A N, Novoselo K S, Booth T J, Stauber T, Peres N M and Geim A K 2008 Fine structure constant defines visual transparency of graphene *Sci.* **320** 1308

[103] Zhu S E, Yuan S and Janssen G C 2014 Optical transmittance of multilayer graphene *Europhys. Lett.* **108** 17007
[104] Wu S, Wan L, Wei L, Talwar D N, He K and Feng Z 2021 Temperature-dependent optical properties of graphene on Si and SiO_2/Si substrates *Crystals* **11** 358
[105] Razzhivina M E, Rukhlenko I D and Tepliakov N V 2023 Chiral optical properties of Möbius graphene nanostrips. *J. Phys. Chem. Lett.* **14** 4426–32
[106] Li Y, Ruan Z, He Y, Li J, Li K, Jiang Y, Xu X, Yuan Y and Lin K 2018 In situfabrication of hierarchically porous g-C_3N_4 and understanding on its enhanced photocatalytic activity based on energy absorption *Appl. Catal.* **B 236** 64–75
[107] Shaik A B and Palla P 2021 Optical quantum technologies with hexagonal boron nitride single photon sources *Sci. Rep.* **11** 12285
[108] Lee H Y, Al Ezzi M M, Raghuvanshi N, Chung J Y, Watanabe K, Taniguchi T, Garaj S, Adam S and Gradecak S 2021 Tunable optical properties of thin films controlled by the interface twist angle *Nano Lett.* **21** 2832–9
[109] Lloyd D, Liu X, Christopher J W, Cantley L, Wadehra A, Kim B L, Goldberg B B, Swan A K and Bunch J S 2016 Band gap engineering with ultralarge biaxial strains in suspended monolayer MoS_2 *Nano Lett.* **16** 5836–41
[110] Frisenda R, Drüppel M, Schmidt R, Michaelis de Vasconcellos S, Perez de Lara D, Bratschitsch R, Rohlfing M and Castellanos-Gomez A 2017 Biaxial strain tuning of the optical properties of single-layer transition metal dichalcogenides *npj 2D Mater. Appl.* **1** 10
[111] Araujo F D *et al* 2023 Substrate-induced changes on the optical properties of single-layer WS_2 *Materials* **16** 2591
[112] Busch R T *et al* 2023 Exfoliation procedure-dependent optical properties of solution deposited MoS_2 films *npj 2D Mater. Appl.* **7** 12
[113] Low T, Rodin A S, Carvalho A, Jiang Y, Wang H, Xia F and Neto A C 2014 Tunable optical properties of multilayer black phosphorus thin films *Phys. Rev.* **B 90** 075434
[114] Han M *et al* 2020 Tailoring electronic and optical properties of MXenes through forming solid solutions *JACS* **142** 19110–8
[115] Gao L, Chen H, Kuklin A V, Wageh S, Al-Ghamdi A A, Ågren H and Zhang H 2022 Optical properties of few-layer Ti_3CN MXene: from experimental observations to theoretical calculations *ACS Nano.* **16** 3059–69
[116] Jethwa V P, Patel K, Pathak V M and Solanki G K 2021 Enhanced electrical and optoelectronic performance of SnS crystal by Se doping *J. Alloys Compd.* **883** 160941
[117] Lewis D J, Kevin P, Bakr O, Muryn C A, Malik M A and O'Brien P 2014 Routes to tin chalcogenide materials as thin films or nanoparticles: a potentially important class of semiconductor for sustainable solar energy conversion *Inorg. Chem. Front.* **1** 577–98
[118] Kitazawa N 1997 Excitons in two-dimensional layered perovskite compounds: $(C_6H_5C_2H_4NH_3)_2Pb(Br, I)_4$ and $(C_6H_5C_2H_4NH_3)_2Pb(Cl, Br)_4$ *Mater. Sci. Eng.:* **B 49** 233–8
[119] Kumar S *et al* 2016 Efficient blue electroluminescence using quantum-confined two-dimensional perovskites *ACS Nano.* **10** 9720–9
[120] Cortecchia D, Neutzner S, Srimath Kandada A R, Mosconi E, Meggiolaro D, De Angelis F, Soci C and Petrozza A 2017 Broadband emission in two-dimensional hybrid perovskites: the role of structural deformation *JACS 11* **139** 39–42
[121] Thummavichai K, Xia Y and Zhu Y 2017 Recent progress in chromogenic research of tungsten oxides towards energy-related applications *Prog. Mater Sci.* **88** 281–324

[122] Shi J, Zhang J, Yang L, Qu M, Qi D C and Zhang K H 2021 Wide bandgap oxide semiconductors: from materials physics to optoelectronic devices *Adv. Mater.* **33** 2006230
[123] Peelaers H, Kioupakis E and Van de Walle C G 2019 Limitations of In_2O_3 as a transparent conducting oxide *Appl. Phys. Lett.* **115**
[124] Mohammadzadeh M R *et al* 2023 Unique photoactivated time-resolved response in 2D GeS for selective detection of volatile organic compounds *Adv. Sci.* **10** 2205458
[125] Sutter E, Zhang B, Sun M and Sutter P 2019 Few-layer to multilayer germanium (II) sulfide: synthesis, structure, stability, and optoelectronics *ACS Nano.* **13** 9352–62
[126] Langer R, Błoński P, Hofer C, Lazar P, Mustonen K, Meyer J C, Susi T and Otyepka M 2020 Tailoring electronic and magnetic properties of graphene by phosphorus doping *ACS Appl. Mater. Interfaces* **12** 34074–85
[127] Valencia A M and Caldas M J 2017 Single vacancy defect in graphene: insights into its magnetic properties from theoretical modeling *Phys. Rev.* B **96** 125431
[128] Miao Q, Wang L, Liu Z, Wei B, Xu F and Fei W 2016 Magnetic properties of N-doped graphene with high Curie temperature *Sci. Rep.* **6** 21832
[129] Shen D *et al* 2022 Synthesis of group VIII magnetic transition-metal-doped monolayer $MoSe_2$ *ACS Nano.* **16** 10623–31
[130] Carrasco J A, Abellán G and Coronado E 2018 Influence of morphology in the magnetic properties of layered double hydroxides *J. Mater. Chem.* C **6** 1187–98
[131] Seijas-Da Silva A, Carrasco J A, Vieira B J, Waerenborgh J C, Coronado E and Abellán G 2023 Two-dimensional magnetic behaviour in hybrid NiFe-layered double hydroxides by molecular engineering *Dalton Trans.* **52** 1219–28
[132] Pascariu P, Airinei A, Grigoras M, Fifere N, Sacarescu L, Lupu N and Stoleriu L 2016 Structural, optical and magnetic properties of Ni doped SnO_2 nanoparticles *J. Alloys Compd.* **668** 65–72
[133] Sharma A, Singh A P, Thakur P, Brookes N B, Kumar S, Lee C G, Choudhary R J, Verma K D and Kumar R 2010 Structural, electronic, and magnetic properties of Co doped SnO_2 nanoparticles *J. Appl. Phys.* **107** 093918
[134] Ahmad N, Khan S and Ansari M M 2018 Optical, dielectric and magnetic properties of Mn doped SnO_2 diluted magnetic semiconductors *Ceram. Int.* **44** 15972–80
[135] Fu Y, Sun N, Feng L, Wen S, An Y and Liu J 2017 Local structure and magnetic properties of Fe-doped SnO_2 films *J. Alloys Compd.* **698** 863–7
[136] Selvi E T and Sundar S M 2017 Effect of size on structural, optical and magnetic properties of SnO_2 nanoparticles *Mater. Res. Express* **4** 075903
[137] Du L, Gao B, Xu S and Xu Q 2023 Strong ferromagnetism of g-C_3N_4 achieved by atomic manipulation *Nat. Commun.* **14** 2278
[138] Kundu K, Dutta P, Acharyya P and Biswas K 2021 Mechanochemical synthesis, optical and magnetic properties of Pb-free Ruddlesden–Popper-type layered $Rb_2CuCl_2Br_2$ perovskite *J. Phys. Chem.* C **125** 4720–9
[139] Lee C, Wei X, Kysar J W and Hone J 2008 Measurement of the elastic properties and intrinsic strength of monolayer graphene *Sci.* **321** 385–8
[140] Jiang H, Zheng L, Liu Z and Wang X 2020 Two-dimensional materials: from mechanical properties to flexible mechanical sensors *InfoMat.* **2** 1077–94
[141] Wang N *et al* 2018 Tailoring the thermal and mechanical properties of graphene film by structural engineering *Small* **14** 1801346

[142] Lin F, Xiang Y and Shen H S 2017 Temperature dependent mechanical properties of graphene reinforced polymer nanocomposites—a molecular dynamics simulation *Compos. Part B: Eng.* **111** 261–9

[143] Qu L H, Deng Z Y, Yu J, Lu X K, Zhong C G, Zhou P X, Lu T S, Zhang J M and Fu X L 2020 Mechanical and electronic properties of graphitic carbon nitride (g-C_3N_4) under biaxial strain *Vacuum* **176** 109358

[144] Bertolazzi S, Brivio J and Kis A 2011 Stretching and breaking of ultrathin MoS_2 *ACS Nano.* **5** 9703–9

[145] Datye I M, Daus A, Grady R W, Brenner K, Vaziri S and Pop E 2022 Strain-enhanced mobility of monolayer MoS_2 *Nano Lett.* **22** 8052–9

[146] Wan Y J, Li X M, Zhu P L, Sun R, Wong C P and Liao W H 2020 Lightweight, flexible MXene/polymer film with simultaneously excellent mechanical property and high-performance electromagnetic interference shielding *Compos. Part A: Appl. Sci. and Manuf.* **130** 105764

[147] Falin A *et al* 2021 Mechanical properties of atomically thin tungsten dichalcogenides: WS_2, WSe_2, and WTe_2 *ACS Nano.* **15** 2600–10

[148] Puebla S, D'Agosta R, Sanchez-Santolino G, Frisenda R, Munuera C and Castellanos-Gomez A 2021 In-plane anisotropic optical and mechanical properties of two-dimensional MoO_3 *npj 2D Mater. Appl.* **5** 37

[149] Wang J Y, Li Y, Zhan Z Y, Li T, Zhen L and Xu C Y 2016 Elastic properties of suspended black phosphorus nanosheets *Appl. Phys. Lett.* **108** 013104

[150] Vaquero-Garzon L, Frisenda R and Castellanos-Gomez A 2019 Anisotropic buckling of few-layer black phosphorus *Nanoscale* **11** 12080–6

[151] Li J, Zhao Z, Ma Y and Qu Y 2017 Graphene and their hybrid electrocatalysts for water splitting *ChemCatChem.* **9** 1554–68

[152] Dileepkumar V G, Balaji K R, Vishwanatha R, Basavaraja B M, Ashoka S, Al-Akraa I M, Santosh M S and Rtimi. S 2022 $CoSe_2$ grafted on 2D gC_3N_4: a promising material for wastewater treatment, electrocatalysis and energy storage *Chem. Eng. J.* **446** 137023

[153] Fu Q, Yang L, Wang W, Han A, Huang J, Du P, Fan Z, Zhang J and Xiang B 2015 Synthesis and enhanced electrochemical catalytic performance of monolayer $WS_{2(1-x)}Se_{2x}$ with a tunable band gap *Adv. Mater.* **27** 4732–8

[154] Luxa J, Mazánek V, Pumera M, Lazar L, Sedmidubský D, Callisti M, Polcar T and Sofer Z 2017 2H → 1T phase engineering of layered tantalum disulfides in electrocatalysis: oxygen reduction reaction *Chem. Eur. J.* **23** 8082–91

[155] Bai S, Yang M, Jiang J, He X, Zou J, Xiong Z, Liao G and Liu S 2021 Recent advances of MXenes as electrocatalysts for hydrogen evolution reaction *npj 2D Mater. Appl.* **5** 78

[156] Thirumal V, Yuvakkumar R, Kumar P S, Ravi G, Arun A, Guduru R K and Velauthapillai D 2023 Heterostructured two-dimensional materials of MXene and graphene by hydrothermal method for efficient hydrogen production and HER activities *Int. J. Hydrogen Energy* **48** 6478–87

[157] Zhang P, Fan C, Wang R, Xu C, Cheng J, Wang L, Lu Y and Luo P 2019 Pd/MXene ($Ti_3C_2T_x$)/reduced graphene oxide hybrid catalyst for methanol electrooxidation *Nanotechnology* **31** 09LT01

[158] Chaudhary K, Zulfiqar S, Somaily H H, Aadil M, Warsi M F and Shahid M 2022 Rationally designed multifunctional Ti_3C_2 MXene@ graphene composite aerogel integrated with bimetallic selenides for enhanced supercapacitor performance and overall water splitting *Electrochim. Acta* **431** 141103

[159] Zhao S *et al* 2016 Ultrathin metal–organic framework nanosheets for electrocatalytic oxygen evolution *Nat. Energy* **1** 1–10
[160] Zhu D, Qiao M, Liu J, Tao T and Guo C 2020 Engineering pristine 2D metal–organic framework nanosheets for electrocatalysis *J. Mater. Chem.* A **8** 8143–70
[161] Song J, Chen J L, Xu Z and Lin R Y 2022 Metal–organic framework-derived 2D layered double hydroxide ultrathin nanosheets for efficient electrocatalytic hydrogen evolution reaction *Chem. Commun.* **58** 10655–8
[162] Qu L, Liu Y, Baek J B and Dai L 2010 Nitrogen-doped graphene as efficient metal-free electrocatalyst for oxygen reduction in fuel cells *ACS Nano.* **4** 1321–6
[163] Kagkoura A, Tzanidis I, Dracopoulos V, Tagmatarchis N and Tasis D 2019 Template synthesis of defect-rich MoS_2-based assemblies as electrocatalytic platforms for hydrogen evolution reaction *Chem. Commun.* **55** 2078–81
[164] Kwon I S, Kwak I H, Zewdie G M, Lee S J, Kim J Y, Yoo S J, Kim J G, Park J and Kang H S 2022 $MoSe_2$–VSe_2–$NbSe_2$ ternary alloy nanosheets to boost electrocatalytic hydrogen evolution reaction *Adv. Mater.* **34** 2205524
[165] Zhang T, Liu Y, Yu J, Ye Q, Yang L, Li Y and Fan H J 2022 Biaxially strained MoS_2 nanoshells with controllable layers boost alkaline hydrogen evolution *Adv. Mater.* **34** 2202195
[166] Zhang C, Mahmood N, Yin H, Liu F and Hou Y 2013 Synthesis of phosphorus-doped graphene and its multifunctional applications for oxygen reduction reaction and lithium ion batteries *Adv. Mater.* **25** 4932–7
[167] Yang Z, Yao Z, Li G, Fang G, Nie H, Liu Z, Zhou X, Chen X A and Huang S 2012 Sulfur-doped graphene as an efficient metal-free cathode catalyst for oxygen reduction *ACS Nano.* **6** 205–11
[168] Sun T, Zhang G, Xu D, Lian X, Li H, Chen W and Su C 2019 Defect chemistry in 2D materials for electrocatalysis *Mater. Today Energy* **12** 215–38
[169] Chu K, Liu Y P, Li Y B, Guo Y L and Tian Y 2020 Two-dimensional (2D)/2D interface engineering of a MoS_2/C_3N_4 heterostructure for promoted electrocatalytic nitrogen fixation *ACS Appl. Mater. Interfaces* **12** 7081–90
[170] Chu K, Luo Y, Shen P, Li X, Li Q and Guo Y 2022 Unveiling the synergy of O-vacancy and heterostructure over MoO_{3-x}/MXene for N_2 electroreduction to NH3 *Adv. Energy Mater.* **12** 2103022
[171] Xiao X *et al* 2018 Engineering NiS/Ni_2P heterostructures for efficient electrocatalytic water splitting *ACS Appl. Mater. Interfaces* **10** 4689–96
[172] Sun L, Xu H, Cheng Z, Zheng D, Zhou Q, Yang S and Lin J 2022 A heterostructured WS_2/WSe_2 catalyst by heterojunction engineering towards boosting hydrogen evolution reaction *Chem. Eng. J.* **443** 136348
[173] Jian J, Kang H, Qiao X, Cui K, Liu Y, Li Y, Qin W and Wu X 2022 Cobalt and aluminum co-optimized 1T phase MoS_2 with rich edges for robust hydrogen evolution activity *ACS Sustain. Chem. Eng.* **10** 10203–10
[174] Yuan F, Zhang E, Liu Z, Yang K, Zha Q and Ni Y 2021 Hollow CoS_x nanoparticles grown on FeCo-LDH microtubes for enhanced electrocatalytic performances for the oxygen evolution reaction *ACS Appl. Energy Mater.* **4** 12211–23
[175] Jiang L, Duan J, Zhu J, Chen S and Antonietti M 2020 Iron-cluster-directed synthesis of 2D/2D Fe–N–C/MXene superlattice-like heterostructure with enhanced oxygen reduction electrocatalysis *ACS Nano.* **14** 2436–44

[176] Zhang Q, Sun M, Yuan C Y, Sun Q W, Huang B, Dong H and Zhang Y W 2023 Strong electronic coupling effects at the heterojunction interface of SnO_2 nanodots and g-C_3N_4 for enhanced CO_2 electroreduction *ACS Catal.* **13** 7055–66

[177] Gopalakrishnan A, Durai L, Ma J, Kong C Y and Badhulika S 2021 Vertically aligned few-layer crumpled MoS_2 hybrid nanostructure on porous Ni foam toward promising binder-free methanol electro-oxidation application *Energy Fuels* **35** 10169–80

Chapter 2

Engineering of 2D materials

K Namsheer†, K A Sree Raj† and Chandra Sekhar Rout

Researchers are focusing more and more on making efficient energy conversion/storage systems because of the rising need for modern clean energy technology. However, there are several technical obstacles that need to be effectively overcome to impede the development of energy systems. 2D materials like transition metal dichalcogenides, Mxene, phosphorene, Mbene have been widely explored in catalytic, energy, optoelectronics, etc applications due to their ultrahigh carrier mobility, large specific surface area, and great mechanical flexibility. Several inherent demerits in 2D materials have been shown to limit their future use in practical applications. To overcome these intrinsic properties researchers have explored various strategies including defects, alloying, doping, stress/strain, morphology, edge and heterostructuring, etc to tune the physicochemical properties of 2D materials. This chapter concentrates on the controllable strategies for engineering physicochemical properties in 2D materials, along with numerous examples.

2.1 Introduction

Since their inception, 2D materials have promised to become a revolutionary candidate in nano and material technologies [1]. 2D materials showed bright prospects in innumerable application fields including energy storage [2] and conversion [3], catalysis [4], electronics [5], biological [6] etc. A thorough understanding of 2D materials, their classification and their properties were discussed in the previous chapter. The unprecedented properties and exquisite possibilities of these materials conjured a widespread research appeal [7]. The exotic physicochemical properties of 2D materials are predominantly contributed by their structure and electronic confinement [8]. Even though 2D materials exhibit exceptional electronic [9], mechanical [10], optical [11], and magnetic properties [12], their practical usage is challenging due to their intrinsic deprivations. Inherently, 2D materials suffer

† NK and KA were equally contributed.

doi:10.1088/978-0-7503-5719-7ch2

© IOP Publishing Ltd 2024

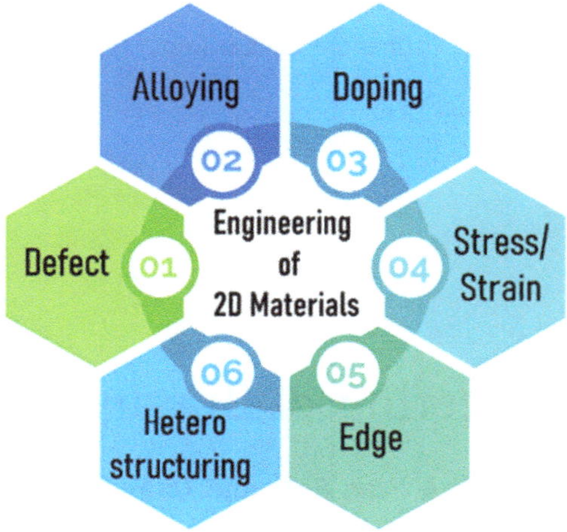

Figure 2.1. Strategies for engineering the 2D materials.

from shortcomings such as restacking, poor structural and chemical stability and unreactive sites, preventing their usage in commercial applications [13]. Researchers are devoted to developing various strategies to engineer various properties of 2D materials to utilize these wonder materials to their full potential. These strategies involve engineering numerous aspects of these materials using a particularly developed toolbox. This chapter aims to provide a detailed and concise analysis of some of these engineering techniques such as defect [14], alloying [15], doping [16], strain and stress engineering [17], morphology [18], edge modification [19] and heterostructures [13] (figure 2.1).

The defect engineering of 2D materials is the easiest and most successful method to tune the physicochemical properties of 2D materials. The 2D materials possess intrinsic defects and doping defects, which according to studies have a noticeable effect on the electrical/electronic transport characteristics [20]. Moreover, adding heteroatom dopants results in vacancies and encourages the growth of active sites, which enhances the capacity for interfacial adsorption [21, 22]. Similarly, to modulate the characteristics of electric transport is also made possible by alloying semiconductors with different charge–carrier polarities. Alloying is a promising method for adjusting the energy band structure of semiconducting 2D materials [15]. Alloying the 2D materials with a foreign guest molecule will customize the optical/electrical/electronic properties [23]. In recent years, 2D transition metal dichalcogenides (TMDs) alloying has become incredibly popular. Most commonly, the alloying of 2D TDMS can be fulfilled by metal replacements or by dichalcogenide replacements [24]. Doping engineering appears to be an effective method for tuning the electronics and optoelectronics properties of 2D materials, which allows control of electrical conduction and the modulation of charge carriers [21]. To achieve predictable and efficient properties, highly customizable doping techniques like surface charge transfer, intercalation, and field-effect modulation methods, and standard substitutional doping procedures with structural distortion have been widely used in recent years [16].

Here, in this chapter, we have summarized various studies focusing on engineering of 2D materials, including defects, alloying, doping, stress/strain, morphology, edge and heterostructuring, etc.

2.2 Strategies for engineering 2D materials

2.2.1 Defect engineering

Crystalline materials possess some degree of imperfection, in accordance with the second law of thermodynamics [25]. In crystallography, defects occur frequently and can take many different forms. Specifically, 2D materials have high surface energy and more unsaturated active sites on their crystal lattice than their bulk counterparts, these will generate intrinsic defects like vacancies, distortions, and lattice dislocations [26]. Generally, the defects in 2D materials can be divided into four types: point defects [27], line defects [28], planar defects [29], and volume defects [20], based on the extension pattern of the atomic disorder. The literature has shown that the presence of defects in 2D materials makes a huge impact on the electronic properties as well as optical properties as well as significantly improving the interfacial adsorption. Therefore, rational creation of defects on 2D materials plays a great role in tuning the carrier transmission efficiency and band structure for the energy conversion applications [30].

A defect is generally described as the change in the periodicity or symmetry of crystal lattice. Point defects arise due to the presence of vacancy or by heteroatom doping. The point defect is an arrangement disturbance that occurs close to the lattice node, typically between one and several lattice constants [31]. The point defects can be again classified into extrinsic and intrinsic defects. The intrinsic defect is caused by the energy fluctuation due to atomic vibrations, and the presence of vacancy defects including Schottky and Frenkel defects [32]. Extrinsic defects (impurity defects) are typically generated by heteroatom doping and do not alter the matrix lattice, but cause lattice deformation [33]. Anion and cation vacancies are mainly generated due to the absence or deficiency of atoms in the lattice plane, which causes charge redistribution at the bandgap level due to the presence of an empty electronic state. Vacancy defects are obvious in the case of 2D materials and researched well, including sulfur vacancies or selenium vacancies in TMDs, carbon vacancies in graphene, oxygen vacancies in transition metal oxides or layered double hydroxides, phosphorous vacancies in black phosphorous [30].

Among the family of 2D materials, TMDs are well studied for energy conversion applications [34]. The chalcogen (S, Se) vacancies directly alter the periodicity of the atomic arrangement in TMDs and make a great impact on the electronic structure of 2D TMDs. Additionally, different vacancies in the basal plane of TMDs can control the dynamic process of local ion diffusion and electron transmission [35]. To activate and optimize basal planes for electrochemical energy conversion applications, a variety of defect engineering-based procedures have been researched [36]. Furthermore, heteroatom substitution in TMDs is also

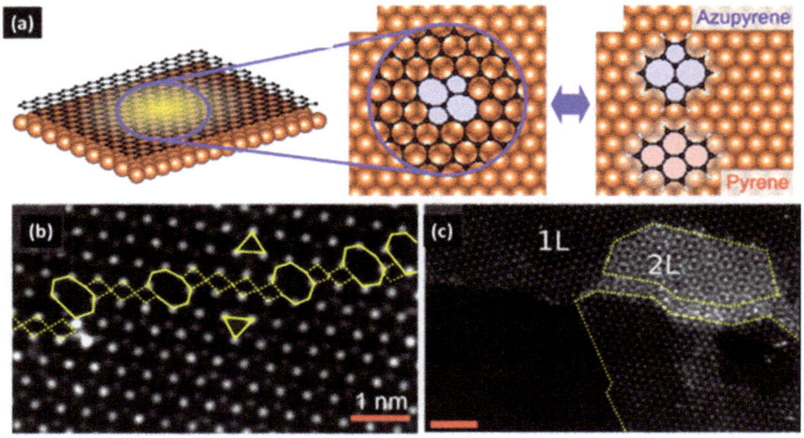

Figure 2.2. (a) Structures of graphene with Stone–Wales defect, reprinted with permission from [39], copyright (2022) American Chemical Society. (b and c) ADF-STEM image of MoS_2 showing line defect and moiré pattern in bilayer MoS_2, reprinted with permission from [43], copyright (2021) American Chemical Society.

considered as a point defect that drastically alters the basal plane of 2D materials [16]. Impurities with unique chemical compositions are typically found in 2D materials and can be introduced during synthesis unintentionally or as precursors. These impurities can significantly improve or impair the physicochemical qualities. The heteroatom doping like transition metal or non-metal elements with monolayer TMDs has recently gained a huge research interest. Recently Li *et al* produced Co-doped 1T-MoS_2 monolayers, with a distorted crystal structure due to the deliberate inclusion of Co and Co–S bonds formation [37]. Similarly, to increase the catalytic activity, Peng and colleagues also created single-atomic Ru doped Mo_2CT_x MXene nanosheets through a reaction in solution method [38]. In graphene, some topological VC defects, such as Stone–Wales defects, can typically be produced by altering the bonding state surrounding the vacancy during defect rebuilding (figure 2.2(a)) [39].

The crystal exhibits line defects spanning its length, primarily caused by dislocations, where crystal planes become misaligned [40]. It might occur because of an additional atomic plane being added somewhere in the crystal structure. Dislocations typically result in movement, stacking, and entanglement when subjected to external stress, which causes crystal distortion in the adjacent area [41]. Direct synthesis of sub-nanometre-wide one-dimensional (1D) MoS_2 channels embedded within WSe_2 monolayers was reported by Han *et al* utilizing a dislocation-catalysed method. The 1D channels create a coherent interface with the embedded 2D matrix because their edges are free of misfit dislocations and dangling bonds. Coherent MoS_2 1D channels are created in 2D superlattices in WSe_2 by periodic dislocation arrays [42]. The periodic atomic arrangements remain inside the domains when a perfect 3D crystal is divided into numerous little ones by interfaces, however, there will be significant atomic misalignments close to the interface between domains, which will result in the production of planar defects. Planar defects are typically grouped into three categories: the stacking fault, twin boundary, and grain boundary. Some vacancies within TMD monolayers will preferentially assemble into line defects because of the low migration barrier [30]. In the MoS_2 monolayer, Bertoldo *et al* discovered zigzag-directed line defect and volume defects (figures 2.2(b) and (c)) [43]. Similarly, volume defects are created by voids, or a crystal's internal surfaces not being formed by enough atoms [44]. They share characteristics with microcracks, due to the bonds that have broken at the surface. Vacancy clusters, which macroscopically resemble voids or holes, can be produced in 2D materials by removing more nearby atoms. Because of the lattice stress brought on by atoms escaping and moving, cracks can occasionally appear and come from the void [30]. Epitaxial growth on metal surfaces makes it simple to create the stacking defects in graphene. These defects typically seem like empty hills with moiré patterns in the graphene's intermittently rippling topography. Such topological stacking flaws in graphene, according to Artaud *et al*, are either encircled by rings made of non-hexagonal carbon polygons or buried in the metal substrate [45].

2.2.2 Alloying

The physicochemical properties of 2D materials can be tuned by the forming alloys by the intentional inclusion of heteroatoms into the crystal lattice. Designing new high-materials for various applications using alloy engineering is another widely utilized and simple process [23]. The electronic structures of the TMDs can be changed after inclusion of guest heteroatoms in different compositions, which can improve electrocatalytic/optical performance [46]. The alloying is a versatile strategy for adjusting semiconductors' energy band structures, which can then be used to customize their overall physicochemical properties. Electric transport capabilities can be modulated by alloying semiconductors with different charge–carrier polarity [47]. Ren *et al* reported that photoluminescence (PL) spectra peak point of the $InAs_xP1_x$ alloy continuously shifts from 860 to 3070 nm as the As content rises from 0% to 100% [48]. Similarly, mercury cadmium telluride alloy with a composition-dependent direct bandgap has also been created for mid-wave to long-wave infrared (3–15 μm) photodetection [49]. This is because alloying enables the on-demand construction of exotic semiconductors in accordance with the practical optoelectronic and catalysis applications [23].

Interestingly, many theoretical studies predicted that alloying can significantly increase the library of 2D materials and introduce a variety of novel physical features beyond the basic limitations of 2D materials. The alloying of 2D TMDs can be divided into four categories: multicomponent replacements, metal replacements, dichalcogenide replacements, and metal and dichalcogenide replacements [24]. When alloying 2D chalcogenides, the structural control during material production, which incorporates thermodynamic stability, is essential. Compared to their bulk equivalent, 2D materials struggle with being unstable and extremely reactive. According to reports, 2D TMDs exhibit bipolar behavior, and extensive band engineering can be accomplished by swapping out alloying elements. Generally, the

TMDs generated from various transition metals are capable of hosting various guest heteroatoms into their crystal lattice resulting in new functionalities compared with the starting material [46]. For instance, Group V (e.g., Nb) and Group VII (e.g., Re) elements are substituted into semiconducting 2D TMDs built of Group VI transition metals (Mo and W) and Group XVI chalcogens (S, Se, and Te) for p-type and n-type doping, respectively. For p- and n-doping, the S, Se, and Te constituents could be switched out for elements from Group XV (such as N and P) and Group XVII (such as F and Cl). TMDs with heavier Group XVI elements (like S to Te) have a lower bandgap [23]. Thus far, various 2D layered material alloys have been exploited, such as, $MoS_{1-2x}Se_{2x}$ [50], $Pb_xSn_{1-x}Se_2$, $Mo_{1-x}Nb_xSe_2$ [51], $WS_{2(1-x)}Te_{2x}$ [52], $Mo_{1-x}W_xS_2$ [53], $MoSe_{2x}Te_{2(1-x)}$ [54], etc. Huang *et al* reported that $Mo_{1-x}W_xSe_2$ alloy with a low W concentration can significantly suppress the deep-level defect states. Recently, by using density functional theory (DFT) analysis, Namsheer *et al* studied the conductivity of the MoSSe alloy. The DFT simulations revealed that the synthesized alloy had a high degree of thermomechanical and structural stability. Additionally, calculations showed that the addition of Se content increases conductivity, which improves electrochemical performance by modulating the density of states [46]. Making multifunctional devices is now a viable option by using TMDs alloy materials. Mukherjee *et al* thoroughly studied the photo response characteristics of $MoS2_xSe_2$ alloys ($x = 0, 0.5, 1$) (figure 2.3(a)). In comparison to their binary

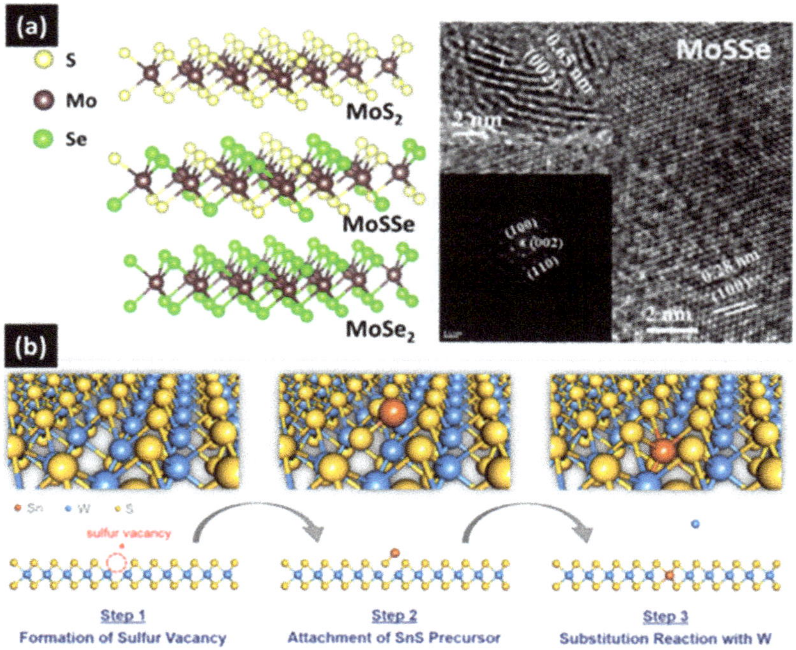

Figure 2.3. DFT derived atomic structure of MoS_2, MoSSe, and $MoSe_2$ nanosheets and TEM image of monolayer MoSSe, reprinted with permission from [55], copyright (2022) American Chemical Society. (b) Alloying in the WS_2 layers, reprinted with permission from [58], copyright (2019) American Chemical Society.

equivalents, the MoSSe hybrid phototransistor stands out as having better opto-electronic characteristics [55]. Similarly, Umrao *et al* reported that low-pressure chemical vapor deposition (CVD) was used to synthesize large-area monolayer $MoS_{2(1x)}Se_{2x}$ alloys. They noticed that the alloy has a hexagonal, crystalline structure. Interestingly. the optical bandgap changes from 1.77 eV to 1.69 eV compared to a pure monolayer of MoS_2 [56]. CVD technique is one of the easiest and most effective methods to synthesize pure TMDS alloys. Fu *et al* have produced monolayer $WS_{2(1-x)}$ Se_{2x} t, and high-resolution scanning transmission electron microscopy has confirmed the crystal structure. By altering the ratio of Se and S, monolayer $WS_{2(1-x)}$ Se_{2x} was made to have an adjustable bandgap. When compared to monolayer WS_2 and WSe_2, as-grown monolayer $WS_{2(1-x)}$ Se_{2x} x has the biggest exchange current density, and better catalytic performance. It validates that TMDs alloys are a good candidate for electrochemical hydrogen production when compared to the bare TMDs [57]. Chang *et al* reported the doping of foreign heteroatoms into single- and double-layer tungsten disulfide (WS_2) crystals. In this study, Sn atoms are energetically swapped to create a substitutional dopant at the W site through a metal exchange process. This research demonstrates that to produce S vacancies, which act as the initial binding sites for metal exchange with SnS precursors, a temperature that is high enough is necessary (figure 2.3(b)) [58].

Alloying of MAX phase was also explored in the last years, the MAX phase is represented as $M_{n+1}AX_n$ where M is transition metal, A is group IIIA or IVA elements and X is C or N [59]. The well-renowned 2D MXenes are derived from MAX phases, and they are explored in various applications including energy storage, energy conversion, catalysis, photovoltaics and biomedical etc. Chiche *et al* introduced 40% Cu into Ti_3AlC_2 by ball milling along with hot pressing technique at a high temperature of 950 °C. They observed that the introduced Cu is well bonded at Al site and the resultant composite alloy shows a low resistance and high mechanical compressive strength [60].

2.2.3 Doping

Doping is a well-known technology which is the process of deliberate introduction of impurities into the materials for tuning their physicochemical properties. Charge transfer doping, intercalation doping, and substitutional doping are common in 2D materials, and substitutional doping is more stable compared to the other [61]. The weak van der Waals (vdW) interactions between 2D layers result in significant interlayer distances that make the intercalation of dopant atoms easier than in traditional semiconductors with bulk crystal structures, which are typically doped by impurity atoms at interstitial sites [61]. Moreover, the 2D materials can be easily doped through surface charge transfer and the effects of an external electrostatic field when they are exfoliated or grown directly into ultrathin thicknesses.

Charge transfer doping: Charge transfer doping to alter the physicochemical behaviour of semiconductors has received a lot of interest. In charge transfer doping, the interaction between the host material and impurities takes place using surface toms, ions, molecules, particles, and by using supporting substrates. But in substitutional doping, the interaction involves incorporating foreign dopant atoms into the lattice. In charge transfer doping the interaction between the host and guest materials is based on acceptor or donor mechanism by the Fermi level. In p-type doping, the transfer of electrons from the host semiconductor to the dopants occurs when the dopant has a lower Fermi level (E_F) than the host semiconductor. On the other hand, n-type doping results from dopants with higher E_F than the semiconductor [62, 63]. The prediction of the charge transfer doping can be done instead using theoretical techniques based on basic principles and DFT. The charge density distribution and the difference between before and after adsorption directly illustrate the direction of charge transfer between the host and surface adsorbents. Also, the final electrical properties are then shown on the electronic density of states that results. These theoretical techniques are very useful for complex compounds [61].

Intercalation doping: 2D materials have larger interlayer spacing between their layers. This interlayer layer spacing acts as a host to take guest materials and results in a change in their physicochemical properties [61, 64]. 2D materials are capable of intercalating foreign ions, atoms, and even molecules into the relatively wide interlayer gap thanks to the vdW interlayer coupling [65]. Prior to now, 2D materials could be exfoliated in solution using this intercalation technique, which was mostly used in electrochemical applications. H and alkali metal ions are mainly intercalated into the interlayer spacing of 2D materials because of their smaller atomic radii [61, 66]. These impurities can alter the electrical structure and lattice parameters of 2D materials by donating electrons to the lattice. Ion intercalation has been extensively exploited for the liquid phase exfoliation of 2D materials by taking advantage of the intercalation-induced interlayer expansion. The 1T phase of MoS_2 is usually prepared by sequentially sonicating in water and n-butyllithium (n-Bu-Li) solution. The resultant MoS_2 structure was deformed from the starting 2H phase lattice and the electronic arrangement was finally destabilized because of the high electron donation from intercalated ions [67]. Intense research has been done on the potential of platinum, ruthenium (Ru), rhodium (Rh), and iridium (Ir) for electrochemical H_2 generation. Among them, Ru has received the most interest for hydrogen evolution reaction (HER) due to its low cost, high catalytic activity, and great stability. On the other hand, Ru has a propensity to aggregate due to its high cohesive energy, which leads to decreased electrocatalytic activity [61]. Therefore, it has been thought that a viable approach enhances its catalytic activity by evenly dispersing and/or anchoring Ru nanoparticles onto a 2D material support with high conductivity. Erdene *et al* experimentally investigated the electrocatalytic activity of Ru attached boron doped Mxene and they found that doping improved the intermediate hydrogen adsorption kinetics and reduced the charge transfer kinetics [68].

Substitutional doping: As discussed above, the cationic and anionic elements can be substitutionally replaced by impurity atoms with similar atomic radii [69]. The valency is one of the main factors to describe weather the doping is n-type or p-type. The substitutional doping principle is like that of semiconductor compounds like CdSe and GaAs, the impurity atoms may go to the M site or X site and act as donor–acceptor according to their valency [24]. For instance, Group V (e.g., Nb) and Group VII (e.g., Re) elements substituting M atoms, group XV (e.g., N) and group XVII (e.g., F) elements substituting X atoms as both p- and n-type dopants, respectively [61]. It is well known that doping of bulk TMDs increases carrier concentration compared to that of monolayer. The doping effect is different in the monolayer state due to less Coulomb interaction [70]. It frequently takes substantial impurity densities in the alloying limit for carrier doping to significantly increase conductivity. Therefore, a detailed understanding of both the doping chemistry and the physical consequences of impurities is necessary to fully realize the promise of 2D TMDs. Recently, Suh *et al* studied the first Nb-doped MoS_2 bulk crystals. The extended x-ray absorption fine structure (EXAFS) examination of the chemical bond environment provided proof of the type of Nb substitutional doping at Mo sites. The 0.5% Nb-doped MoS_2 showed significant p-type electrical conductivity when exfoliated into a few layers. It is noteworthy that such degenerate Nb doping at a rate of 40.1% may change the structural makeup of MoS_2 from the typical 2H to 3R mode stacking [71].

2.2.4 Strain and stress engineering

In 2D materials, it has been successfully accomplished to modify different physical properties by stress and strain engineering. These mechanical modifications can directly alter the phonon structure and energy band structure of 2D materials, resulting in Raman and PL peaks in terms of optical characteristics [72]. In the last few decades, the semiconductor industry has made extensive use of strain engineering. For instance, a silicon transistor's channel may experience uniaxial or biaxial tensile strain, which can significantly increase electron or hole mobility. However, the application of strain modulation is severely constrained because conventional semiconductor bulk single crystals can only endure a relatively small amount of strain [17, 73]. 2D materials exhibit considerable potential for strain engineering since they can bear larger elastic strain without fracture than bulk materials. In addition to optoelectronic applications, the strain engineering of 2D materials is beneficial for high-energy conversion applications. The catalytic efficacy of noble metal nanosheets and TMDs has also been shown to benefit from strain engineering. A single-crystalline Ni_3S_2@MoS_2 core–shell heterostructure with variable layers was created by Zhang *et al* to generate biaxial strain to MoS_2 nano-shells. A bilayer MoS_2 nano-shell electrode was discovered to have exceptional HER activity [74].

2.2.5 Morphology

In the nanoscale, the morphology of a nanostructure can influence its properties; hence modifying the morphology of 2D materials became a key strategy in engineering 2D materials for various functions [75]. There are various synthesis approaches employed to implement this tactic. Engineering the morphology is a customized approach using different toolkits for different 2D materials and applications [76]. Morphological engineering of nanostructure can provide structure–activity information of the nanomaterial irrespective of the application [77]. Researchers believe this is an efficient way to tune various properties of 2D material to improve its activity. Here we are discussing different synthesis procedures of different 2D materials and how they alter the morphology of the nanostructures. With the recent progress in synthesis, approaches like CVD can produce high--quality 2D materials with an effective morphological control. In 2012, Jaramillo and co-workers explored a surface structure engineered nanoporous double-gyroid MoS_2 bicontinuous network as an efficient electrocatalyst using an electrodeposition method. This unique morphology with high surface area and active edge sites showed exceptional hydrogen evolution activity. This work exhibited that the morphological engineering of 2D MoS_2 at the nanoscale can influence atomic scale surface structure changes and eventually improves its catalytic performance [78]. The morphological engineering of MoS_2 was achieved using Mo_2C MXene for an efficient pH independent hydrogen evolution catalyst. The microspherical morphology of MoS_2 is modified using Mo_2C nanoparticles during a CVD process. The morphology modified MoS_2/Mo_2C catalyst showed improved pH independent HER performance. The morphological engineering enhanced the interfacial mass transfer and the formation of surface oxygen on Mo_2C are the main reasons behind the improved HER performance of the catalyst (figures 2.4(a) and (b)) [79]. Vertical graphene (VG) is one of the attractive morphologically engineered graphene complex systems with similar properties of graphene. VG is primarily composed of graphene sheets of few-layer thickness grown vertically on the deposition substrate [80]. The morphology of VG has three different aspects; edges of few-layer graphene, free-standing graphene vertically aligned sheets holding carbon nanosheets and the basal plane of few-layer graphene. There are various vapor deposition methods such as plasma enhanced chemical vapor deposition (PECVD), microwave plasma enhanced chemical vapor deposition (MPCVD), inductively coupled plasma enhanced chemical vapor deposition (ICP-PECVD) etc [81–83]. Due to the unique morphological advantages, VG finds applications in various fields like field emission [84], energy storage and conversion [82], flexible electronics etc

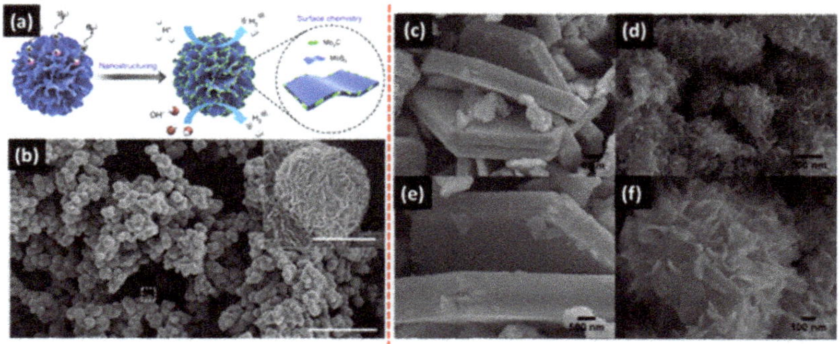

Figure 2.4. (a) Schematic design of MoS_2/Mo_2C formation. (b) SEM image of the synthesized MoS_2/Mo_2C (inset is an enlarged view of the dotted square which is an MoS_2/Mo_2C microsphere; the scale bar is 20 μm). Reproduced from [79] CC BY 4.0. FESEM images of pristine VSe_2 showing large hexagonal sheets like morphology. (c, e) and (d, f) Sulfur doping has transformed the morphology of hexagonal VSe_2 sheets into a garland-like structure. Reproduced from [90] with permission from the Royal Society of Chemistry.

[85]. Cao and group developed MoS_2 pyramid platelets through CVD for electrocatalytic application. The key feature in the formation of edge rich pyramid platelets of MoS_2 is the regulation of the pressure parameter (760 Torr) during the vapour deposition [86]. In another interesting work, hexagonal boron nitrate (h-BN) has been subjected to morphological modification using a simple saline coupling agent. The coupling agent used here is KH550 (3-aminopropyltriethoxysilane) in the presence of a needle-like BN precursor. With the increase of KH550 concentration, the dispersion of BN nanoparticles also improved along with an increment in the crystal growth at room temperature. Here the saline coupling agent acts as a dispersant to change the dispersion state of the BN particle. The study further regulated the morphology of the h-BN using temperature. When the temperature of the system reached 80 °C the BN transformed into hollow spherical spheres [87]. Apart from this, the thermal treatment of h-BN is also considered an effective method in morphological engineering. h-BN thermal treated at 950 °C for 3 h and sonicated for 18 h showed a clear morphological shift from the original. This change in morphology is attributed to the formation of a boron oxide phase in the sample. Thermal treatment is rather a common method to induce changes in nanostructures [88]. The morphological change contributed by the oxide formation might have an unpredictable impact in certain applications. Wu and co-workers created a controllable morphology of WS_2 in a WS_2/graphene hybrid using a one-step hydrothermal process. With the addition of graphene sheets the microporous nanowire spheres of WS_2 are transformed into a honeycomb structure. Here graphene plays a crucial role in transforming the morphology of WS_2 by acting as a growth template for the WS_2 to grow and self-assemble [89]. Similarly, dopants are also used to engineer the morphology of 2D materials. In one such work, sulfur dopants are employed to controllably modify the morphology of metallic vanadium diselenide (VSe_2). The sulfur dopant changed the hexagonal sheet-like morphology of VSe_2 into a garland-like structure. The change in morphology also impacts the supercapacitor performance of the sample. The doped sample showed a considerable enhancement in charge storage compared to VSe_2 hexagonal sheets (figures 2.4(c)–(f)) [90]. All the dopants cannot guarantee a morphological shift, for example, in a similar study cobalt was used as the dopant in VSe_2 but after doping there was no considerable change in the morphology of VSe_2 hexagonal sheets [91]. While speaking in terms of morphology, the ultrathin black phosphorous showed better photosensitization behaviour than its bulk phase. The ultrathin black phosphorous was prepared using liquid exfoliation of the bulk sample. After the exfoliation, the layer-stacking morphology of bulk black phosphorous is changed into ultrathin free-standing nanosheets [92]. The drastic change in morphology from bulk to ultrathin black phosphorous has a significant impact on the enhanced photosensitization performance.

2.2.6 Edge engineering

In 2D materials, when you reduce one dimension into atomic scale, unique material properties emerge. One of the cornerstone features of 2D materials is the formation of 1D edges as part of the discontinuity similar to the surface of bulk materials.

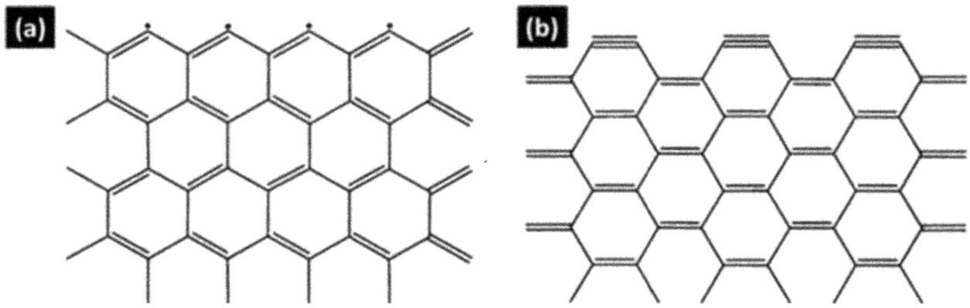

Figure 2.5. (a) Armchair edges and (b) zigzag edges of the graphene structure. Reproduced from [95], copyright (2010) American Chemical Society.

Controlling and tuning the edges of 2D materials gives rise to improved electronic, magnetic, optical and catalytical properties. Edges also play a crucial role in the growth and morphological development of 2D materials. Edges form during the chemical growth of graphene, exfoliation of graphite into graphene or mechanical and chemical approaches on graphene sheets. Therefore, understanding the role of edges and their engineering plays a significant contribution in the development of modern 2D materials-based technologies.

The edge sites in graphene arise when the aromaticity of the honeycomb lattice is broken. That is, the breaking of σ bonds between adjacent carbon atoms of the π conjugation in a graphene lattice creates an edge. Zigzag edge and armchair edge are the two types of edges observed in graphene depending on the lattice crystallographic orientation [93, 94] (figures 2.5(a) and (b)). Each of these edges shows specific electronic properties and chemical reactivity [95]. The broken σ bonds at the edges create radical groups with active and accessible electrons. The difference in conjugation systems in each of the edges creates the difference in their chemical activity [19]. Usually, an edge develops into complex alternated zigzag and armchair sequences without any particular crystallographic direction; this alternated segment is termed as chiral edge [96, 97]. Local defects such as dislocation or imperfection can also be considered as edges in graphene since these defects terminate the honeycomb conjugation of the lattice. In a vacuum, it is observed that the atoms on the edge are metastable σ and π dangling bonds. The dangling bonds formed during the edge formation are unstable and difficult to observe. Zigzag edges are very energetic and undergo planar reconstruction of the benzine ring to lower their energy. The edge atomic structure often affects the energy states of the edge atoms by determining their electronic distribution [98].

The properties of the graphene edges can be engineered by employing various methods. Chemical functionalization is one of the common tools to modify graphene edges. Chemical functionalization at the edges is achieved either by a covalent or non-covalent bond. The functionalization of graphene involves a regioselective binding at the π-conjugated network to attain a stoichiometric edge plane functionalization [98]. Therefore, edge selectivity is an important factor in achieving edge functionalization. Graphene nanoflakes prepared using simple acid oxidation of single-walled carbon nanotubes have modified edges with an attachment of carboxyl group. Here graphene nanoflakes mediate electron transfer with the redox active groups at the edges [99]. Edge reconstruction is another method to engineer the edge sites of graphene [100]. It is revealed that the zigzag edges of graphene nanoribbons reconstruct using two hexagons back-to-back into one pentagon due to the unstable dangling bonds resulting in armchair-like edges [101, 102]. Gómez-Navarro and co-workers showed an electrochemical approach to modify the edges of graphene using palladium nanoparticles. The electrodeposition enables the metal functionalization of graphene monolayers with a controllable density of palladium particles. It is shown that the controllable particle size of palladium is due to its nucleation at vacancies along the edge sites of monolayer graphene islands [103].

In the case of TMDs, which are composed of hexagonal unit cells, the hexagonal lattice is similar to that of graphene. Coincidently, the edge sites in TMDs are also termed as armchair and zigzag as in the case of graphene [104] (figures 2.6(a) and (b)). Theoretical predictions showed that in an ideal scenario armchair edge in TMDs is semiconducting while zigzag edges are metallic. The experimental validation of this theoretical prediction is difficult to achieve due to the complexity of preparing ideal edge preparation and characterization. Generally, it is been considered that all edges in TMDs are either armchair or zigzag or a linear superimposition of both [105]. MoS_2 is one of the well-known TMDs that has been developed in the last few decades. With its interesting properties, MoS_2 can be used in many applications including 2D transistors, lubricants, energy storage and HER catalysts [106]. The excellent HER performance of 2D MoS_2 is derived from its highly active edges while its basal planes are mostly inert [106]. To activate the edge sites of 2D MoS_2, researchers prepared MoS_2 on a high surface area carbon black surface. This edge-modified system showed excellent HER activity with high stability in acidic media. MoS_2 and its selenide counterpart $MoSe_2$ showed better catalytic activity in an acidic medium but their activity in an alkaline medium is

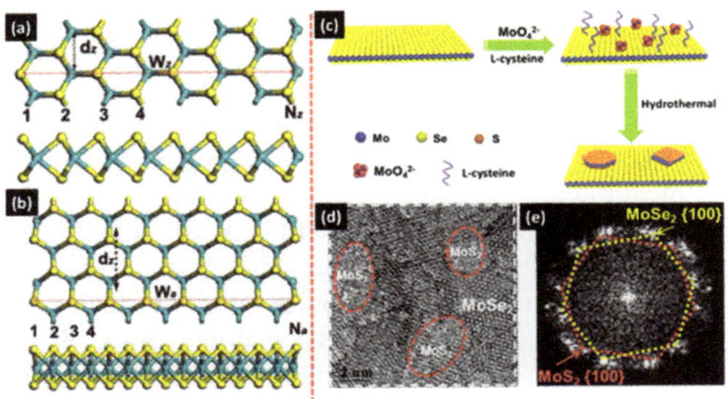

Figure 2.6. Top and side views of MoS_2 structures in (a) 8-ZMoS_2NR (zigzag edge) and (b) 15-AMoS2NR (armchair edge). Reproduced with permission from [104], copyright (2008) American Chemical Society. (c) Schematic of the synthesis of $MoS_2/MoSe_2$. (d) HRTEM image and (e) the corresponding fast fourier transform pattern of $MoS_2/MoSe_2$. Reproduced from [107] with permission from the Royal Society of Chemistry.

inferior. To obtain better HER activity in alkaline medium, the edges of both these materials are engineered by constructing a $MoS_2/MoSe_2$ heterostructure. In the heterostructure, ultrasmall MoS_2 nanoclusters were anchored over the $MoSe_2$ surface in order to prevent the agglomeration of MoS_2 which provides abundant edge sites, resulting in improved water adsorption/desorption capability [107] (figures 2.6(c)–(e)). Engineering the chalcogen edges in TMDs using transition metals is a strategy used by researchers. For example, the Mo edges in MoS_2 are predominantly active in chemical reactions such as HER where S edges remain inert. Transition metals like Fe, Co, Ni or Cu can be doped into these S edges and alter various properties of MoS_2 and making them more active in HER reactions. Doped MoS_2 nanofilms with modified S edges exhibit enhancement in exchange current during the electrocatalytic activity [108]. In a similar fashion, the HER kinetics of metallic VSe_2 has been modified using this edge-modified doping strategy. Co doped at the Se edges activated the basal planes promoted the electron transfer and enhanced the HER activity of the catalyst [109].

Recently, it has been revealed that the edge sites in MXenes grabbed attention due to their role in activities like HER, electrochemical N_2 reduction reaction (NRR) and so on. For example, single-layer Ti_3C_2 MXene has different active edges such as lateral Ti edge, middle Ti edge, C and O [110]. A DFT study conducted by Wang and co-workers showed that the edge site of the middle Ti atom in Ti_3C_2 MXene is more active in NRR due to its high adsorption energy. The NRR activity of Ti_3C_2 edge sites with exposed middle Ti atoms should be better than that of the basal planes having an O termination group [111]. The reduction of MXene sheet size can enhance the NRR activity by providing more exposure to the active edge sites [110]. On the other hand, the top C atomic edge sites in an MXene nanoribbon are saturated by H^* species in acid media, while the hollow site in between the Ti edges can contribute as reaction sites for HER [112]. The HER activity in MXene nanoribbons clearly depends on the composition of MXene and the ribbon width. It has been reported that the armchair nanoribbons of Ti_3C_2 exhibit more favorable HER activity than others [112].

2.2.7 Heterostructure

One of the most promising and widely explored engineering approaches of 2D materials is the construction of heterostructures using other materials. The inherent deficiencies of 2D materials can be ameliorated using heterostructure formation. Particularly, heterostructures of 2D materials with 0D, 1D and 2D materials can be an effective solution for the wide application of 2D materials. Heterostructures of 2D materials are highly investigated by the research community owing to their unique property alterations and improved activities in applications such as electrocatalysis, energy storage, electrochemical CO_2 reduction, photocatalysis and so on. In this section, we are exploring the formation and impact of heterostructures of 2D materials using 0D, 1D and 2D materials.

2.2.7.1 2D/0D heterostructures

0D materials are a novel class of materials possess many advantages in terms of large surface area, many active edges, multiple defect sites, fast electron movement and electron reservoir properties [113, 114]. These intriguing properties and their uniform structure make them an important candidate in many of the next-generation applications. Carbon quantum dots (CQDs) are considered as a promising candidate in the class of 0D materials. A heterostructure made up of 0D material and 2D material can provide unprecedented properties depending on the application. Jia *et al* constructed a CQD/graphene heterostructure using a simple *in situ* solvothermal treatment of exfoliated graphene sheets for electrocatalytic application. The prepared heterostructure showed high electrocatalytic activity and stability owing to the intimate contact between the CQD and the graphene which prevents aggregation and fastens the electron transportation. Also, the improved edge sites in the heterostructure enhance the active sites for the electrocatalytic performance [115]. Graphitic carbon nitride (g-C_3N_4) is a widely popular 2D material that shows excellent stability, nontoxic nature and cost effectiveness in mass production [116]. The photoelectrocatalytic performance of g-C_3N_4 can be improved by constructing a heterostructure with N-doped graphene quantum dot (NGQD). The deamination coupling between these two materials in the heterostructure creates a specific π (NGQD)-p (lone pair in ternary N)-π (g-C_3N_4) interaction. This uniquely constructed heterostructure effectively enhances the charge separation and visible light absorption for the photoelectrocatalytic reactions [117]. Similarly, NGQD is hybridized with graphene to prepare an electrocatalyst for oxygen reduction reaction (ORR). The hydrothermally prepared NGQD/graphene heterostructure combines the advantages of large surface area, rich active edges, high electronic conductivity and N-doped active sites to provide a high performance electrocatalytic activity with good electrochemical stability and resistance to methanol crossover [118].

2.2.7.2 2D/1D heterostructures

The research community's attraction behind 1D materials is mostly due to their unique properties such as high aspect ratio, numerous highly coordinated atomic sites, stability and excellent flexibility [119]. Compared to 0D materials, 1D and 2D materials show better electrocatalytic performance. The agglomeration, high surface free energy and small contact area with the substrate truncates the electrocatalytic performance of 0D materials [120]. The synergistic interaction between these two structures enhances the catalytic activity of the heterostructure [120]. The 2D/1D heterostructure possesses many advantages for electrocatalytic activities such as: (1) the growth of nano arrays on the substrate enhances the number of active sites; (2) the well aligned nano arrays on the substrate can provide well defined electronic pathways; (3) the optimal interspace between both the nanostructures elevates the diffusion of electrolytes and the release of gas bubbles [119]. In this regard, it is significant to explore the 2D/1D heterostructure formation for various electrocatalytic applications. In an attempt to create a 2D/1D heterostructure of N-doped graphene/CNT, researchers grew graphene nanosheets on the inner surface of CNTs. The N-graphene/CNT hybrid structure was prepared in a convenient experimental set-up in which a powder mixture of $FeCl_3$ and cyanamide was annealed at 900 °C in Ar gas. During the process, the Fe^{3+} reduces to Fe nanocrystals which assist the gases released from cyanamide decomposition into

N-graphene/CNT. The unique morphology and defect structure and improved surface area of this heterostructure contributed towards its high electrocatalytic performance [121]. Wang and co-workers explored a controllable synthesis approach of MoS_2/CNT heterostructure for HER using a simple hydrothermal method. An efficient interlayer expansion of MoS_2 sheets can be observed by varying the CNT loading. The synergistic effect between MoS_2 and CNT can be accountable for the improved HER activity of the hybrid structure [122]. In our recent work, we have explored the selective growth of MoSSe nanosheets on defect rich carbon nanotubes for electrocatalytic applications. The 2D/1D architecture of MoSSe/CNT exhibited improved HER performance over pristine MoSSe and CNT. The selective growth of MoSSe on CNT provides enhanced active sites and fast electronic pathways for electrocatalytic activity (figures 2.7(a)–(g)) [123].

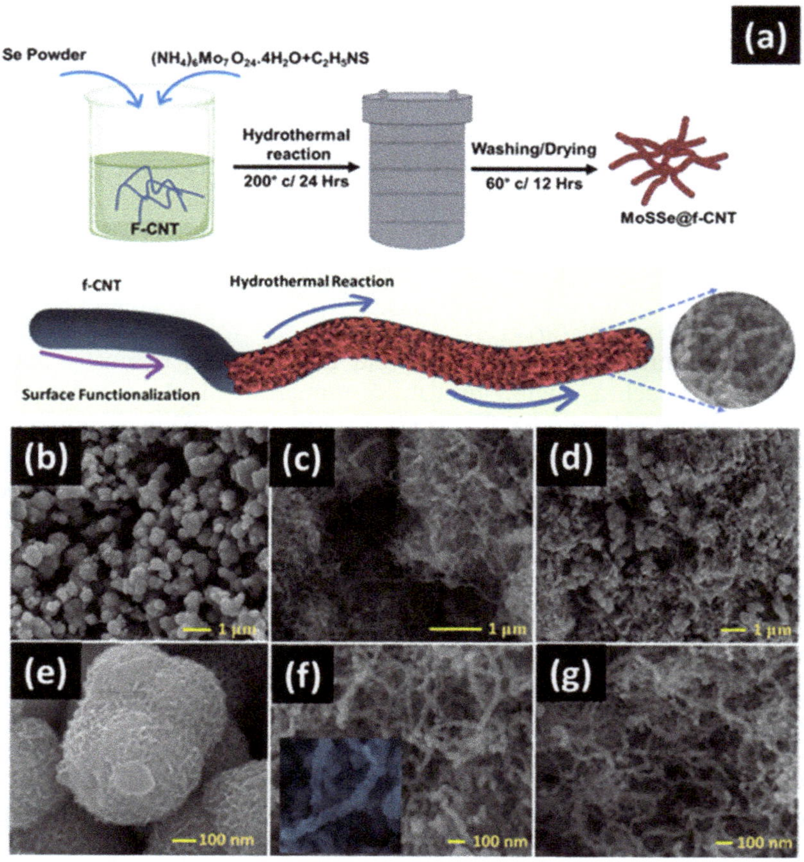

Figure 2.7. (a) Schematic representation of the synthesis of MoSSe/CNT heterostructure, FESEM images of (b, e) MoSSe, (c, f) MoSSe/CNT (low CNT concentration), (d, g) MoSSe/CNT (high concentration of CNT). Reproduced with permission from [123], copyright (2022) American Chemical Society.

2.2.4.3 2D/2D heterostructure

Similar to 2D/1D combination, 2D/2D heterostructures are also a widely appealing research area, especially for electrocatalytic applications. One of the key advantages of 2D/2D heterostructure is the presence of slit-shaped and nanospaced ionic diffusion channels which provide shorter ionic pathways in the heterostructure. Also, the interfacial properties and nondisplaced phase change during the electrochemical reactions provide excellent electrocatalytic activity and stability to 2D/2D heterostructures. Liu *et al* prepared ultrathin MoS_2 nanoplatelets inlaid on graphene sheets using a unique thermal synthesis route. The prepared 2D/2D structure showed a large surface area with a huge number of active sites with excellent collective properties of electron transportation and hydrogen ion trapping. This particular catalyst showed exceptional electrocatalytic activity towards HER. The strong connection between the MoS_2 nanoplatelets and graphene nanosheets provides structural and electrochemical stability to the hybrid structure (figures 2.8(a)–(c)) [124]. In another approach, a simple hydrothermal method was used to prepare a MoS_2/graphene heterostructure. During the hydrothermal process, an *in situ* growth of few-layer MoS_2 is taking place on the graphene sheet substrate. With the integrated sheet-on-sheet advantage this 2D/2D structure exhibited enhanced electrocatalytic activity [125]. An MXene/MoS_2 2D/2D heterostructure with precise design and orientation was prepared using the CVD method. This heterostructure exhibited a metallic nature in theoretical calculation. The unique 2D/2D structure provides enhanced active sites for catalytic activities which leads to excellent HER kinetics [126]. Seh and co-workers developed an interesting heterostructure of Mo_2CT_x MXene and 2H-MoS_2 to circumvent the oxidation of the MXene for efficient HER activity. The 2D/2D Mo_2CT_x/MoS_2 structure was formed by the *in situ* two-step sulfidation of Mo_2CT_x MXene. Strong epitaxial coupling between Mo_2CT_x and MoS_2 at the interface of the heterostructure can be achieved by avoiding the excessive covering that usually hinders the active interface and delivers

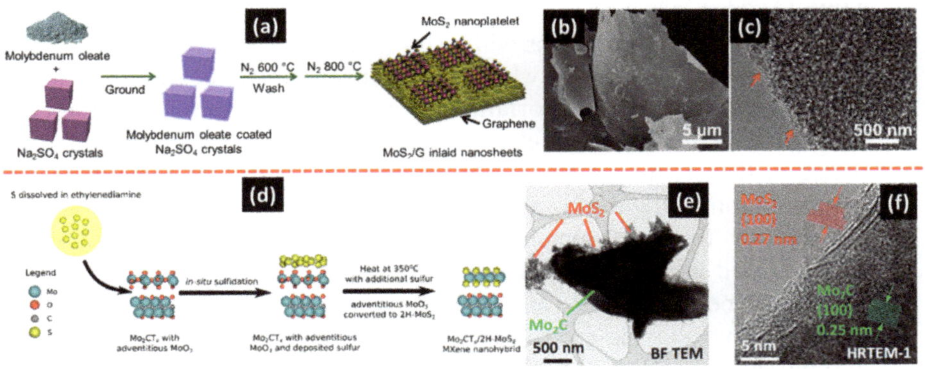

Figure 2.8. (a) Synthesis of MoS_2/graphene heterostructure, (b) FESEM and (c) TEM images of the MoS_2/graphene. Reproduced with permission from [124], copyright (2016) American Chemical Society; (d) schematic of the synthesis of Mo_2CT_x/MoS_2 heterostructure, (e) bright field TEM and (f) HRTEM images of Mo_2CT_x/MoS_2 heterostructure. Reproduced with permission from [127], copyright (2020) American Chemical Society.

superior HER performance (figure 2.8(d)–(f)) [127]. Work done by Shinde *et al* provides a new perspective on the formation of a 2D/2D heterostructure derived from semimetallic $MoTe_2$ and Ti_3C_2 MXene using a one-pot hydrothermal method. Petal clusters of $MoTe_2$ grown on MXene create a remarkable electrocatalyst with excellent durability [3].

2.3 Conclusion and future perspectives

In summary, in this chapter we mainly focused on the various engineering strategies for tuning the physicochemical properties 2D materials including defects, alloying, doping, stress/strain, morphology, edge and heterostructuring, etc. Four general categories of defects, namely point, line, planar, and volume defects were systematically introduced. Similarly, alloying of 2D materials by metal replacement and dichalcogenide replacements were discussed. The various doping processes to engineer 2D materials like charge transfer doping, intercalation doping and substitutional doping are discussed in detail in this chapter. Morphological modification as an engineering tool for 2D materials is attracting great attention for electrochemical reactions. The recent progress in vapour deposition techniques has accelerated this strategy. Morphology-engineered 2D materials with active edge sites are an effective candidate for various electrochemical activities. Edge engineering predominantly revolves around zigzag and armchair edges. The fundamental aspects and the properties of these edges are discussed above. Some of the contrasting physicochemical properties of these edges are attracting more attention from researchers. Heterostructuring is one of most well-established engineering strategies for all nanostructured materials. Nano heterostructures of 2D materials show abundant electrochemical properties. They provide better electrical and chemical conductivity with an excellent stability and high number of active sites due to the synergistic interaction.

2.3.1 Future perspectives

1. Diverse defect configurations will result in different electronic structures and charge distribution states, which will impact on applications. The defect concentration is quite important; a low defect concentration results in a noticeable improvement in electrochemical performance, whereas an abundance of defects may cause structural instability. In this context, real-time material monitoring during synthesis for controlled defect formation is made possible by the development of more delicate synthesis procedures and sophisticated *in situ* characterization tools.
2. Designing new 2D materials with novel structures or phases is greatly facilitated by alloying in 2D materials. It significantly alters the chemical and physical properties. In this situation, choosing the necessary 2D materials with the right structure or phase and the right physical and chemical qualities becomes a challenging issue. Theoretical analysis and machine learning can be used in this situation. There have been a few attempts to use machine learning for the synthesis of 2D or single-phase TMDs with a single component. It can

be used in the future to design alloy 2D materials with a particular composition and set of physical and chemical characteristics.

3. To preserve the crystalline integrity of host materials, more focus should be placed on explaining the dopant interaction with different 2D materials in the case of substitutional doping, charge transfer doping, and intercalation doping.
4. More sophisticated *in situ* operando characterization tools should be explored to understand the effect of stress and strain engineering effect on 2D materials.
5. With the development of sophisticated vapour deposition techniques, it is possible to create atomically thin 2D materials with desired morphologies depending on the type of electrochemical application. With these advancements in synthesis technologies, fabrication of diverse, multifunctional 2D nanostructures and their hybrids using one pot in vapour phase is awaiting exploration.
6. In the case of heterostructure formation, attaining the full potential of the heterostructure is an issue needing to be looked into. Implementation of new or modified synthesis strategies for the formation of high-quality 2D materials-based heterostructure is a necessary step in the realm of 2D materials. A lack of comprehensive understanding in electrochemical activity in the heterostructure also hinders their possible commercialization.
7. With the advancement in edge engineering realization nanoribbons of various 2D materials for multiscale application is likely. These materials with modified edges specially tuned for desired applications will be a future goal for researchers. The atomic characterization of edges of 2D materials after modification is one of the crucial challenges that needed to overcome.

Acknowledgments

The authors gratefully acknowledge financial assistance from the SERB Core Research Grant (Grant No. CRG/2022/000897), Department of Science and Technology (DST/NM/NT/2019/205(G)), and Minor Research Project Grant, Jain University (JU/MRP/CNMS/29/2023).

References

[1] Lemme M C, Akinwande D, Huyghebaert C and Stampfer C 2022 2D materials for future heterogeneous electronics *Nat. Commun.* **13** 1392

[2] Raj KA S, Mane P, Radhakrishnan S, Chakraborty B and Rout C S 2022 Heterostructured metallic 1T-$VSe_2/Ti_3C_2T_x$ MXene nanosheets for energy storage *ACS Appl. Nano Mater.* **5** 4423–36

[3] Shinde P V, Mane P, Late D J, Chakraborty B and Rout C S 2021 Promising 2D/2D $MoTe_2/Ti_3C_2T_x$ hybrid materials for boosted hydrogen evolution reaction *ACS Appl. Energy Mater.* **4** 11886–97

[4] Ru Fan F, Wang R, Zhang H and Wu W 2021 Emerging beyond-graphene elemental 2D materials for energy and catalysis applications *Chem. Soc. Rev.* **50** 10983–1031

[5] Pang Y, Yang Z, Yang Y and Ren T-L 2020 Wearable electronics based on 2D materials for human physiological information detection *Small* **16** 1901124

[6] Huh W, Lee D and Lee C-H 2020 Memristors based on 2D materials as an artificial synapse for neuromorphic electronics *Adv. Mater.* **32** 2002092

[7] Molaei M J, Younas M and Rezakazemi M 2021 A comprehensive review on recent advances in two-dimensional (2D) hexagonal boron nitride *ACS Appl. Electron. Mater.* **3** 5165–87

[8] Nasrin K, Sudharshan V, Subramani K and Sathish M 2022 Insights into 2D/2D MXene heterostructures for improved synergy in structure toward next-generation supercapacitors: a review *Adv. Funct. Mater.* **32** 2110267

[9] Naumis G G 2020 Electronic properties of two-dimensional materials *Synthesis, Modeling, and Characterization of 2D Materials, and Their Heterostructures Micro and Nano Technologies* ed E-H Yang, D Datta, J Ding and G Hader (Amsterdam: Elsevier) pp 77–109 ch 5

[10] Jiang H, Zheng L, Liu Z and Wang X 2020 Two-dimensional materials: from mechanical properties to flexible mechanical sensors *InfoMat* **2** 1077–94

[11] Ma Q, Ren G, Xu K and Ou J Z 2021 Tunable optical properties of 2D materials and their applications *Adv. Opt. Mater.* **9** 2001313

[12] Zhang S, Xu R, Luo N and Zou X 2021 Two-dimensional magnetic materials: structures, properties and external controls *Nanoscale* **13** 1398–424

[13] Chakraborty S K, Kundu B, Nayak B, Dash S P and Sahoo P K 2022 Challenges and opportunities in 2D heterostructures for electronic and optoelectronic devices *iScience* **25** 103942

[14] Liang Q, Zhang Q, Zhao X, Liu M and Wee A T S 2021 Defect engineering of two-dimensional transition-metal dichalcogenides: applications, challenges, and opportunities *ACS Nano.* **15** 2165–81

[15] Wang T, Park M, Yu Q, Zhang J and Yang Y 2020 Stability and synthesis of 2D metals and alloys: a review *Mater. Today Adv.* **8** 100092

[16] Zhu H, Gan X, McCreary A, Lv R, Lin Z and Terrones M 2020 Heteroatom doping of two-dimensional materials: from graphene to chalcogenides *Nano Today* **30** 100829

[17] Yang S, Chen Y and Jiang C 2021 Strain engineering of two-dimensional materials: methods, properties, and applications *InfoMat* **3** 397–420

[18] Chowdhury T, Sadler E C and Kempa T J 2020 Progress and prospects in transition-metal dichalcogenide research beyond 2D *Chem. Rev.* **120** 12563–91

[19] Acik M and Chabal Y J 2011 Nature of graphene edges: a review *Jpn. J. Appl. Phys.* **50** 070101

[20] Jiang J, Xu T, Lu J, Sun L and Ni Z 2019 Defect engineering in 2D materials: precise manipulation and improved functionalities *Research* **2019** 4641739

[21] Komsa H-P, Kotakoski J, Kurasch S, Lehtinen O, Kaiser U and Krasheninnikov A V 2012 Two-dimensional transition metal dichalcogenides under electron irradiation: defect production and doping *Phys. Rev. Lett.* **109** 035503

[22] Dolui K, Rungger I, Das Pemmaraju C and Sanvito S 2013 Possible doping strategies for MoS_2 monolayers: an *ab initio* study *Phys. Rev.* B **88** 075420

[23] Yao J and Yang G 2022 2D layered material alloys: synthesis and application in electronic and optoelectronic devices *Adv. Sci.* **9** 2103036

[24] Singh A K, Kumbhakar P, Krishnamoorthy A, Nakano A, Sadasivuni K K, Vashishta P, Roy A K, Kochat V and Tiwary C S 2021 Review of strategies toward the development of alloy two-dimensional (2D) transition metal dichalcogenides *iScience* **24** 103532

[25] Banhart F, Kotakoski J and Krasheninnikov A V 2011 Structural defects in graphene *ACS Nano.* **5** 26–41

[26] Tang T, Wang Z and Guan J 2022 A review of defect engineering in two-dimensional materials for electrocatalytic hydrogen evolution reaction *Chin. J. Catal.* **43** 636–78

[27] Bertoldo F, Ali S, Manti S and Thygesen K S 2022 Quantum point defects in 2D materials—the QPOD database *NPJ Comput. Mater.* **8** 16

[28] Qin H, Sorkin V, Pei Q-X, Liu Y and Zhang Y-W 2020 Failure in two-dimensional materials: defect sensitivity and failure criteria *J. Appl. Mech.* **87**

[29] Williams D B and Carter C B 1996 Planar defects *Transmission Electron Microscopy: A Textbook for Materials Science* ed D B Williams and C B Carter (Boston, MA: Springer) pp 379–99

[30] Liu F and Fan Z 2023 Defect engineering of two-dimensional materials for advanced energy conversion and storage *Chem. Soc. Rev.* **52** 1723–72

[31] Meggiolaro D, Mosconi E and De Angelis F 2019 Formation of surface defects dominates ion migration in lead-halide perovskites *ACS Energy Lett.* **4** 779–85

[32] Vala M D, Bhatt M, Kansara S and Sonvane Y 2023 Schottky and Frenkel defect on SbS_2 monolayer: first principles calculations *J. Phys.: Conf. Ser.* **2518** 012011

[33] Xie L *et al* 2020 Effect of pore structure and doping species on charge storage mechanisms in porous carbon-based supercapacitors *Mater. Chem. Front.* **4** 2610–34

[34] Upadhyay S N, Satrughna J A K and Pakhira S 2021 Recent advancements of two-dimensional transition metal dichalcogenides and their applications in electrocatalysis and energy storage *Emergent Mater.* **4** 951–70

[35] Wang J, Liu J, Luo J, Liang P, Chao D, Lai L, Lin J and Shen Z 2015 MoS_2 architectures supported on graphene foam/carbon nanotube hybrid films: highly integrated frameworks with ideal contact for superior lithium storage *J. Mater. Chem.* A **3** 17534–43

[36] Zhang Y, Zhang Y, Zhang H, Bai L, Hao L, Ma T and Huang H 2021 Defect engineering in metal sulfides for energy conversion and storage *Coord. Chem. Rev.* **448** 214147

[37] Li X, Qian T, Zai J, He K, Feng Z and Qian X 2018 Co stabilized metallic 1Td MoS_2 monolayers: bottom-up synthesis and enhanced capacitance with ultra-long cycling stability *Mater. Today Energy* **7** 10–7

[38] Peng W, Luo M, Xu X, Jiang K, Peng M, Chen D, Chan T-S and Tan Y 2020 Spontaneous atomic ruthenium doping in Mo_2CTX MXene defects enhances electrocatalytic activity for the nitrogen reduction reaction *Adv. Energy Mater.* **10** 2001364

[39] Klein B P *et al* 2022 Topological Stone–Wales defects enhance bonding and electronic coupling at the graphene/metal interface *ACS Nano.* **16** 11979–87

[40] Khossossi N, Singh D, Ainane A and Ahuja R 2020 Recent progress of defect chemistry on 2D materials for advanced battery anodes *Chem.—Asian J.* **15** 3390–404

[41] Li X, Hu Q, Wang H, Chen M, Hao X, Ma Y, Liu J, Tang K, Abudula A and Guan G 2021 Charge induced crystal distortion and morphology remodeling: formation of Mn-CoP nanowire @ Mn-CoOOH nanosheet electrocatalyst with rich edge dislocation defects *Appl. Catal.* B **292** 120172

[42] Han Y, Li M-Y, Jung G-S, Marsalis M A, Qin Z, Buehler M J, Li L-J and Muller D A 2018 Sub-nanometre channels embedded in two-dimensional materials *Nat. Mater.* **17** 129–33

[43] Bertoldo F *et al* 2021 Intrinsic defects in MoS_2 grown by pulsed laser deposition: from monolayers to bilayers *ACS Nano.* **15** 2858–68

[44] Ran L, Hou J, Cao S, Li Z, Zhang Y, Wu Y, Zhang B, Zhai P and Sun L 2020 Defect engineering of photocatalysts for solar energy conversion *Sol. RRL* **4** 1900487

[45] Artaud A *et al* 2020 Depressions by stacking faults in nanorippled graphene on metals *2D Mater.* **7** 025016

[46] Namsheer K, Thomas S, Sharma A, Thomas S A, See Raj K A, Kumar V, Gagliardi A, Aravind A and Rout C S 2022 Rational design of selenium inserted 1T/2H mixed-phase molybdenum disulfide for bifunctional energy storage and pollutant degradation applications *Nanotechnology*

[47] Cui Y, Zhou Z, Li T, Wang K, Li J and Wei Z 2019 Versatile crystal structures and (opto) electronic applications of the 2D metal mono-, di-, and tri-chalcogenide nanosheets *Adv. Funct. Mater.* **29** 1900040

[48] Ren P *et al* 2014 Band-selective infrared photodetectors with complete-composition-range $InAs_xP_{1-x}$ alloy nanowires *Adv. Mater.* **26** 7444–9

[49] Rogalski A 2005 HgCdTe infrared detector material: history, status and outlook *Rep. Prog. Phys.* **68** 2267

[50] Hou K, Huang Z, Liu S, Liao G, Qiao H, Li H and Qi X 2020 A hydrothermally synthesized $MoS_{2(1-x)}Se_{2x}$ alloy with deep-shallow level conversion for enhanced performance of photodetectors *Nanoscale Adv.* **2** 2185–91

[51] Wang S *et al* 2020 Phase-dependent band gap engineering in alloys of metal-semiconductor transition metal dichalcogenides *Adv. Funct. Mater.* **30** 2004912

[52] Wang Z *et al* 2020 2H/1T′ phase $WS_{2(1-x)}Te_{2x}$ alloys grown by chemical vapor deposition with tunable band structures *Appl. Surf. Sci.* **504** 144371

[53] Patra A, Shaikh M, Ghosh S, Late D J and Rout C S 2022 $MoWS_2$ nanosheets incorporated nanocarbons for high-energy-density pseudocapacitive negatrode material and hydrogen evolution reaction *Sustain. Energy Fuels* **6** 2941–54

[54] Binwal D C, Kaur M, Pramoda K and Rao C N R 2020 HER activity of nanosheets of 2D solid solutions of $MoSe_2$ with MoS_2 and $MoTe_2$ *Bull. Mater. Sci.* **43** 313

[55] Mukherjee S, Bhattacharya D, Ray S K and Pal A N 2022 High-performance broad-band photodetection based on graphene–$MoS_{2x}Se_{2(1-x)}$ alloy engineered phototransistors *ACS Appl. Mater. Interfaces* **14** 34875–83

[56] Umrao S, Jeon J, Jeon S M, Choi Y J and Lee S 2017 A homogeneous atomic layer $MoS_{2(1-x)}Se_{2x}$ alloy prepared by low-pressure chemical vapor deposition, and its properties *Nanoscale* **9** 594–603

[57] Fu Q, Yang L, Wang W, Han A, Huang J, Du P, Fan Z, Zhang J and Xiang B 2015 Synthesis and enhanced electrochemical catalytic performance of monolayer $WS_{2(1-x)}Se_{2x}$ with a tunable band gap *Adv. Mater.* **27** 4732–8

[58] Chang R-J, Sheng Y, Ryu G H, Mkhize N, Chen T, Lu Y, Chen J, Lee J K, Bhaskaran H and Warner J H 2019 Postgrowth substitutional tin doping of 2D WS_2 crystals using chemical vapor deposition *ACS Appl. Mater. Interfaces* **11** 24279–88

[59] Ghosh N C and Harimkar S P 2012 Consolidation and synthesis of MAX phases by spark plasma sintering (SPS): a review *Advances in Science and Technology of $M_{n+1}AX_n$ Phases* ed I M Low (Woodhead Publishing) ch 3 pp 47–80

[60] Nechiche M, Gauthier-Brunet V, Mauchamp V, Joulain A, Cabioc'h T, Milhet X, Chartier P and Dubois S 2017 Synthesis and characterization of a new $(Ti_{1-\varepsilon},Cu_{\varepsilon})_3(Al,Cu)C_2$ MAX phase solid solution *J. Eur. Ceram. Soc.* **37** 459–66

[61] Luo P, Zhuge F, Zhang Q, Chen Y, Lv L, Huang Y, Li H and Zhai T 2018 Doping engineering and functionalization of two-dimensional metal chalcogenides *Nanoscale Horiz.* **4** 26–51

[62] Chen M, Nam H, Wi S, Priessnitz G, Gunawan I M and Liang X 2014 Multibit data storage states formed in plasma-treated MoS_2 transistors *ACS Nano.* **8** 4023–32

[63] Kim Y, Jhon Y I, Park J, Kim C, Lee S and Jhon Y M 2016 Plasma functionalization for cyclic transition between neutral and charged excitons in monolayer MoS_2 *Sci. Rep.* **6** 21405

[64] Lukatskaya M R, Mashtalir O, Ren C E, Dall'Agnese Y, Rozier P, Taberna P L, Naguib M, Simon P, Barsoum M W and Gogotsi Y 2013 Cation intercalation and high volumetric capacitance of two-dimensional titanium carbide *Sci.* **341** 1502–5

[65] Wan J, Lacey S D, Dai J, Bao W, Fuhrer M S and Hu L 2016 Tuning two-dimensional nanomaterials by intercalation: materials, properties and applications *Chem. Soc. Rev.* **45** 6742–65

[66] Acerce M, Voiry D and Chhowalla M 2015 Metallic 1T phase MoS_2 nanosheets as supercapacitor electrode materials *Nat. Nanotech.* **10** 313–8

[67] Fan X, Xu P, Zhou D, Sun Y, Li Y C, Nguyen M A T, Terrones M and Mallouk T E 2015 Fast and efficient preparation of exfoliated 2H MoS_2 nanosheets by sonication-assisted lithium intercalation and infrared laser-induced 1T to 2H phase reversion *Nano Lett.* **15** 5956–60

[68] Bat-Erdene M, Batmunkh M, Sainbileg B, Hayashi M, Bati A S R, Qin J, Zhao H, Zhong Y L and Shapter J G 2021 Highly dispersed Ru nanoparticles on boron-doped $Ti_3C_2T_x$ (MXene) nanosheets for synergistic enhancement of electrocatalytic hydrogen evolution *Small* **17** 2102218

[69] Barakat F, Laref A, AlSalhi M S and Faraji S 2020 The impact of anion elements on the engineering of the electronic and optical characteristics of the two dimensional monolayer janus MoSSe for nanoelectronic device applications *Res. Phys.* **18** 103284

[70] Loh L, Zhang Z, Bosman M and Eda G 2021 Substitutional doping in 2D transition metal dichalcogenides *Nano Res.* **14** 1668–81

[71] Suh J *et al* 2014 Doping against the native propensity of MoS_2: degenerate hole doping by cation substitution *Nano Lett.* **14** 6976–82

[72] Dai Z, Liu L and Zhang Z 2019 Strain engineering of 2D materials: issues and opportunities at the interface *Adv. Mater.* **31** 1805417

[73] Lee J H, Jang W S, Han S W and Baik H K 2014 Efficient hydrogen evolution by mechanically strained MoS_2 nanosheets *Langmuir* **30** 9866–73

[74] Yang Y, Zhang K, Lin H, Li X, Chan H C, Yang L and Gao Q 2017 MoS_2–Ni_3S_2 heteronanorods as efficient and stable bifunctional electrocatalysts for overall water splitting *ACS Catal.* **7** 2357–66

[75] Bhimanapati G R *et al* 2015 Recent advances in two-dimensional materials beyond graphene *ACS Nano.* **9** 11509–39

[76] Vaarkamp M, Miller J T, Modica F S and Koningsberger D C 1996 On the relation between particle morphology, structure of the metal-support interface, and catalytic properties of Pt/γ-Al_2O_3 *J. Catal.* **163** 294–305

[77] Ebrahimi S and Yarmand B 2019 Morphology engineering and growth mechanism of ZnS nanostructures synthesized by solvothermal process *J. Nanopart. Res.* **21** 264
[78] Kibsgaard J, Chen Z, Reinecke B N and Jaramillo T F 2012 Engineering the surface structure of MoS_2 to preferentially expose active edge sites for electrocatalysis *Nat. Mater.* **11** 963–9
[79] Luo Y, Tang L, Khan U, Yu Q, Cheng H-M, Zou X and Liu B 2019 Morphology and surface chemistry engineering toward pH-universal catalysts for hydrogen evolution at high current density *Nat. Commun.* **10** 269
[80] Zheng W, Zhao X and Fu W 2021 Review of vertical graphene and its applications *ACS Appl. Mater. Interfaces* **13** 9561–79
[81] Bo Z, Mao S, Han Z J, Cen K, Chen J and Ostrikov K K 2015 Emerging energy and environmental applications of vertically-oriented graphenes *Chem. Soc. Rev.* **44** 2108–21
[82] K N, Polaki S R and Rout C S 2022 Molybdenum sulfo-selenides grown on surface engineered vertically aligned graphitic petal arrays for solid-state supercapacitors *J. Energy Storage* **52** 105007
[83] Shinde P V, Samal R and Rout C S 1959 Vertically aligned graphene-analogous low-dimensional materials: a review on emerging trends, recent developments, and future perspectives *Adv. Mater. Interfaces* **9** 2101959
[84] Jiang L, Yang T, Liu F, Dong J, Yao Z, Shen C, Deng S, Xu N, Liu Y and Gao H-J 2013 Controlled synthesis of large-scale, uniform, vertically standing graphene for high-performance field emitters *Adv. Mater.* **25** 250–5
[85] Deng C *et al* 2019 Ultrasensitive and highly stretchable multifunctional strain sensors with timbre-recognition ability based on vertical graphene *Adv. Funct. Mater.* **29** 1907151
[86] Yu Y, Huang S-Y, Li Y, Steinmann S N, Yang W and Cao L 2014 Layer-dependent electrocatalysis of MoS_2 for hydrogen evolution *Nano Lett.* **14** 553–8
[87] Ma X, Lee N-H, Oh H-J, Jung S-C, Lee W-J and Kim S-J 2011 Morphology control of hexagonal boron nitride by a silane coupling agent *J. Cryst. Growth* **316** 185–90
[88] Garro Mena L and Hohn K L 2021 Modification of hexagonal boron nitride by thermal treatment *J. Mater. Sci.* **56** 7298–307
[89] Song Y, Liao J, Chen C, Yang J, Chen J, Gong F, Wang S, Xu Z and Wu M 2019 Controllable morphologies and electrochemical performances of self-assembled nano-honeycomb WS_2 anodes modified by graphene doping for lithium and sodium ion batteries *Carbon* **142** 697–706
[90] Sree Raj K A, Pramoda K and Rout C S 2023 Assembling a high-performance asymmetric supercapacitor based on pseudocapacitive S-doped VSe_2/CNT hybrid and 2D borocarbonitride nanosheets *J. Mater. Chem.* C **11** 2565–73
[91] Sree Raj K A, Adhikari S, Radhakrishnan S, Johari P and Rout C S 2022 Effect of cobalt doping on the enhanced energy storage performance of 2D vanadium diselenide: experimental and theoretical investigations *Nanotechnology* **33** 295703
[92] Wang H, Yang X, Shao W, Chen S, Xie J, Zhang X, Wang J and Xie Y 2015 Ultrathin black phosphorus nanosheets for efficient singlet oxygen generation *J. Am. Chem. Soc.* **137** 11376–82
[93] Son Y-W, Cohen M L and Louie S G 2006 Energy gaps in graphene nanoribbons *Phys. Rev. Lett.* **97** 216803

[94] Aurbach D, Markovsky B, Weissman I, Levi E and Ein-Eli Y 1999 On the correlation between surface chemistry and performance of graphite negative electrodes for Li ion batteries *Electrochim. Acta* **45** 67–86

[95] Xu K and Ye P D 2010 Theoretical study of atomic layer deposition reaction mechanism and kinetics for aluminum oxide formation at graphene nanoribbon open edges *J. Phys. Chem.* C **114** 10505–11

[96] Jiang D, Sumpter B G and Dai S 2007 Unique chemical reactivity of a graphene nanoribbon's zigzag edge *J. Chem. Phys.* **126** 134701

[97] Sheka E F and Chernozatonskii L A 2010 Chemical reactivity and magnetism of graphene *Int. J. Quantum Chem.* **110** 1938–46

[98] Bellunato A, Arjmandi Tash H, Cesa Y and Schneider G F 2016 Chemistry at the edge of graphene *Chem. Phys. Chem.* **17** 785–801

[99] Avlasevich Y, Li C and Müllen K 2010 Synthesis and applications of core-enlarged perylene dyes *J. Mater. Chem.* **20** 3814–26

[100] Wassmann T, Seitsonen A P, Saitta A M, Lazzeri M and Mauri F 2008 Structure, stability, edge states, and aromaticity of graphene ribbons *Phys. Rev. Lett.* **101** 096402

[101] Koskinen P, Malola S and Häkkinen H 2008 Self-passivating edge reconstructions of graphene *Phys. Rev. Lett.* **101** 115502

[102] Huang B, Liu M, Su N, Wu J, Duan W, Gu B and Liu F 2009 Quantum manifestations of graphene edge stress and edge instability: a first-principles study *Phys. Rev. Lett.* **102** 166404

[103] Sundaram R S, Gómez-Navarro C, Balasubramanian K, Burghard M and Kern K 2008 Electrochemical modification of graphene *Adv. Mater.* **20** 3050–3

[104] Li Y, Zhou Z, Zhang S and Chen Z 2008 MoS_2 nanoribbons: high stability and unusual electronic and magnetic properties *J. Am. Chem. Soc.* **130** 16739–44

[105] Davelou D, Kopidakis G, Kaxiras E and Remediakis I N 2017 Nanoribbon edges of transition-metal dichalcogenides: stability and electronic properties *Phys. Rev.* **B 96** 165436

[106] Benck J D, Hellstern T R, Kibsgaard J, Chakthranont P and Jaramillo T F 2014 Catalyzing the hydrogen evolution reaction (HER) with molybdenum sulfide nanomaterials *ACS Catal.* **4** 3957–71

[107] Zhou Q, Zhao G, Rui K, Chen Y, Xu X, Dou S X and Sun W 2019 Engineering additional edge sites on molybdenum dichalcogenides toward accelerated alkaline hydrogen evolution kinetics *Nanoscale* **11** 717–24

[108] Wang H, Tsai C, Kong D, Chan K, Abild-Pedersen F, Nørskov J K and Cui Y 2015 Transition-metal doped edge sites in vertically aligned MoS_2 catalysts for enhanced hydrogen evolution *Nano Res.* **8** 566–75

[109] Zhu Q, Shao M, Yu S H, Wang X, Tang Z, Chen B, Cheng H, Lu Z, Chua D and Pan H 2019 One-pot synthesis of Co-doped VSe_2 nanosheets for enhanced hydrogen evolution reaction *ACS Appl. Energy Mater.* **2** 644–53

[110] Huang L, Gu X and Zheng G 2019 Tuning active sites of MXene for efficient electrocatalytic N_2 fixation *Chem.* **5** 15–7

[111] Luo Y, Chen G-F, Ding L, Chen X, Ding L-X and Wang H 2019 Efficient electrocatalytic N_2 fixation with MXene under ambient conditions *Joule* **3** 279–89

[112] Yang X, Gao N, Zhou S and Zhao J 2018 MXene nanoribbons as electrocatalysts for the hydrogen evolution reaction with fast kinetics *Phys. Chem. Chem. Phys.* **20** 19390–7

[113] Lim S Y, Shen W and Gao Z 2014 Carbon quantum dots and their applications *Chem. Soc. Rev.* **44** 362–81

[114] Li Y, Zhao Y, Cheng H, Hu Y, Shi G, Dai L and Qu L 2012 Nitrogen-doped graphene quantum dots with oxygen-rich functional groups *J. Am. Chem. Soc.* **134** 15–8

[115] Zhao M, Zhang J, Xiao H, Hu T, Jia J and Wu H 2019 Facile *in situ* synthesis of a carbon quantum dot/graphene heterostructure as an efficient metal-free electrocatalyst for overall water splitting *Chem. Commun.* **55** 1635–8

[116] Zhao Z, Sun Y and Dong F 2014 Graphitic carbon nitride based nanocomposites: a review *Nanoscale* **7** 15–37

[117] Mou Z, Lu C, Yu K, Wu H, Zhang H, Sun J, Zhu M and Cynthia Goh M 2019 Chemical interaction in nitrogen-doped graphene quantum dots/graphitic carbon nitride heterostructures with enhanced photocatalytic H_2 evolution *Energy Technol.* **7** 1800589

[118] Fan M, Zhu C, Yang J and Sun D 2016 Facile self-assembly N-doped graphene quantum dots/graphene for oxygen reduction reaction *Electrochim. Acta* **216** 102–9

[119] Wang P, Jia T and Wang B 2020 A critical review: 1D/2D nanostructured self-supported electrodes for electrochemical water splitting *J. Power Sources* **474** 228621

[120] Meng W, He H, Yang L, Jiang Q, Yuliarto B, Yamauchi Y, Xu X and Huang H 2022 1D–2D hybridization: nanoarchitectonics for grain boundary-rich platinum nanowires coupled with MXene nanosheets as efficient methanol oxidation electrocatalysts *Chem. Eng. J.* **450** 137932

[121] Wen Z, Ci S, Hou Y and Chen J 2014 Facile one-pot, one-step synthesis of a carbon nanoarchitecture for an advanced multifunctonal electrocatalyst *Angew. Chem. Int. Ed.* **53** 6496–500

[122] Xu W, Dong X, Wang Y, Zheng N, Zheng B, Lin Q and Zhao Y 2020 Controllable synthesis of MoS_2/carbon nanotube hybrids with enlarged interlayer spacings for efficient electrocatalytic hydrogen evolution *ChemistrySelect* **5** 13603–8

[123] Kuniyil N, Koottumvathukkal Anil Raj S R and Rout C S 2022 Selective growth of molybdenum sulfo-selenides on a defect-rich carbon nanotube skeleton for efficient energy storage and hydrogen generation applications *Energy Fuels* **36** 13346–55

[124] Ma L *et al* 2016 *In situ* thermal synthesis of inlaid ultrathin MoS_2/graphene nanosheets as electrocatalysts for the hydrogen evolution reaction *Chem. Mater.* **28** 5733–42

[125] Zhang X, Zhang Q, Sun Y, Zhang P, Gao X, Zhang W and Guo J 2016 MoS_2-graphene hybrid nanosheets constructed 3D architectures with improved electrochemical performance for lithium-ion batteries and hydrogen evolution *Electrochim. Acta* **189** 224–30

[126] Zhang R, Sun Y, Jiao F, Li L, Geng D and Hu W 2023 MXene-MoS_2 nanocomposites via chemical vapor deposition with enhanced electrocatalytic activity for hydrogen evolution *Nano Res.* **16** 8937–44

[127] Lim K R G, Handoko A D, Johnson L R, Meng X, Lin M, Subramanian G S, Anasori B, Gogotsi Y, Vojvodic A and Seh Z W 2020 2H-MoS_2 on Mo_2CT_x MXene nanohybrid for efficient and durable electrocatalytic hydrogen evolution *ACS Nano.* **14** 16140–55

Chapter 3

Engineered 2D materials for hydrogen evolution reaction (HER)

Pratik V Shinde, Komal N Patil, Vitthal M Shinde and Chandra Sekhar Rout

As one of the cleanest renewable resources, hydrogen fuel should be among the best options to replace fossil fuels in the production of future energy. One effective method for producing efficient, environmentally friendly, and long-lasting hydrogen is electrochemical water splitting. Herein, the performance of engineered 2D materials, on electrocatalytic HER is systematically discussed. The most recent developments and mechanistic understandings of HER activity are presented first. Then, the impact of several engineered techniques, including heterostructure formation, edge engineering, morphological tuning, doping, alloying, strain, defect, and doping, on electrocatalytic activity is briefly discussed. In conclusion, the main issues that this rapidly growing field is currently facing are outlined, along with potential solutions.

3.1 Introduction

The need for alternative, sustainable, and clean energy sources is a result of the current and impending energy crisis as well as concerns about climate change. It has become a prior need to find alternative energy resources that can substitute or limit the use of conventional non-renewable energy supplies and fulfill the need for energy. Due to its higher gravimetric energy density (140 MJ kg^{-1}), hydrogen (H_2) is regarded as an ideal energy source [1]. Typically, coal gasification and steam reforming technologies are used to produce large-scale H_2. Their main concern is the emission of CO_2, which makes them an unsuitable choice for green hydrogen generation. Additionally, steam reforming necessitates the use of methane and other low-molecular-weight hydrocarbons, both of which are short in supply [2]. This implies that current methods are not an appropriate answer for all of our future hydrogen needs. Therefore, the generation of hydrogen through the splitting of water molecules is a fascinating way toward a carbon-neutral future [3, 4].

doi:10.1088/978-0-7503-5719-7ch3

© IOP Publishing Ltd 2024

However, the large overpotential and unfavorable kinetics of the hydrogen evolution reaction (HER) prevent it from being widely used in applications. There is a need for an electrocatalyst to achieve higher efficiency in HER.

To date, platinum (Pt) is regarded as the most electroactive HER catalyst due to its low binding energy for the reaction [5]. As a result, the reaction in its presence proceeds at low overpotential values close to zero. The widespread implementation of platinum in water electrolyzers is restricted due to a low natural abundance and high cost. There is a need for an alternative catalyst that is less expensive but still effective for large-scale technology. Two-dimensional (2D) materials have drawn a lot of attention as alternatives to the scarce and expensive platinum-based catalysts because of their inherent activity, tunable electronic properties, cost-effectiveness, and abundance reserves [3, 6, 7]. The family of 2D materials has grown from the first discovered material, graphene, to several classes, including transition metal dichalcogenides (TMDs), perovskites, MXenes, Xenes, metal oxides, and so on [8, 9]. There is a lot of room for improvement in the production of non-precious electrocatalysts for water splitting. By implementing defects, alloying, doping, strain and stress engineering, morphology tuning, edge engineering, and heterostructure with an optimistic perspective, 2D materials can be improved for effective HER [10–16].

Herein, the 2D materials-based electrocatalysts for HER are thoroughly represented to explain the underlying science so that their structural changes can improve their catalytic performance. Investigation of recent studies in structure engineering, including defects, alloying, doping, strain and stress engineering, morphology tuning, edge engineering, and heterostructure fabrication is deeply discussed. These studies reveal the most cutting-edge approaches for making catalysts based on 2D materials that are both affordable and highly effective. The perspectives, difficulties, and upcoming study directions of HER electrocatalysis are also provided to assist in the ongoing research for 2D materials electrocatalysts.

3.2 HER-mechanism and insights

Theoretically, the splitting of water molecules takes place at a thermodynamic voltage of 1.23 V, which corresponds to energy of 237.2 kJ mol^{-1}, at 25 °C and 1 atm [17]. However, due to complex electron transfer, this process actually needs more voltage than 1.23 V. This additional potential or overpotential (η), is brought on by unfavorable factors like activation energy, bubble or electrode resistance, ion and gas diffusion, and electrolyte diffusion blockage [18]. With suitable catalysts, the overpotential could be significantly reduced while the reaction speed and efficiency can be increased.

The hydrogen evolution reaction is one of the most investigated electrocatalytic processes. The reaction takes place at the cathodic side in the electrochemical water-splitting reaction. HER can be performed in acidic or alkaline solutions, but it is also investigated in intermediate pH and buffered solutions. Under various pH conditions, the HER produces molecular hydrogen from water using various pathways [19].

The reactions that take place in acidic media are shown below:
At the anode:

$$2H_2O_{(l)} \rightarrow O_{2(g)} + 4H^+ + 4e^- \quad E^\circ = 1.23 \text{ V versus SHE} \tag{3.1}$$

At the cathode:

$$4H^+ + 4e^- \rightarrow 2H_{2(g)}\ E^\circ = 0 \text{ V versus SHE} \tag{3.2}$$

Overall reaction:

$$2H_2O_{(l)} \rightarrow 2H_{2(g)} + O_2(g)\ \Delta E^\circ = 1.23 \text{ V} \tag{3.3}$$

$$\Delta G^\circ = +\ 237.2 \text{ kJ mol}^{-1}$$

The reactions that take place in alkaline media are shown below:
At anode:

$$4OH^- \rightarrow O_{2(g)} + 2H_2O + 4e^-\ E^\circ = 0.40 \text{ V versus SHE} \tag{3.4}$$

At the cathode:

$$4H_2O + 4e^- \rightarrow 2H_{2(g)} + 4\ OH^-\ E^\circ = -0.83 \text{ V versus SHE} \tag{3.5}$$

Overall reaction:

$$2H_2O_{(l)} \rightarrow 2H_{2(g)} + O_{2(g)}\ \Delta E^\circ = 1.23 \text{ V} \tag{3.6}$$

$$\Delta G^\circ = +\ 237.2 \text{ kJ mol}^{-1}$$

For water-splitting reactions, regardless of the electrolyte medium, the ΔG° is positive (endothermic reaction). There are two steps in a possible reaction pathway for the HER [20].

In the acidic medium, the first step is discharging protons on the catalyst surface to form adsorbed hydrogen (H_{ads}). This step is referred to as the Volmer reaction and is represented as:

$$H^+_{(aq)} + e^- \rightarrow H_{ads} \tag{3.7}$$

The Tafel slope for this reaction is shown as

$$b_{1,V} = \frac{2.303RT}{\beta F}. \tag{3.8}$$

where R is the ideal gas constant, β is the symmetry factor (equal to 0.5), T is the absolute temperature, and F is the Faraday constant.

Based on the coverage of H_{ads}, the second step is the evolution of the H_2 molecular process which is either through the recombination of two adsorbed protons (Tafel reaction) or a second proton/electron transfer (Heyrovsky reaction).

If the coverage of H_{ads} is sufficient, then adjacent H_{ads} join together and the evolution of the H_2 molecule takes place. The step is called the Tafel reaction and its Tafel slope is represented as follows:

$$H_{ads} + H_{ads} \rightarrow H_{2(g)} \tag{3.9}$$

$$b_{2,T} = \frac{2.303RT}{2F}. \tag{3.10}$$

Adsorbed hydrogen atoms will preferably combine with a proton and an electron simultaneously to evolve a molecule of H_2 in the case of low H_{ads} coverage. This is the Heyrovsky reaction, its equation, and the Tafel slope expressed as follows:

$$H_{ads} + H^{+}_{(aq)} + e^{-} \rightarrow H_{2(g)} \tag{3.11}$$

$$b_{2,H} = \frac{2.303RT}{(1 + \beta)F} \tag{3.12}$$

The HER could occur via either the Volmer–Heyrovsky or the Volmer–Tafel mechanism and the Volmer step is inevitable in both steps (figure 3.1). The Tafel slope values for Volmer, Tafel, and Heyrovsky are 118, 29, and 39 mV/dec, respectively [20]. The Tafel slope of 29 mV/dec, is referred to as the Tafel-Heyrovsky mechanism, and the rate-limiting step would be the electrochemical desorption step. In practical water electrolysis, the smaller Tafel slope is thought to be advantageous because a strongly enhanced HER rate can be attained with a moderate increase in applied potential.

In an alkaline medium, adsorbed hydrogen (H_{ads}) is primarily formed by the discharge of water. This step is termed the Volmer reaction:

$$H_2O + e^{-} \rightarrow H_{ads} + OH^{-}_{(aq)} \tag{3.13}$$

This step is followed by the Heyrovsky reaction as follows:

$$H_{ads} + H_2O + e^{-} \rightarrow H_{2(g)} + OH^{-}_{(aq)} \tag{3.14}$$

Or Tafel's reaction is as follows:

$$H_{ads} + H_{ads} \rightarrow H_{2(g)} \tag{3.15}$$

Whenever the pH is alkaline, there is almost no chance of finding a free proton in the solution, so the Volmer step must always rely on a water dissociation reaction [2]. The Volmer step affects the overall reaction rate because it demands a separate water dissociation step. In an alkaline medium, anodic OER provides a proton to the cathode via the deprotonation of hydroxide ions. Since the alkaline solution contains plenty of OH^{-} ions, the newly formed protons rejoin them and affect

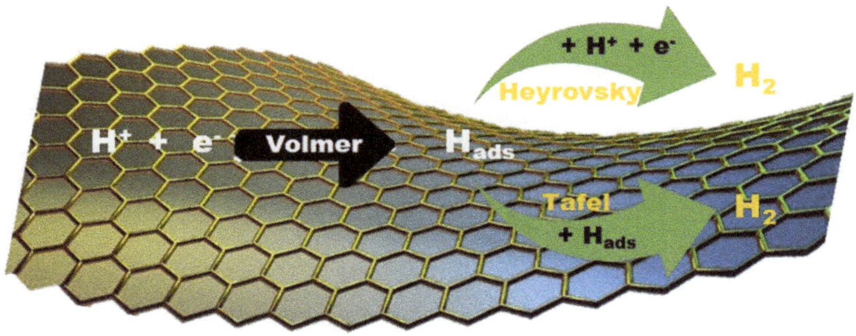

Figure 3.1. The schematic of mechanism paths for hydrogen evolution activity.

HER's ability to move further [20]. Based on these findings from the mechanism, alkaline HER is likely to require more energy than acidic HER.

An ideal electrocatalyst is one that has an appropriate balance of adsorption and desorption energies. Adsorption that is too strong results in a Heyrovsky/Tafel reaction, whereas adsorption that is too weak results in a Volmer reaction as the rate-determining step. As a result, the hydrogen adsorption energy (ΔG_H) of the electrocatalyst for HER should be close to zero [21]. The optimized free energy of hydrogen adsorption (ΔG_H) is a requirement for an ideal HER catalyst [22, 23]. A catalyst's excellent HER performance in an alkaline solution should be dependent on the low activation energies for water dissociation (ΔG_B) [24].

3.3 Different engineered approaches to enhance HER performance

A hydrogen-producing electrocatalyst must have effective catalytic properties such as feasible surface structure, superior electronic conductivity, and plenty of active sites. When considering an ideal electrocatalyst, the material should possess characteristics like high activity, facile upscaling, low overpotential, and good stability. As a result, researchers opted to engineer 2D materials in a controlled manner in order to enhance electrocatalytic HER performance.

3.3.1 Defect

2D catalysts typically do not have catalytic activity on their thermodynamically stable basal surface. Therefore, in order to control the electrochemical activity of 2D materials, defect engineering is crucial. In addition to tuning electronic structure, defect sites also increase the ratio of functional active centres for increased HER activity [25]. Jia and their group reported 2D graphene material possessing carbon defects (DG) for HER [26]. In an acidic medium, the current density of 10 mA cm^{-2} was obtained for −0.42 V for graphene, −0.35 V for N-doped graphene, and −0.15 V for DG. Zuo *et al* prepared $PdTe_2$-based catalysts with three different types of vacancies (*d*-$PdTe_x$), including single Pd, Te defect site, and double Te defect sites [10]. The dashed lines in figure 3.2(a) transmission electron microscopy (TEM) image highlights the porous positions in *d*-$PdTe_x$. The porous structure that can be

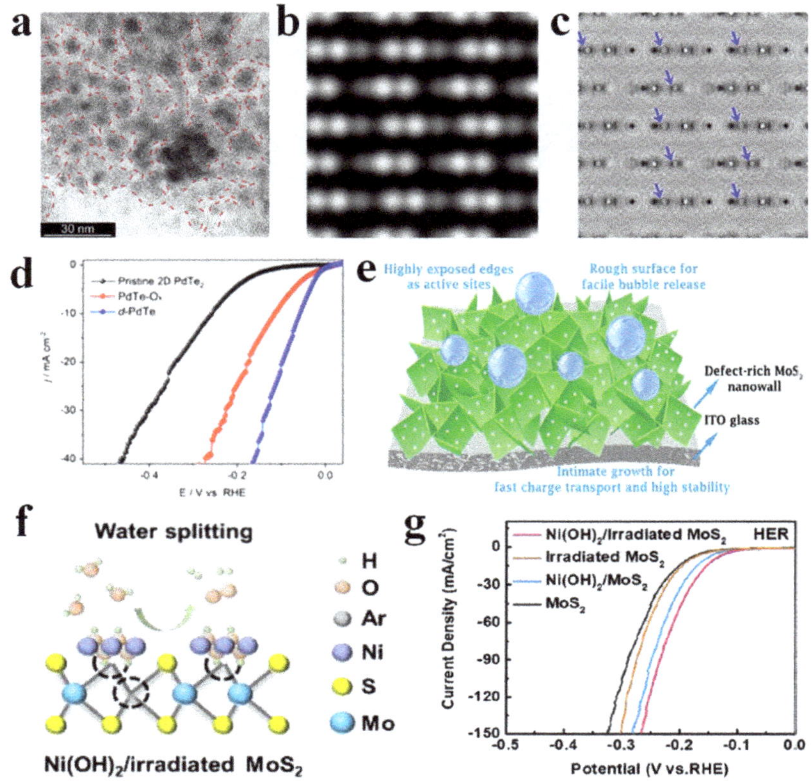

Figure 3.2. (a) The HRTEM image at the atomic scale in the panel emphasizes the porous positions of d-PdTe$_x$. (b and c) The simulated STEM and exit-wave images of d-PdTe$_x$ full of site defects. Arrows in the colour blue indicate the Te defects. (d) LSV polarization curves of the three catalysts in 0.5 M H_2SO_4 solution at a scan rate of 5 mV s^{-1}. Reprinted with permission from [10]. Copyright (2023) by the American Chemical Society. (e) Schematic representation of the defect-rich MoS_2 nanowall for HER. The abundant defects are labeled by grey dots. Reprinted from [28]. Copyright (2017) with permission from Springer Nature. (f) Defect-mediated metal ion adsorption for the formation of $Ni(OH)_2/MoS_2$ heterostructure. (g) Linear sweep voltammetry polarization curves for HER. Reprinted with permission from [29]. Copyright (2020) by the American Chemical Society.

accessed in three dimensions might make it easier to maximize the use of the defect sites and enhance the material's intrinsic activity. The simulated STEM and exit-wave images of the *d*-PdTe$_x$ in figures 3.2(b) and (c) show full of site defects and blue arrows pointed to the existence of Te defect sites. The *d*-PdTe$_x$ shows a current density of 10 mA cm^{-2} with a 76 mV overpotential, which is lower than the original $PdTe_2$ (259 mV). The performance of the HER is enhanced by local electron delocalization of unsaturated Pd sites close to the defects, which optimizes the adsorption of the Gibbs free energy of intermediate hydrogen. Li *et al* designed MoS_2 nano mesh with evenly distributed holes and flaws prepared by a combination of ball-milling and ultrasonic techniques [27]. The MoS_2 nano mesh exhibits a current density of 10 mA cm^{-2} with an overpotential of 160 mV and a Tafel slope of

46 mV/dec. The monolayer, holey, and defective structure of the catalyst, increased the number of active edge sites, improved conductivity, and made hydrogen adsorption ability stronger.

The defect-rich MoS_2 nanowall with an optimized thickness presented a low onset overpotential of 85 mV and a high current density of 310.6 mA cm^{-2} at an overpotential of 300 mV [28]. To achieve a 10 mA cm^{-2} cathodic current, a low potential of 95 mV is required. As shown in figure 3.2(e), the vertically aligned defect-rich MoS_2 nanosheets expose more active edge sites and allow facile ion penetration. The nanowall's extremely rough surface gives it the ability to easily release gas bubbles, which ensures that empty active sites will later be exposed for continuous hydrogen evolution. The heterostructure $Ni(OH)_2/MoS_2$ prepared by defect-mediated metal ion adsorption delivered the lowest overpotential of 139 mV at the current density of 10 mA cm^{-2} and Tafel slope of 45 mV/dec, shown in figures 3.2(f) and (g) [29]. The high barrier for H–OH adsorption on defect-free MoS_2 limits hydrogen generation, preventing the formation of H_{ads} intermediates. The introduction of defects in the MoS_2 basal plane efficiently lowers the Gibbs free energy for adsorption of hydrogen. In MoS_2, the Fermi level's surrounding gap states enable hydrogen to bind directly to exposed atoms of Mo, which causes defects like S-vacancies to create new catalytic sites in the basal plane [30]. In order to create vacancies and become interstitials by settling in nearby areas, a part of the Mo atoms in MoS_2 spontaneously leave their positions in the lattice and form the Frenkel-defected monolayer MoS_2 catalyst (FD-MoS_2), as shown in figure 3.3(a) [31]. To reach the current density of 10 mA cm^{-2}, FD-MoS_2 requires a lower

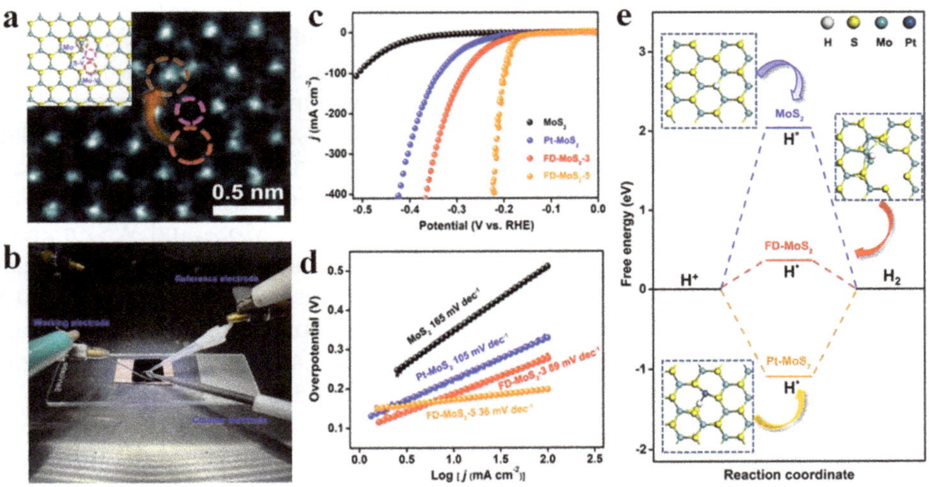

Figure 3.3. (a) Atomic-resolution HAADF-STEM images and in inset equivalent atomic models of FD-MoS_2. (b) Optical microscopy image of microelectrochemical devices. Here, Ag/AgCl is the reference electrode, graphite carbon is the counter electrode. (c) Polarization curves in 0.5 M H_2SO_4. (d) Corresponding Tafel plots. (e) Free energy diagrams of HER. Reprinted from [31]. Copyright (2022) with permission from Springer Nature.

overpotential of 164 mV as compared to pristine monolayer MoS_2 (358mV) and Pt-single-atom doped MoS_2 (211 mV) (figures 3.3(b)–(d)). Theoretical research demonstrates that the presence of Frenkel defects in MoS_2 alters the H adsorption site and delivers a moderate amount of H* adsorption energy for FD-MoS_2 (figure 3.3(e)).

In acidic media, the S-defect-rich Re/ReS_2/carbon cloth shows an overpotential of 42 mV at 10 mA cm^{-2} with a Tafel slope of 36 mV/dec [32]. While, in alkaline conditions, the electrocatalyst reaches a current density of 10 mA cm^{-2} at an overpotential of 44 mV and a Tafel slope of 53 mV/dec. Sulfur defects optimize the adsorption-free energy of hydrogen at active sites and improve charge transfer kinetics. Structural defects like Se-vacancy, Se_2-vacancy, MoSe antisite point defect, and Se_{Se2} adatom are favoured as catalytic centres for hydrogen generation [33]. The $MoSe_2$ nanosheets required 300 mV overpotential to produce a current of 10 mA cm^{-2} and exhibited a Tafel slop of 90 mV/dec. SnS_2 with S vacancy concentrations (Vs-SnS_2) display an onset potential of 141 mV and a Tafel slope of 74 mV/dec [34]. Without S atoms, surface charge modulation is triggered, which improves electronic conductivity. The charge density redistribution and lattice distortion caused by the under-coordinated Sn atoms next to the S vacancy are advantageous for hydrogen binding in HER.

3.3.2 Alloying

An alloy is a type of material prepared by two or more types of metal or a metal and nonmetal. Alloying has the potential to generate significant defect sites such as phase boundaries and substitutional doping sites [35–38]. Such effect of alloying is beneficial for performance improvement in HER activity. The electric construction of TMDs can be tailored with a modulated bandgap and defect states within the gap by adjusting the alloy composition [39, 40]. As shown in figure 3.4(a), for hydrogen evolution, Gan *et al* prepared pyrolytic carbon film with $Mo_xW_{1-x}S_2$ nanoflakes embedded as free-standing flexible electrodes [41]. The $Mo_{0.37}W_{0.63}S_2$/C electrocatalyst shows an overpotential of 0.137 V at current densities of 10 mA cm^{-2} and a Tafel slope of 53 mV/dec (figure 3.4(b)). The alloying of MoS_2 and WS_2 materials activates the 'inert' basal plane and improves the electrocatalytic HER performance of $Mo_xW_{1-x}S_2$. Furthermore, the curved $Mo_xW_{1-x}S_2$ structure induces strain at the surface layer and boosts catalytic activity. Kwak *et al* prepared $Re_{1-x}Mo_xSe_2$ alloy nanosheets with the range of x from 0 to 100% (figure 3.4(c)) [11]. With an increment in x, the transition of phase takes place from the order of 1T″ (triclinic) → 1T′ (monoclinic) → 2H (hexagonal). The 10% Mo substitution sample shows an overpotential of 77 mV for 10 mA cm^{-2} and a Tafel slope of 42 mV/dec, in 0.5 M H_2SO_4 solution (figure 3.4(d)). The overpotential is 107 mV and 188 mV for x is equal to 0 and 100%, respectively. Due to the addition of 1T′ phase $MoSe_2$, the alloy samples are more metallic than the semiconducting $ReSe_2$. At $x = 0\%$, 12.5%, 50%, and 100%, the Gibbs free energy values are 1.74, 0.76, 0.84, and 1.88 eV, respectively. As a result, the alloy compositions show a noticeably lower activation barrier for HER. When WSe_2 and VSe_2 are combined, a 2H-to-1T phase transition

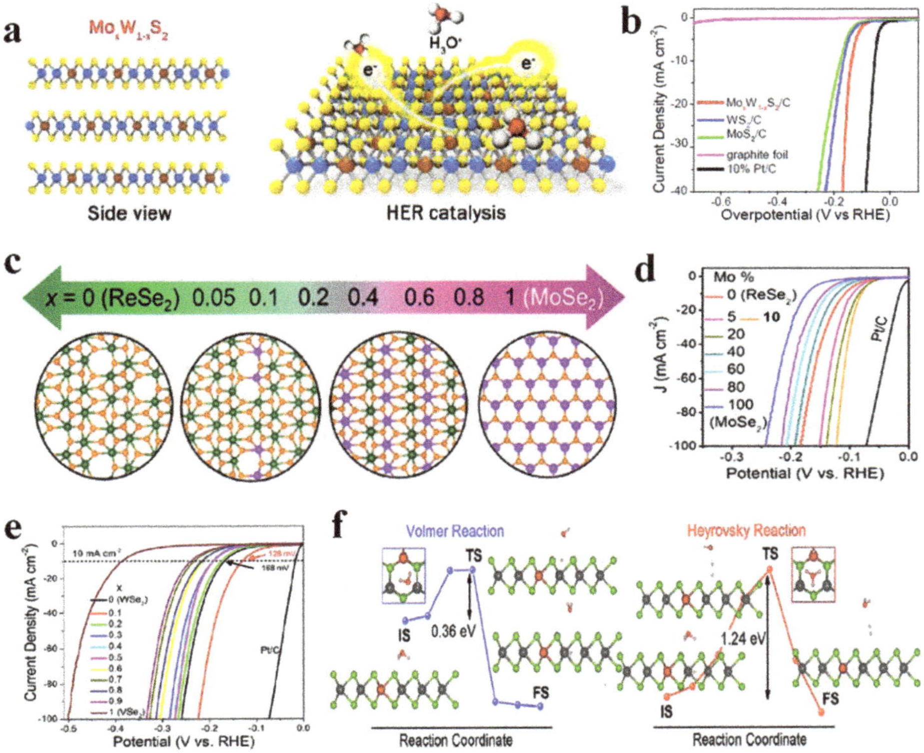

Figure 3.4. (a) Schematic of $Mo_xW_{1-x}S_2$ alloy and its function as HER electrocatalyst. The S, Mo, W, H, and O atoms are indicated here by the yellow, blue, orange, white, and red balls, respectively. (b) LSV polarization curves at a scan rate of 5 mV s^{-1} in 0.5 M H_2SO_4 solution. Reprinted from [41], Copyright (2018), with permission from Elsevier. (c) Schematic diagram for the ratio between $NaReO_4$ and Na_2MoO_4 varied to produce samples with $x = 0\%$, 5%, 10%, 20%, 40%, 60%, 80%, and 100%. (d) LSV curves at scan rate of 2 mV s^{-1} for $Re_{1-x}Mo_xSe_2$ in 0.5 M H_2SO_4. Reprinted with permission from [11]. Copyright (2020) by the American Chemical Society. (e) LSV curves at scan rate of 2 mV s^{-1} for $W_{1-x}V_xSe_2$ in 0.5 M H_2SO_4. (f) Activation barrier of the Volmer reaction and Heyrovsky reaction in the C model at $x = 0.11$. The initial (IS), transitional (TS), and final (FS) state structures are displayed. In each reaction step, the TS is shown from the top. The O atom of the H_2O molecule is represented by a small red ball. Reprinted with permission from [42]. Copyright (2022) by the American Chemical Society.

occurs at $x = 0.7$, which changes the material from semiconducting to metallic [42]. The lowest value of overpotential for 10 mA cm^{-2} is 128 mV, which is observed for the $x = 0.1$ sample (figure 3.4(e)). The Volmer reaction and Heyrovski reaction's activation barriers are 0.36 and 1.24 eV, respectively, for the heteroatom sites (2W, V). The enhanced HER performance of the alloy phase was supported by the Gibbs free energy of H adsorption on the basal planes (Se or hole sites) and the activation barriers along the Volmer–Heyrovsky reaction pathway (figure 3.4(f)).

Du *et al* attached a series of Ni/Co alloys on Nb-doped $Ti_3C_2T_x$ MXene nanohybrids and achieve the current density of 10 mA cm^{-2} with the overpotential of 43.4 mV in 1.0 M KOH solution [43]. The interaction between Pt nanoclusters and Nb_2CT_x MXene substrates, forming Pt_3Nb alloy, which shows an overpotential of 5 mV and 46 mV at 10 mA cm^{-2} and 100 mA cm^{-2}, respectively [44]. The improved metal–support interaction as well as the evenly distributed Pt_3Nb alloy phase with small size and greater intrinsic HER activity can be credited for the superior performance. High-entropy alloys are composed of five or more elements rather than the traditional multicomponent alloys of one or two elements [45]. In addition to optimizing the kinetic barrier for the adsorption–desorption of reaction intermediates, the continuously tunable adsorption energies of these alloys also boost the catalytic performance [46, 47]. With the advantage of layered morphology and high-entropy, $Co_{0.6}(VMnNiZn)_{0.4}PS_3$ nanosheet alloys exhibit an overpotential of 65.9 mV at the current density of 10 mA cm^{-2} and a Tafel slope of 65.5 mV/dec [48]. The alkaline HER performance is improved by the high-entropy strategy by increasing the richness of active sites and ideal adsorption for reaction intermediates. In 1.0 M KOH electrolyte, high entropy CoNiCuMgZn alloy embedded onto ultrathin 2D holey graphene, reach the current density of 10 mA cm^{-2} with the overpotential of 158 mV, showing a Tafel slope value of 36.1 mV/dec, and exhibit electrochemical active surface area of 19.72 mF cm^{-2} [45]. While in an acidic medium (0.5 M H_2SO_4) electrocatalyst displays an overpotential of 131 mV for 10 mA cm^{-2} and a Tafel slope of 32.6 mV/dec. Theoretical calculations showed that the combination of alloys and extremely thin graphene nanosheets produced a value for hydrogen adsorption Gibbs free energy that was very close to zero and enhanced the catalytic kinetics. A maximum molar configurational entropy in high entropy alloys contributes to the minimization of Gibbs free energy, allowing them to be chemically stable in severe electrochemical environments [49–52].

3.3.3 Doping

The water absorption and reduction processes of the active part in the water-splitting process control the hydrogen evolution activity. As a result, investigating the factors influencing active site adjustment and developing catalysts with rich abundance and relatively simple active sites is extremely important. 2D materials are thought to be the best platforms for providing active sites and catalysis relationships because of their planar and large surface area. The electrical structure of materials can be tuned by doping them with exotic atoms, which enhances the catalytic activity of those materials [53, 54]. To present, one of the most efficient strategies to further enhance the HER performance of TMDs is metal doping modification [55–59]. As shown in figures 3.5(a)–(c), the Co-doped MoS_2 material achieved overpotentials of 67 and 155 mV at 10 mA cm^{-2} in 1.0 M KOH and 0.5 M H_2SO_4, respectively [55]. Theoretical calculations suggest that the edged Co site reveals Gibbs free energy for hydrogen adsorption value of −0.04 eV probably because of the advantageous charge transfer between Co and adjacent S/Mo sites. This value is lower than the calculated value of the Pt site at its (111) plane. Cui-Hua *et al* synthesized Pt–Te co-doped MoS_2 nanosheets, which exhibit an overpotential of 52 mV and 79 mV to reach 10 mA cm^{-2} current density in 1.0 M KOH and 0.5 M H_2SO_4, respectively [56]. The synergistic interaction between Pt, Te, and MoS_2

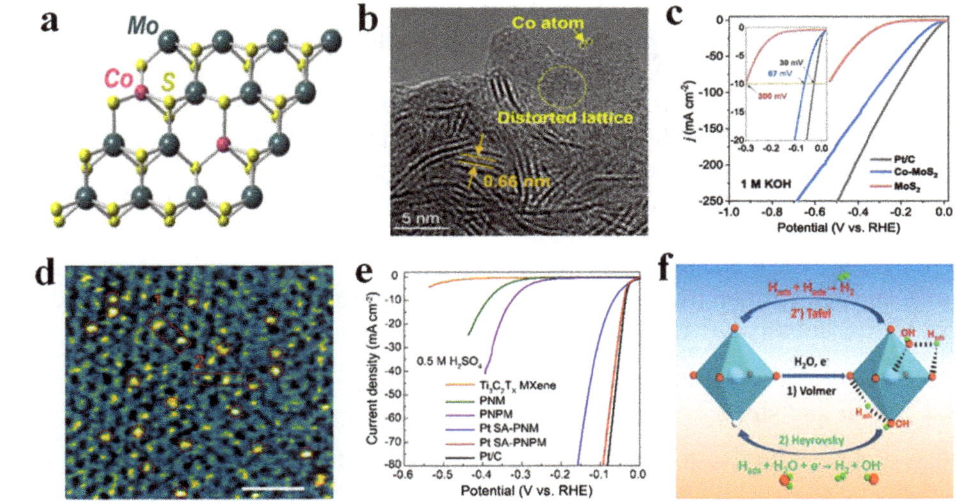

Figure 3.5. (a) Schematic of the proposed structure of Co(S_4)–MoS_2. (b) HRTEM image of Co–MoS_2. (c) LSV curves in 1.0 M KOH. Reproduced from [55], Copyright (2022), with permission from Elsevier. (d) The magnified HAADF-STEM image of Pt SA-PNPM, display noticeable isolated Pt atoms (red circles). (e) LSV curves in 0.5 M H_2SO_4 electrolyte. Reprinted with permission from [12]. Copyright (2023) by the American Chemical Society. (f) Possible HER mechanism for the SFMON-450 in alkaline media. Reprinted from [62], Copyright (2021), with permission from Elsevier.

increases the intrinsic HER activity. In order to expose more active sites and increase the material's conductivity, Sun *et al* doped P in WS_2, which resulted in the formation of a 1 T-phase enriched WS_2P catalyst [57]. The optimized electrocatalyst surpasses the performance of 2H WS_2 and shows an overpotential of 125 mV to reach the current density of 10 mA cm^{-2}, a small Tafel slope of 73.73 mV/dec, and great stability under acidic conditions.

Peng *et al* prepared a series of single-metal sites (Pt, Ir, Ru, Pd, and Au) on 3D porous N, P co-doped $Ti_3C_2T_x$ MXene (signified as M1 SA-PNPM, M1 = Pt, Ir, Ru, Pd, and Au) by an electrostatic gelation process and high-temperature pyrolysis [12]. The 3D porous structure exposes the active surface sites and facilitates charge–mass transfer for HER. As a result, over a wide pH range, the Pt SA-PNPM catalyst exhibits activity that is about 20 times more than that of the conventional Pt/C catalyst. The S-doped multilayer niobium carbide ($Nb_4C_3T_x$) electrocatalyst displays an overpotential of 118 mV@10 mA cm^{-2}, a Tafel slope of 104 mV/dec, and stable performance as long as 24 h in 1.0 M KOH solution [60]. The doping of S prevents the restacking of multilayer $Nb_4C_3T_x$. Therefore, the catalyst's porous structure and increased interlayer spacing are favorable for charge transfer. Lin *et al* prepared a catalyst for HER in which Ir single atoms are confined in porous heteroatom (N, S) co-doped $Ti_3C_2T_x$ [61]. Through a well-coordinated Ir–N and Ir–S interaction, the N, S co-doped $Ti_3C_2T_x$ support can capture electrons from single Ir atoms, causing charge redistribution in the interfacial region of the $Ti_3C_2T_x$ support. The optimized catalyst shows an overpotential of 57.7 mV at a current density of 10 mA cm^{-2} and a Tafel slope of 25 mV/dec. The N-doped $Sr_2Fe1.5Mo_{0.5}O_{6-\delta}$ (SFMON) perovskite shows an overpotential of 251 mV at the current density of 10 mA cm^{-2}, a Tafel slope of 138 mV/dec, and long stability of 40 h [62]. This superior performance is attributed to the addition of a proper amount of N^{3-} into the perovskite lattice increased the content of oxygen vacancy as well as high-valence Fe species. The high-valence B-site transition metal centre in the BO_6 octahedron can be preferentially adsorbed with the generated OH^- from water reduction in the Volmer process (figure 3.5(f)). Hydrogen atoms are more likely to anchor on the neighboring O site because of the electrostatic affinity. The presence of oxygen vacancies on the catalyst surface promotes the adsorption and breakdown of water for subsequent H_2 evolution.

3.3.4 Strain and stress

Because 2D materials are highly flexible and strain sensitive, it is possible to alter their physicochemical and optoelectronic properties by applying strain and stress engineering techniques [63–65]. The deformation of a solid given by stress is known as strain. Elastic and reversible deformation occurs when stress is removed, while a plastic strain is irreversible. The nanocone-like moiré superlattices (MSLs) WS_2 reveal an overpotential of 60 mV at a current density of 10 mA cm^{-2} and a Tafel slope of 40 mV/dec [13]. Twisting strain helps to activate the nanobelts' basal plane by altering the electrical structure of catalytic active sites and promoting mass transfer (figures 3.6(a)–(d)). The highly active under-coordinated sites in WS_2 MSLs are more active than those in WS_2 non-twisted bilayers along the edges, polymorphs interface, and strained metallic phase surface. Both metastable and temperature-sensitive chemically exfoliated MoS_2 (ce-MoS_2) show a Tafel slope of 64 mV/dec, exchange current densities of 10.4 μA cm^{-2}, high cathodic current densities of >100 mA cm^{-2} at less than −300 mV, and an overpotential of −191 mV @10 mA cm^{-2} [66]. The hierarchically strained morphology improves electronic coupling between active sites and current collecting substrates. The ability to simultaneously modulate active site density and intrinsic HER activity is made possible by the synergistic combination of high strain load resulting from capillarity-induced self-crumpling and sulfur vacancies intrinsic to chemical exfoliation. Liao *et al* prepared single crystal MoO_2 sheets with vertical and horizontal orientations, in which strain was produced by extrusion deformation of two adjacent MoO_2 sheets [67]. The intensity-enhanced noise was seen at the local strained region compared to the unstrained regions in the same frame, revealing the beneficial impact of compressive strain on hydrogen generation.

Jiang *et al* designed strain-tunable sulfur vacancies around single-atom Ru sites on nanoporous MoS_2 (Ru/np-MoS_2) for alkaline HER (figure 3.6(e)) [68]. The synergistic effect between sulfur vacancies and Ru sites is boosted by altering the strain of this system, which influences the catalytic behavior of active sites (figures 3.6(f)–(h)). The strained sulfur vacancies enhanced reactant density and the trigger hydrogen evolution process on Ru sites. The resultant electrocatalyst shows an overpotential of 30 mV at 10 mA cm^{-2} current density, a Tafel slope of 31 mV/dec, and long durability. In 1T′ $Re_xMo_{1-x}S_2$–2H MoS_2 lateral

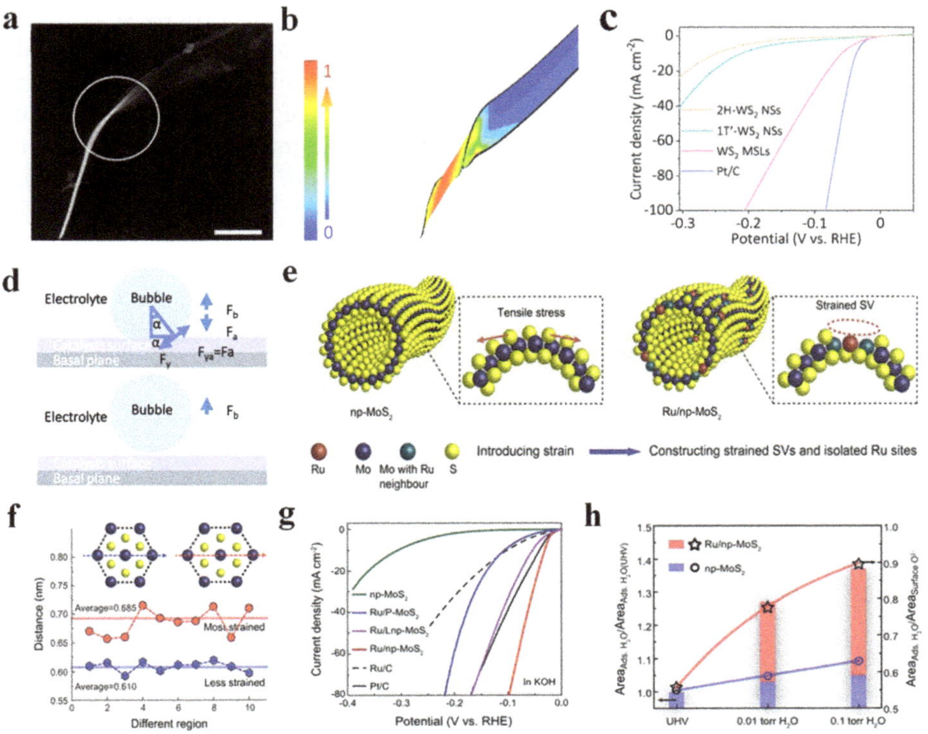

Figure 3.6. (a) STEM image shows a WS_2 nanobelt with a single screw. Scale bar-300 nm. (b) Calculations of strain in a nanocone using finite elements. The relative scale of the strain distribution is displayed by the color bar. (c) LSV curves with a scan rate of 10 mV s^{-1} in Ar-bubbled 0.5 M H_2SO_4. (d) Stress examination of one single bubble on the surface of the catalyst. Reproduced with permission from [13]. Copyright (2021) Springer Nature. (e) Illustration of the construction Ru/np-MoS_2. (f) The confirmed strain of Ru/np-MoS_2 using intensity line profiles at different regions. (g) Polarization curves. (h) Comparison of water absorptive capability for np-MoS_2 and Ru/np-MoS_2. Reproduced with permission from [68]. Copyright (2021) Springer Nature.

heterostructure, around 50% Re atoms of ReS_2 were substituted by Mo atoms, which induced ≈7% of the tensile strain in ReS_2 lattices [69]. The resultant catalyst exhibits an overpotential of 84 mV at 10 mA cm^{-2} and a Tafel slope of 58 mV/dec. According to Kelvin probe force microscopy, the difference in work function between 2H MoS_2 and 1T′ $Re_xMo_{1-x}S_2$ is 40 meV, which makes it easier for electrons to transfer from 2H MoS_2 to 1T′ $Re_xMo_{1-x}S_2$ at the heterojunction.

3.3.5 Morphology

Another appealing strategy is to enhance the electrocatalytic HER performance by carefully controlling the structural morphology of the materials. By selenizing Pt film, Hao's group prepared $PtSe_2$ nanowall films with morphology controlled at the centimeters level [70]. At 550 °C, the Pt atoms reorganized into an ordered distribution, which is used to generate $PtSe_2$ nanowalls with orderly structure. The nanowall structure was developed by adjusting the growth temperature, and the original thickness of the Pt film was used to tailor the nanowall thickness. The well-ordered $PtSe_2$ nanowall films depict an overpotential of 300 mV at −10 mA cm^{-2} and a Tafel slope of 52 mV/dec. The well-ordered $PtSe_2$ nanowall films expose more edge active sites at the nanowall structure. As shown in figures 3.7(a)–(d), the

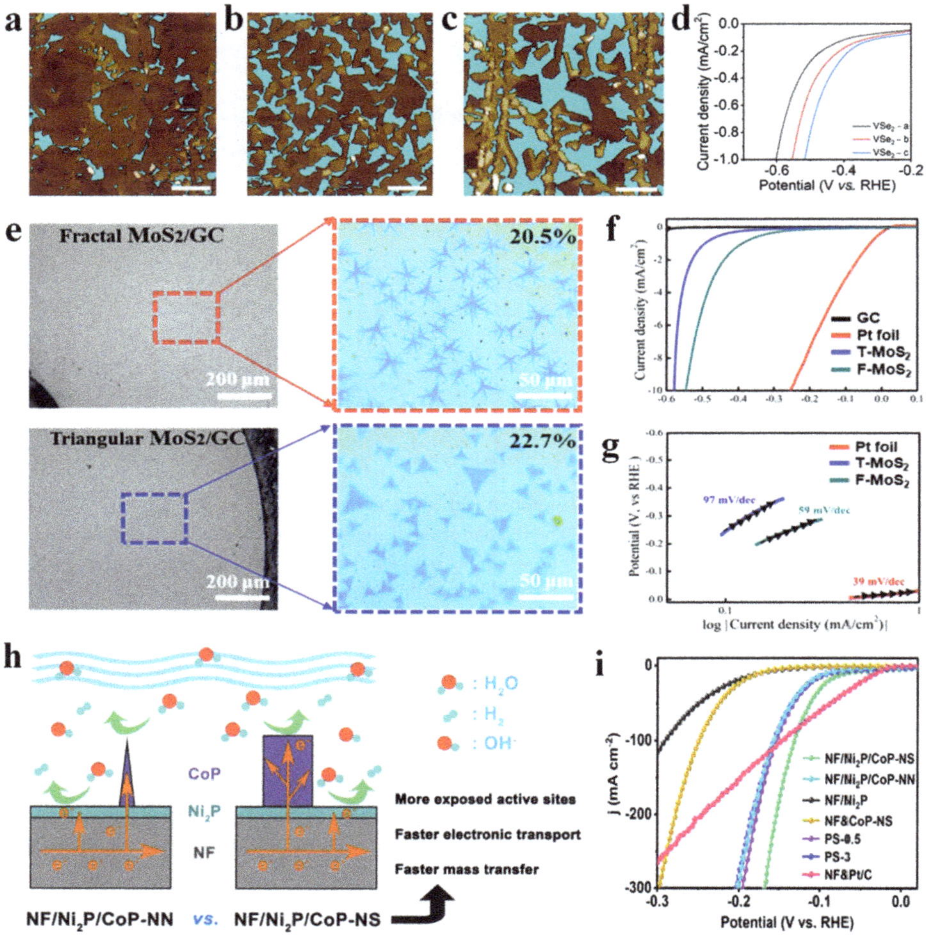

Figure 3.7. AFM images of VSe_2 flakes grown on HOPG with various surface coverages of (a) 89.7% (b) 70.5% and (c) 65.1% (scale bars: 400 nm). (d) Polarization curves of VSe_2 flakes in (a)–(c). Reprinted with permission from [14]. Copyright (2022) by the American Chemical Society. (e) Optical microscopy images of different kinds of MoS_2 transferred to the GC electrode. (f) Polarization curves of the two MoS_2 samples. (g) The Tafel slope diagram. Reprinted with permission from [71]. Copyright (2022) by the American Chemical Society. (h) LSV curves in 1.0 M KOH. (i) Schematic illustration of the comparison between NF/Ni_2P/CoP-NN and NF/Ni_2P/CoP-NS catalyzed hydrogen evolution. Reprinted from [72], Copyright (2022), with permission from Elsevier.

controlled synthesized VSe_2 2D flakes on highly oriented pyrolytic graphite (HOPG) show an overpotential of 543 mV at a current density of 1 mA cm^{-2} [14]. The nanobelt structures have a higher edge density and can offer more catalytic active sites than the triangular flakes' shapes. More catalytic site densities are present in the plentiful dendritic edge nanostructures than in the monolayer MoS_2 produced under thermodynamic equilibrium circumstances [71]. The Tafel slope of dendritic monolayer MoS_2 was demonstrated by HER performance to be significantly lower (59 mV/dec) than that of the triangular sample (97 mV/dec) (figures 3.7(e)–(g)). Zhou *et al* studied HER activity in an alkaline medium by using CoP nanoarrays and engineered them with different morphologies [72]. The performance of Nickel foam/Ni_2P/CoP-nanosheet (NF/Ni_2P/CoP-NS) is far better than NF/Ni_2P/CoP with a nanoneedle-like morphology (figures 3.7(h) and (i)). Because of the abundant exposed active sites, high conductivity, and mass transfer capabilities, the NF/Ni_2P/CoP-NS electrocatalyst exhibits an overpotential of 64 mV at the current density of 10 mA cm^{-2} and a Tafel slope of 58.2 mV/dec.

3.3.6 Edge

Edge engineering is a reliable method for enhancing active edge sites, modifying the electronic structure, and enhancing the electrocatalytic activity for hydrogen evolution [73]. In simple words, edges are the catalytically active sites of 2D materials. By decreasing antimonene nanosheets to the size of ultrasmall antimonene nanodots (AMNDs), Wang *et al* used active edge sites engineering to reveal many catalytically active edge sites [74]. The ionic liquid-modified AMNDs show an overpotential of 116 mV at 10 mA cm^{-2}, a Tafel slope of 104 mV/dec, and long stability in 1 M KOH. AMNDs' edges, which serve as catalytically active sites, are most fully exposed when their size is reduced. The basal surfaces of MoS_2 flakes are catalytically inert, and HER activity is strongly correlated with their edge sites [15]. The edge length of triangular monolayer MoS_2 flakes changes from nanometre to micrometer with changes in growth temperature or proximity to the precursor substrate. As shown in figures 3.8(a) and (b), the nanosized triangular MoS_2 flakes on Au foil show an exchange current density of 38.1 μA cm^{-2} and a Tafel slope of 61 mV/dec. $MoSe_2$ nanoflowers showed an onset potential of 170 mV and a Tafel slope value of 61 mV/dec in 0.5 M H_2SO_4 electrolyte [73]. The vertical orientation of $MoSe_2$ nanosheets, which reduces the adsorption energy of hydrogen on catalytically active edge sites, is what caused the low overpotential for $MoSe_2$ nanoflowers. The greater coverage of hydrogen ions on edge sites improves catalytic activity.

Kumatani *et al* showed that the edge structure of graphene and nitrogen/phosphorous co-doping synergistically improve HER activity (figures 3.8(c) and (d)) [75]. Theoretically, nitrogen/phosphorus dopants placed on the edges improve the contrast between positive and negative charges on atom sites, decreasing the Gibbs free energy of the HER process. The catalyst shows an overpotential of 344 mV at 10 mA cm^{-2}, a turnover frequency of 0.64 H_2/s, an exchange current density of 12.4 μA cm^{-2}, and a Tafel slope of 118 mV/dec. The large-area and edge-rich $PtSe_2$ shows a maximum cathodic current density of 227 mA cm^{-2} for HER

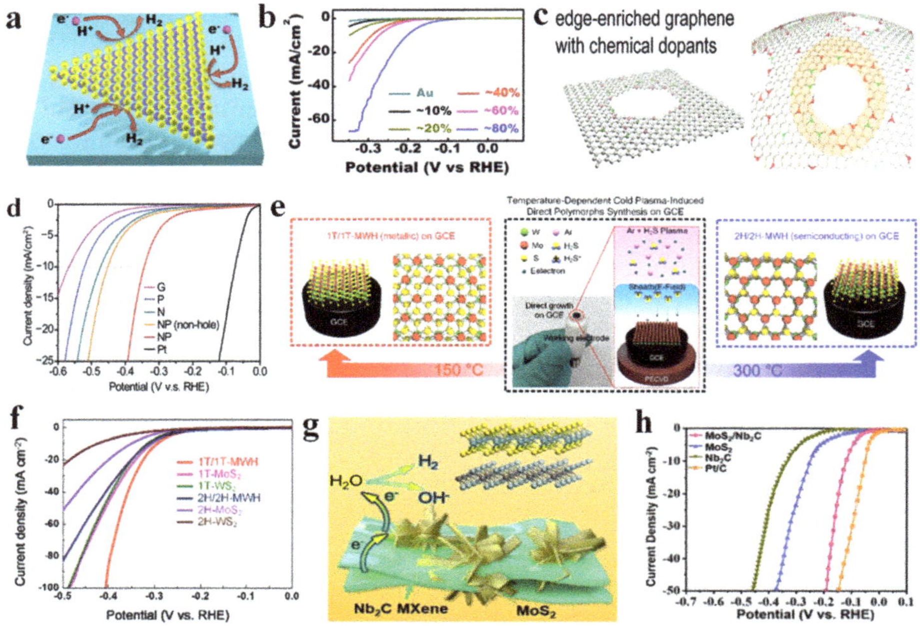

Figure 3.8. (a) Schematic represent the edges of monolayer MoS_2 functioning as catalytically active sites for hydrogen evolution. (b) Polarization curves of as-grown monolayer MoS_2 on Au foils samples. Reprinted with permission from [15]. Copyright (2014) by the American Chemical Society. (c) Schematic of edge-enriched graphene with chemical dopants. (d) LSV of graphene samples with and without edges in 0.5 M H_2SO_4 medium. Reproduced from [75]. CC BY 4.0. (e) Temperature-dependent cold plasma approach shown schematically for polymorph production directly on GCE. (f) Polarization curves of all samples. Reprinted with permission from [79]. Copyright (2023) by the American Chemical Society. (g) Schematic diagram of HER mechanism of the MoS_2/Nb_2C hybrids in 1 M KOH. (h) LSV curves of all samples at 5 mV s^{-1}. Reprinted with permission from [16]. Copyright (2022) by the American Chemical Society.

activity [76]. As the density of the edge rises, so does the current density, and the performance of hydrogen generation eventually improves.

3.3.7 Heterostructuring

More exposed edge sites for electrochemical reactions were made possible by heterostructure interfaces. Through the blending of various compositions, homogeneous heterointerfaces could promote charge transmission and amplify the synergetic effect [77]. On the other hand, in heterogeneous structures lattice mismatch leads to a lattice defect and additional edge active sites, which can help electrocatalytic reactions. The WS_2/WSe_2 heterojunction changes the electrical structure and exposes additional electrochemically active HER sites [77]. With significant stability, the WS_2/WSe_2 catalyst shows an overpotential of 121 mV at 10 mA cm^{-2} and a small Tafel slope of 74.08 mV/dec. The combination of WS_2 and WSe_2 stimulates the redistribution of charge density at the heterointerface, which is advantageous for establishing effective electrocatalytic activity. The development of vertical heterostructures considerably improves HER performance because of synergistic effects such as a low energy barrier produced by orbital overlap or strongly conducting support materials [37, 78]. Seok *et al* tailored 1T-MoS_2/1T-WS_2 vertical heterostructures (1T/1T-MWH) or 2H-MoS_2/2H-WS_2 vertical heterostructures (2H/2H-MWH) and studied for hydrogen evolution in a 0.5 M H_2SO_4 solution [79]. On the glassy carbon electrode (GCE), different MWH polymorphic heterostructures were synthesized based on the substrate temperature (figure 3.8(e)). The 1T/1T-MWH heterointerfaces exhibited the lowest overpotential of 294 mV @10 mA cm^{-2} current density and a Tafel slope of 99 mV/dec, which is better than 2H phase materials (figure 3.8(f)). Due to its metallic 1T phase and heterostructures with alloy structures at the heterointerface, 1T/1T-MWH exhibits the best electrocatalytic performance. In an acidic medium (0.5 M H_2SO_4), WSe_2/MoS_2 heterostructure shows the overpotential of 116 mV to reach a current density of 10 mA cm^{-2} and the Tafel slope of 76 mV/dec [80]. Their electrochemically active surface area helps to enhance HER activity.

The $PB_{0.94}C$-based double/simple perovskite heterostructure ($PB_{0.94}C$-DSPH) shows an overpotential of 364 mV at a current density of 500 mA cm^{-2} and a Tafel slope of 56.5 mV/dec, in 1.0 M KOH solution [81]. Such high performance is attributed to the large electrochemical surface area, more hydrophilic surface, efficient charge transfer at high current densities, and high conductivity. Under alkaline conditions, the MoS_2/Nb_2C catalyst delivers an overpotential of 117 mV at 10 mA cm^{-2} and a Tafel slope of 65.1 mV/dec and good stability (figures 3.8(g) and (h)) [16]. The combination of MoS_2 and Nb_2C creates a conductive network that benefits the HER process. In addition to these results, numerous other reports have demonstrated that creating heterostructures with 2D materials is an effective strategy for accelerating the HER catalytic reaction kinetics [82–85].

The performance of strategically engineered various 2D materials is presented in table 3.1.

3.4 Conclusion

In the future, HER will be an essential aspect of sustainable energy. Accordingly, this chapter summarizes the significant efforts put forth in the development of highly active catalysts based on 2D materials for the cost-effective production of hydrogen. Although 2D materials hold the potential for better electrocatalytic performance towards HER, finding designs that have favorable morphological and structural control continues to be challenging. The ability to transport charge and the density of active sites could both be significantly altered by controlling the structure–performance relationship of electrocatalysts. Through precise manipulation of atomic defects, the intrinsic properties of 2D materials must be modified. The engineered defects in electrocatalysts enhance exposure to numerous active sites, improve the adsorption behavior of H_2O, reduce the free energy of hydrogen adsorption, and rise electron mobility. High entropy alloys' electrocatalytic performance is undoubtedly diminished by their low electronic conductivity and

Table 3.1. The performance of engineered 2D materials toward hydrogen evolution.

Engineered technique	Material	Electrolyte	Overpotential	Tafel slope	References
Defect	*d*-$PdTe_x$	0.5 M H_2SO_4	76 mV @10 mA cm^{-2}	118 mV/dec	[10]
	MoS_2 nano mesh	0.5 M H_2SO_4	160 mV @10 mA cm^{-2}	46 mV/dec	[27]
	MoS_2 nanowall	0.5 M H_2SO_4	95 mV @10 mA cm^{-2}	78 mV/dec	[28]
	$Ni(OH)_2/MoS_2$	1.0 M KOH	139 mV @10 mA cm^{-2}	45 mV/dec	[29]
	FD-MoS_2	0.5 M H_2SO_4	164 mV @10 mA cm^{-2}	36 mV/dec	[31]
	S-defect-rich Re/ReS_2/carbon cloth	0.5 M H_2SO_4	42 mV @10 mA cm^{-2}	36 mV/dec	[32]
	$MoSe_2$ nanosheets	1.0 M H_2SO_4	300 mV @10 mA cm^{-2}	90 mV/dec	[33]
	SnS_2	0.5 M H_2SO_4	141 mV @10 mA cm^{-2}	74 mV/dec	[34]
Alloy	$Mo_xW_{1-x}S_2$ nanoflakes	0.5 M H_2SO_4	137 mV @10 mA cm^{-2}	53 mV/dec	[41]
	$Re_{1-x}Mo_xSe_2$	0.5 M H_2SO_4	77 mV @10 mA cm^{-2}	42 mV/dec	[11]
	$W_{1-x}V_xSe_2$	0.5 M H_2SO_4	128 mV @10 mA cm^{-2}	80 mV/dec	[42]
	Ni/Co alloy@Nb-doped $Ti_3C_2T_x$ MXene	1.0 M KOH	43.4 mV @10 mA cm^{-2}	116 mV/dec	[43]
	Pt/Nb_2CT_x	0.5 M H_2SO_4	5 mV @10 mA cm^{-2}	34.66 mV/dec	[44]
	CoNiCuMgZn@Graphene	1.0 M KOH	158 mV @10 mA cm^{-2}	36.1 mV/dec	[45]
	$Co_{0.6}(VMnNiZn)_{0.4}PS_3$	1.0 M KOH	65.9 mV @10 mA cm^{-2}	65.5 mV/dec	[48]
Doping	Co-doped MoS_2	1.0 M KOH	67 mV @10 mA cm^{-2}	67 mV/dec	[55]
	Pt–Te co-doped MoS_2	1.0 M KOH	52 mV @10 mA cm^{-2}	62.27 mV/dec	[56]
	P-doped MoS_2	0.5 M H_2SO_4	125 mV @10 mA cm^{-2}	73.73 mV/dec	[57]
	S-doped $Nb_4C_3T_x$	1.0 M KOH	118 mV @10 mA cm^{-2}	104 mV/dec	[60]
	Ir–N, S co-doped $Ti_3C_2T_x$	0.5 M H_2SO_4	58 mV @10 mA cm^{-2}	25.1 mV/dec	[61]
	N-doped $Sr_2Fe1.5Mo_{0.5}O_{6-\delta}$	1.0 M KOH	251 mV @10 mA cm^{-2}	138 mV/dec	[62]

Strain and stress	WS_2	0.5 M H_2SO_4	60 mV @10 mA cm^{-2}	40 mV/dec	[13]
	ce-MoS_2	0.5 M H_2SO_4	191 mV @10 mA cm^{-2}	64 mV/dec	[66]
	Ru/np-MoS_2	1.0 M KOH	30 mV @10 mA cm^{-2}	31 mV/dec	[68]
	1T′ $Re_xMo_{1-x}S_2$–2H MoS_2	0.5 M H_2SO_4	84 mV @10 mA cm^{-2}	58 mV/dec	[69]
Morphology	$PtSe_2$ nanowall	0.5 M H_2SO_4	300 mV @10 mA cm^{-2}	52 mV/dec	[70]
	Nickel foam/Ni_2P/CoP-nanosheet	1.0 M KOH	64 mV @10 mA cm^{-2}	58.2 mV/dec	[72]
Edge	Antimonene	1.0 M KOH	116 mV @10 mA cm^{-2}	104 mV/dec	[74]
	$MoSe_2$ on carbon cloth	0.5 M H_2SO_4	220 mV @10 mA cm^{-2}	61 mV/dec	[73]
	Nitrogen/phosphorus co-doped graphene	0.5 M H_2SO_4	344 mV @10 mA cm^{-2}	118 mV/dec	[75]
Heterostructuring	WS_2/WSe_2 heterojunction	0.5 M H_2SO_4	121 mV @10 mA cm^{-2}	74.08 mV/dec	[77]
	1T-MoS_2/1T-WS_2	0.5 M H_2SO_4	294 mV @10 mA cm^{-2}	99 mV/dec	[79]
	WSe_2/MoS_2	0.5 M H_2SO_4	116 mV @10 mA cm^{-2}	76 mV/dec	[80]
	MoS_2/Nb_2C	1.0 M KOH	117 mV @10 mA cm^{-2}	65.1 mV/dec	[16]

micron-scale grain sizes. The introduction of conductive substrates like graphene offers a fresh approach to the conductivity problem. It is possible to use controlled multi-element doping engineering to modify 2D materials that function synergistically well as HER catalysts. Under experimental conditions, it is challenging to precisely control the strain state because it is constrained by the mechanical characteristics of the substrate and the nanomaterial. There is a need to further develop methods for regulating strain engineering in an optimistic way. It is necessary to expand the range of materials available for strain and stress engineering. The current research concentrated on graphene and mostly 2D TMDs. The addition of strain with defect, doping, or other engineering will help to further regulate the catalytic activity of atomic sites. A powerful platform for modifying atomic defects, alloying, and doping is provided by the large surface area in the plane of 2D materials. Therefore, the engineering of 2D materials is one significant approach for performance improvement by modulating the electronic structure and electrical conductivity. There is a need to design tailored morphologies of 2D materials with plenty of active edge sites. By adjusting the shape and thickness of layered materials, active edge sites can be given better catalytic characteristics. Therefore, there is still plenty of room for improvement in engineering approaches like doping, alloying, and defects, including the challenging fabrication process, formation mechanism, and less control over concentration and proportion. This chapter provided guidance for improving hydrogen evolution through the use of various 2D material engineering techniques.

Acknowledgments

The authors gratefully acknowledge financial assistance from the SERB Core Research Grant (Grant No. CRG/2022/000897), Department of Science and Technology (DST/NM/NT/2019/205(G)), and Minor Research Project Grant, Jain University (JU/MRP/CNMS/29/2023).

References

[1] Kadam S R, Bar-Ziv R and Bar-Sadan M 2022 A cobalt-doped WS_2/WO_3 nanocomposite electrocatalyst for the hydrogen evolution reaction in acidic and alkaline media *New J. Chem.* **46** 20102–7

[2] Anantharaj S, Noda S, Jothi V R, Yi S, Driess M and Menezes P W 2021 Strategies and perspectives to catch the missing pieces in energy-efficient hydrogen evolution reaction in alkaline media *Angew. Chem. Int. Ed.* **60** 18981–9006

[3] Shinde P V, Mane P, Late D J, Chakraborty B and Rout C S 2021 Promising 2D/2D $MoTe_2$/$Ti_3C_2T_x$ hybrid materials for boosted hydrogen evolution reaction *ACS Appl. Energy Mater.* **4** 11886–97

[4] Shinde P V, Gavali D S, Thapa R, Singh M K and Rout C S 2021 Ternary VS_2/ZnS/CdS hybrids as efficient electrocatalyst for hydrogen evolution reaction: experimental and theoretical insights *AIP Adv.* **11** 105010

[5] García-Miranda Ferrari A, Brownson D A and Banks C E 2019 Investigating the integrity of graphene towards the electrochemical hydrogen evolution reaction (HER) *Sci. Rep.* **49** 15961

[6] Shinde P V, Mane P, Chakraborty B and Rout C S 2021 Spinel $NiFe_2O_4$ nanoparticles decorated 2D Ti_3C_2 MXene sheets for efficient water splitting: experiments and theories *J. Colloid Interface Sci.* **602** 232–41
[7] Shinde P V, Babu S, Mishra S K, Late D, Rout C S and Singh M K 2021 Tuning the synergistic effects of MoS_2 and spinel $NiFe_2O_4$ nanostructures for high performance energy storage and conversion applications *Sustain. Energy Fuels* **5** 3906–17
[8] Shinde P V, Samal R and Rout C S 2022 Vertically aligned graphene-analogous low-dimensional materials: a review on emerging trends, recent developments, and future perspectives *Adv. Mater. Interfaces* **9** 2101959
[9] Shinde P V and Singh M K 2019 Synthesis, characterization, and properties of graphene analogs of 2D material *Fundamentals and Sensing Applications of 2D Materials* **Vol 1** (Woodhead Publishing) pp 91–143
[10] Zuo Y *et al* 2023 Defect engineering in two-dimensional layered $PdTe_2$ for enhanced hydrogen evolution reaction *ACS Catal.* **613** 2601–9
[11] Kwak I H, Kwon I S, Debela T T, Abbas H G, Park Y C, Seo J, Ahn J P, Lee J H, Park J and Kang H S 2020 Phase evolution of $Re_{1-x}Mo_xSe_2$ alloy nanosheets and their enhanced catalytic activity toward hydrogen evolution reaction *ACS Nano.* **14** 11995–2005
[12] Peng W, Han J, Lu Y R, Luo M, Chan T S, Peng M and Tan Y 2023 A general strategy for engineering single-metal sites on 3D porous N, P Co-doped $Ti_3C_2T_X$ MXene *ACS nano* **16** 4116–25
[13] Xie L, Wang W, Zhao W, Liu S, Huang W and Zhao Q 2021 WS_2 moiré superlattices derived from mechanical flexibility for hydrogen evolution reaction *Nat. Commun.* **12** 5070
[14] Zhang X *et al* 2022 Morphology-controlled electrocatalytic performance of two-dimensional VSe_2 nanoflakes for hydrogen evolution reactions *ACS Appl. Nano Mater.* **5** 2087–93
[15] Shi J *et al* 2014 Controllable growth and transfer of monolayer MoS_2 on Au foils and its potential application in hydrogen evolution reaction *ACS Nano.* **8** 10196–204
[16] Zong H, Hu L, Gong S, Yu K and Zhu Z 2022 Flower-petal-like Nb_2C MXene combined with MoS_2 as bifunctional catalysts towards enhanced lithium-sulfur batteries and hydrogen evolution *Electrochim. Acta* **404** 139781
[17] Bae S Y, Mahmood J, Jeon I Y and Baek J B 2020 Recent advances in ruthenium-based electrocatalysts for the hydrogen evolution reaction *Nanoscale Horiz.* **5** 43–56
[18] Zeng M and Li Y 2015 Recent advances in heterogeneous electrocatalysts for the hydrogen evolution reaction *J. Mater. Chem.* A **3** 14942–62
[19] Peera S G, Koutavarapu R, Chao L, Singh L, Murugadoss G and Rajeshkhanna G 2020 2D MXene nanomaterials as electrocatalysts for hydrogen evolution reaction (HER): a review *Micromachines* **13** 1499
[20] Anantharaj S, Ede S R, Sakthikumar K, Karthick K, Mishra S and Kundu S 2016 Recent trends and perspectives in electrochemical water splitting with an emphasis on sulfide, selenide, and phosphide catalysts of Fe, Co, and Ni: a review *ACS Catal.* **6** 8069–97
[21] Li C and Baek J B 2019 Recent advances in noble metal (Pt, Ru, and Ir)-based electrocatalysts for efficient hydrogen evolution reaction *ACS Omega* **5** 31–40
[22] Liu S, Zhang Q, Bao J, Li Y, Dai Z and Gu L 2017 Significantly enhanced hydrogen evolution activity of freestanding pd-ru distorted icosahedral clusters with less than 600 atoms *Chem.–A Eur. J.* **23** 18203–7

[23] Mahmood J, Li F, Jung S M, Okyay M S, Ahmad I, Kim S J, Park N, Jeong H Y and Baek J B 2017 An efficient and ph-universal ruthenium-based catalyst for the hydrogen evolution reaction *Nat. Nanotechnol.* **12** 441–6

[24] Chen C H, Wu D, Li Z, Zhang R, Kuai C G, Zhao X R, Dong C K, Qiao S Z, Liu H and Du X W 2019 Ruthenium-based single-atom alloy with high electrocatalytic activity for hydrogen evolution *Adv. Energy Mater.* **9** 1803913

[25] Xie J, Yang X and Xie Y 2020 Defect engineering in two-dimensional electrocatalysts for hydrogen evolution *Nanoscale* **12** 4283–94

[26] Jia Y, Zhang L, Du A, Gao G, Chen J, Yan X, Brown C L and Yao X 2016 Defect graphene as a trifunctional catalyst for electrochemical reactions *Adv. Mater.* **28** 9532–8

[27] Li Y, Yin K, Wang L, Lu X, Zhang Y, Liu Y, Yan D, Song Y and Luo S 2018 Engineering MoS_2 nanomesh with holes and lattice defects for highly active hydrogen evolution reaction *Appl. Catal.* B **239** 537–44

[28] Xie J, Qu H, Xin J, Zhang X, Cui G, Zang X, Bao J, Tang B and Xie Y 2017 Defect-rich MoS_2 nanowall catalyst for efficient hydrogen evolution reaction *Nano Res.* **10** 1178–88

[29] He Z, Liu Q, Zhu Y, Tan T, Cao L, Zhao S and Chen Y 2020 Defect-Mediated adsorption of metal ions for constructing Ni hydroxide/MoS_2 heterostructures as high-performance water-splitting electrocatalysts *ACS Appl. Energy Mater.* **3** 7039–47

[30] Li H *et al* 2016 Activating and optimizing MoS_2 basal planes for hydrogen evolution through the formation of strained sulphur vacancies *Nat. Mater.* **15** 48–53

[31] Xu J *et al* 2022 Frenkel-defected monolayer MoS_2 catalysts for efficient hydrogen evolution *Nat. Commun.* **13** 2193

[32] Pang Q Q, Niu Z L, Yi S S, Zhang S, Liu Z Y and Yue X Z 2020 Hydrogen-etched bifunctional sulfur-defect-rich ReS_2/CC electrocatalyst for highly efficient HER and OER *Small* **16** 2003007

[33] Truong Q D *et al* 2020 Defect-rich exfoliated $MoSe_2$ nanosheets by supercritical fluid process as an attractive catalyst for hydrogen evolution in water *Appl. Surf. Sci.* **505** 144537

[34] Shao G, Xiang H, Huang M, Zong Y, Luo J, Feng Y, Xue X X, Xu J, Liu S and Zhou Z 2022 Vacancies in 2D SnS_2 accelerating hydrogen evolution reaction *Sci. China Mater.* **65** 1833–41

[35] Xie L M 2015 Two-dimensional transition metal dichalcogenide alloys: preparation, characterization and applications *Nanoscale* **7** 18392–401

[36] Gong Q, Cheng L, Liu C, Zhang M, Feng Q, Ye H, Zeng M, Xie L, Liu Z and Li Y 2015 Ultrathin $MoS_{2(1-x)}Se_{2x}$ alloy nanoflakes for electrocatalytic hydrogen evolution reaction *ACS Catal.* **5** 2213–9

[37] Lei Y *et al* 2017 Low-temperature synthesis of heterostructures of transition metal dichalcogenide alloys ($W_xMo_{1-x}S_2$) and graphene with superior catalytic performance for hydrogen evolution *ACS Nano.* **11** 5103–12

[38] Kiran V, Mukherjee D, Jenjeti R N and Sampath S 2014 Active guests in the MoS_2/$MoSe_2$ host lattice: efficient hydrogen evolution using few-layer alloys of $MoS_{2(1-x)}Se_{2x}$ *Nanoscale* **6** 12856–63

[39] Chen Y, Xi J, Dumcenco D O, Liu Z, Suenaga K, Wang D, Shuai Z, Huang Y S and Xie L 2013 Tunable band gap photoluminescence from atomically thin transition-metal dichalcogenide alloys *ACS Nano.* **7** 4610–6

[40] Yao J, Zheng Z and Yang G 2016 Promoting the performance of layered-material photodetectors by alloy engineering *ACS Appl. Mater. Interfaces* **8** 12915–24

[41] Gan X, Lv R, Wang X, Zhang Z, Fujisawa K, Lei Y, Huang Z H, Terrones M and Kang F 2018 Pyrolytic carbon supported alloying metal dichalcogenides as free-standing electrodes for efficient hydrogen evolution *Carbon* **132** 512–9

[42] Kwon I S, Kwak I H, Zewdie G M, Lee S J, Kim J Y, Yoo S J, Kim J G, Park J and Kang H S 2022 WSe_2–VSe_2 alloyed nanosheets to enhance the catalytic performance of hydrogen evolution reaction *ACS Nano.* **16** 12569–79

[43] Du C F, Sun X, Yu H, Liang Q, Dinh K N, Zheng Y, Luo Y, Wang Z and Yan Q 2019 Synergy of Nb doping and surface alloy enhanced on water–alkali electrocatalytic hydrogen generation performance in Ti-based MXene *Adv. Sci.* **6** 1900116

[44] Fan X *et al* 2021 Mechanochemical synthesis of Pt/Nb_2CT_x MXene composites for enhanced electrocatalytic hydrogen evolution *Materials* **14** 2426

[45] Feng D, Dong Y, Nie P, Zhang L and Qiao Z A 2022 CoNiCuMgZn high entropy alloy nanoparticles embedded onto graphene sheets via anchoring and alloying strategy as efficient electrocatalysts for hydrogen evolution reaction *Chem. Eng. J.* **430** 132883

[46] Batchelor T A, Pedersen J K, Winther S H, Castelli I E, Jacobsen K W and Rossmeisl J 2019 High-entropy alloys as a discovery platform for electrocatalysis *Joule* **3** 834–45

[47] Koo W T, Millstone J E, Weiss P S and Kim I D 2020 The design and science of poly-elemental nanoparticles *ACS Nano.* **14** 6407–13

[48] Wang R, Huang J, Zhang X, Han J, Zhang Z, Gao T, Xu L, Liu S, Xu P and Song B 2022 Two-dimensional high-entropy metal phosphorus trichalcogenides for enhanced hydrogen evolution reaction *ACS Nano.* **16** 3593–603

[49] Sun Y and Dai S High-entropy materials for catalysis: a new frontier *Sci. Adv.* **7** eabg1600

[50] Chen L, Zhou Z, Tan Z, He D, Bobzin K, Zhao L, Öte M and Königstein T 2018 High temperature oxidation behaviour of $Al_{0.6}CrFeCoNi$ and $Al_{0.6}CrFeCoNiSi_{0.3}$ high entropy alloys *J. Alloys Compd.* **764** 845–52

[51] Leng Y, Zhang Z, Chen H, Du S, Liu J, Nie S, Dong Y, Zhang P and Dai S 2021 Overcoming the phase separation within high-entropy metal carbide by poly (ionic liquid) *Chem. Commun.* **57** 3676–9

[52] Gwalani B, Pohan R M, Lee J, Lee B, Banerjee R, Ryu H J and Hong S H 2018 High-entropy alloy strengthened by *in situ* formation of entropy-stabilized nano-dispersoids *Sci. Rep.* **8** 14085

[53] Wei H, Si J, Zeng L, Lyu S, Zhang Z, Suo Y and Hou Y 2023 Electrochemically exfoliated Ni-doped MoS_2 nanosheets for highly efficient hydrogen evolution and Zn-H_2O battery *Chin. Chem. Lett.* **34** 107144

[54] Bolar S, Shit S, Kumar J S, Murmu N C, Ganesh R S, Inokawa H and Kuila T 2019 Optimization of active surface area of flower like MoS_2 using V-doping towards enhanced hydrogen evolution reaction in acidic and basic medium *Appl. Catal.* B **254** 432–42

[55] Li Z, Li C, Chen J, Xing X, Wang Y, Zhang Y, Yang M and Zhang G 2022 Confined synthesis of MoS_2 with rich co-doped edges for enhanced hydrogen evolution performance *J. Energy Chem.* **70** 18–26

[56] An C H, Kang W, Deng Q B and Hu N 2022 Pt and Te co-doped ultrathin mos_2 nanosheets for enhanced hydrogen evolution reaction with wide Ph range *Rare Met.* **41** 378–84

[57] Sun L, Gao M, Jing Z, Cheng Z, Zheng D, Xu H, Zhou Q and Lin J 2022 1 T-phase enriched P doped WS_2 nanosphere for highly efficient electrochemical hydrogen evolution reaction *Chem. Eng. J.* **429** 132187

[58] Xu J, Yu B, Zhao H, Cao S, Song L, Xing K, Zhou R and Lu X 2020 Oxygen-doped VS4 microspheres with abundant sulfur vacancies as a superior electrocatalyst for the hydrogen evolution reaction *ACS Sustain. Chem. Eng.* **8** 15055–64
[59] Rong J, Ye Y, Cao J, Liu X, Fan H, Yang S, Wei M, Yang L, Yang J and Chen Y 2022 Restructuring electronic structure via W-doped 1T MoS_2 for enhancing hydrogen evolution reaction *Appl. Surf. Sci.* **579** 152216
[60] Wu F, Xu C, Yang X, Yang L and Yin S 2022 S-doped multilayer niobium carbide ($Nb_4C_3T_x$) electrocatalyst for efficient hydrogen evolution in alkaline solutions *Int. J. Hydrogen Energy* **47** 17233–40
[61] Lin W, Lu Y R, Peng W, Luo M, Chan T S and Tan Y 2022 Atomic bridging modulation of Ir–N, S co-doped MXene for accelerating hydrogen evolution *J. Mater. Chem.* A **10** 9878–85
[62] Gao J, Zhang Y, Wang X, Jia L, Jiang H, Huang M and Toghan A 2021 Nitrogen-doped $Sr_2Fe_{1.5}Mo_{0.5}O_{6-\delta}$ perovskite as an efficient and stable catalyst for hydrogen evolution reaction *Mater. Today Energy* **20** 100695
[63] Dai Z, Liu L and Zhang Z 2019 Strain engineering of 2D materials: issues and opportunities at the interface *Adv. Mater.* **31** 1805417
[64] Deng S, Sumant A V and Berry V 2018 Strain engineering in two-dimensional nanomaterials beyond graphene *Nano Today* **22** 14–35
[65] Du J, Yu H, Liu B, Hong M, Liao Q, Zhang Z and Zhang Y 2021 Strain engineering in 2D material-based flexible optoelectronics *Small Methods* **5** 2000919
[66] Chen Y C *et al* 2017 Structurally deformed MoS_2 for electrochemically stable, thermally resistant, and highly efficient hydrogen evolution reaction *Adv. Mater.* **29** 1703863
[67] Liao M, Zhu Q, Li S, Li Q, Tao Z and Fu Y 2022 In-situ imaging of strain-induced enhancement of hydrogen evolution activity on the extruded MoO_2 sheets *Nano Res.* **5** 1–8
[68] Jiang K, Luo M, Liu Z, Peng M, Chen D, Lu Y R, Chan T S, de Groot F M and Tan Y 2021 Rational strain engineering of single-atom ruthenium on nanoporous MoS_2 for highly efficient hydrogen evolution *Nat. Commun.* **12** 1687
[69] Nguyen H T *et al* 2023 $1T'Re_xMo_{1-x}S_2$–2H MoS_2 lateral heterojunction for enhanced hydrogen evolution reaction performance *Adv. Funct. Mater.* **33** 2209572
[70] Hao R, Feng Q L, Wang X J, Zhang Y C and Li K S 2022 Morphology-controlled growth of large-area $PtSe_2$ films for enhanced hydrogen evolution reaction *Rare Met.* **41** 1314–22
[71] Wang S *et al* 2022 Morphology regulation of MoS_2 nanosheet-based domain boundaries for the hydrogen evolution reaction *ACS Appl. Nano Mater.* **5** 2273–9
[72] Zhou P, Li R, Lv J, Huang X, Lu Y and Wang G 2022 Synthesis of CoP nanoarrays by morphological engineering for efficient electrochemical hydrogen production *Electrochim. Acta* **426** 140768
[73] Masurkar N, Thangavel N K and Arava L M 2018 CVD-grown $MoSe_2$ nanoflowers with dual active sites for efficient electrochemical hydrogen evolution reaction *ACS Appl. Mater. Interfaces* **10** 27771–9
[74] Wang J, Wang C, Song Y, Sha W, Wang Z, Cao H, Zhao M, Liu P and Guo J 2022 Ionic liquid modified active edge-rich antimonene nanodots for highly efficient electrocatalytic hydrogen evolution reaction *ChemCatChem.* **14** e202101765
[75] Kumatani A *et al* 2019 Chemical dopants on edge of holey graphene accelerate electrochemical hydrogen evolution reaction *Adv. Sci.* **6** 1900119
[76] Lin S *et al* 2017 Tunable active edge sites in $PtSe_2$ films towards hydrogen evolution reaction *Nano Energy* **42** 26–33

[77] Sun L, Xu H, Cheng Z, Zheng D, Zhou Q, Yang S and Lin J 2022 A heterostructured WS_2/WSe_2 catalyst by heterojunction engineering towards boosting hydrogen evolution reaction *Chem. Eng. J.* **443** 136348

[78] Kim H U *et al* 2019 Low-temperature wafer-scale growth of MoS_2-graphene heterostructures *Appl. Surf. Sci.* **470** 129–34

[79] Seok H *et al* 2023 Tailoring polymorphic heterostructures of MoS_2–WS_2 (1T/1T, 2H/2H) for efficient hydrogen evolution reaction *ACS Sustain. Chem. Eng.* **11** 568–77

[80] Vikraman D, Hussain S, Truong L, Karuppasamy K, Kim H J, Maiyalagan T, Chun S H, Jung J and Kim H S 2019 Fabrication of MoS_2/WSe_2 heterostructures as electrocatalyst for enhanced hydrogen evolution reaction *Appl. Surf. Sci.* **480** 611–20

[81] Liu Y, Dou Y, Li S, Xia T, Xie Y, Wang Y, Zhang W, Wang J, Huo L and Zhao H 2021 Synergistic interaction of double/simple perovskite heterostructure for efficient hydrogen evolution reaction at high current density *Small Methods* **5** 2000701

[82] Chen Y, Meng G, Yang T, Chen C, Chang Z, Kong F, Tian H, Cui X, Hou X and Shi J 2022 Interfacial engineering of co-doped 1T-MoS_2 coupled with V_2C MXene for efficient electrocatalytic hydrogen evolution *Chem. Eng. J.* **450** 138157

[83] Lim K R, Handoko A D, Johnson L R, Meng X, Lin M, Subramanian G S, Anasori B, Gogotsi Y, Vojvodic A and Seh Z W 2020 2H-MoS_2 on Mo_2CT_x Mxene nanohybrid for efficient and durable electrocatalytic hydrogen evolution *ACS Nano.* **14** 16140–55

[84] Li W, Liu D, Yang N, Wang J, Huang M, Liu L, Peng X, Wang G, Yu X F and Chu P K 2019 Molybdenum diselenide–black phosphorus heterostructures for electrocatalytic hydrogen evolution *Appl. Surf. Sci.* **467** 328–34

[85] Li S, Zang W, Liu X, Pennycook S J, Kou Z, Yang C, Guan C and Wang J 2019 Heterojunction engineering of $MoSe_2$/MoS_2 with electronic modulation towards synergetic hydrogen evolution reaction and supercapacitance performance *Chem. Eng. J.* **359** 1419–26

Chapter 4

Engineered 2D materials for oxygen evolution reaction (OER)

Pratik V Shinde, Komal N Patil, Vitthal M Shinde and Chandra Sekhar Rout

The development of affordable, highly effective electrocatalysts to accelerate the sluggish oxygen-evolution reaction (OER) is one of the most crucial concerns for efficient water splitting. Therefore, in this chapter, recent research progress in engineered 2D material electrocatalysts for improved OER performance is summarized. Initially, we have outlined the sequential events that took place in an alkaline and an acidic medium in order to provide knowledge about the evolution of oxygen. The key points related to these mechanisms were also addressed. Next, various material designing strategies for improving OER performance, including defects, doping, alloying, strain, edge engineering, morphological tuning, and heterostructure generation used for enhancing OER performance are described. These approaches will contribute to increasing the availability of active sites and their exposure to catalytic reactions, which will help to improve the performance of 2D materials. Finally, we highlight a number of challenges and possibilities that may present design opportunities for 2D material electrocatalysts of the future.

4.1 Introduction

Currently, nonrenewable energy sources like fossil fuels, coal, and natural gas are used to meet the majority of the world's energy needs. Their future use is, however, constrained by their limited sources on the earth and the greenhouse gas emissions associated with them. The production of sustainable energy in perspective of the expanding human population is one of the most important issues of the twenty-first century. Hydrogen energy has received a lot of attention as the ideal renewable energy due to its high energy density and environmental friendliness [1]. Presently, the main sources for producing hydrogen are natural gases and fossil fuels. The main drawbacks of these technologies, however, are their dependence on nonrenewable energy sources and the release of carbon dioxide into the atmosphere [2]. In order to

doi:10.1088/978-0-7503-5719-7ch4

© IOP Publishing Ltd 2024

get around all of these limitations, researchers are working to develop hydrogen using renewable technologies, like water splitting [3]. One of the most intriguing and cost-effective ways to conserve renewable energy is to split water into hydrogen and oxygen.

The electrocatalytic water splitting involves two half-cell reactions, such as the hydrogen evolution reaction (HER) at the cathode, and the oxygen evolution reaction (OER) at the anode [4]. The OER is kinetically less advantageous than the HER because the cathodic HER involves a two-electron transfer while the anodic OER involves a four-electron transfer [5]. As a result, considerable effort has been expended in the search for effective electrocatalysts for the OER. The most commonly used catalysts today are ruthenium (Ru) and iridium (Ir), but their high cost and scarcity have prompted researchers to investigate cost-effective OER catalysts. Moreover, in acidic electrolytes with high applied potentials, both RuO_2 and IrO_2 further oxidize and then dissolve, resulting in unstable performance during water electrolysis [6].

It is crucial for clean energy systems to develop effective and affordable electrocatalysts for the OER. Two-dimensional (2D) materials are potentially effective catalysts for OER reactions among the numerous classes of low-dimensional electrocatalysts that are currently in demand [7, 8]. 2D materials offer excellent electronic conductivity, admirable structural stability, worthy surface area, and suitable active edges for the role of electrocatalysts [7]. As a result, 2D materials are regarded as one of the most auspicious electrocatalysts for water splitting. 2D materials like graphene, MXenes, metal oxides, metal–organic frameworks (MOF), etc have recently been identified as promising OER electrocatalysts [1, 9, 10]. The majority of pure 2D nanomaterials, however, have very poor catalytic performance. As a result, various optimization strategies have been established to build efficient electrocatalysts based on graphene and its analogs. Some of the efficient strategies to increase the catalytic activities of 2D materials are alloying, doping, defect engineering, strain/stress engineering, modification of morphology, edge engineering, and designing of the heterostructure [11–17].

In this chapter, we provide an overview of recent advancements in 2D materials-based OER electrocatalysts, to clarify the correlations between engineered structures and their catalytic activity. We illustrated how 2D materials-based catalysts outperformed traditional Ru/Ir catalysts in terms of OER kinetics, overpotential, Tafel slope, and stability. For a better comprehension of OER, the mechanisms or pathways involved in OER are also briefly presented. The chapter describes how the fabrication of heterostructures, doping, alloying, defecting, and engineering of edges on 2D materials affect their performance in OER. Finally, this chapter summarizes the conclusion based on current outputs and also presents future strategies for improving the performance of OER.

4.2 OER-mechanism and insights

The overall reaction of the splitting of water molecules is signified as follows [7]:

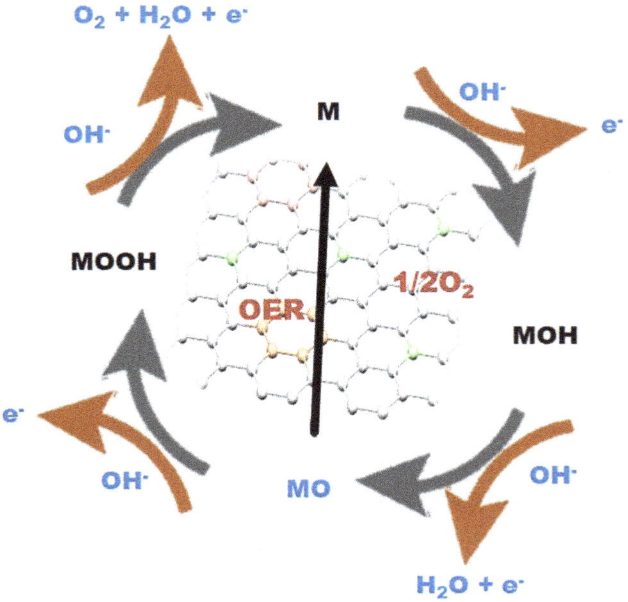

Figure 4.1. Schematic illustration of the evolution routes of the oxygen molecule in an alkaline medium.

$$H_2O \rightarrow H_2 + \frac{1}{2}O_2 \tag{4.1}$$

In acidic conditions, the following reactions take place on electrodes

$$H_2O \rightarrow 2H^+ + \frac{1}{2}O_2 + 2e^- \text{ (on anode)} \tag{4.2}$$

$$2H^+ + 2e^- \rightarrow H_2 \text{ (on cathode)} \tag{4.3}$$

while in alkaline conditions, the following reactions take place on electrodes

$$2OH^- \rightarrow H_2O + \frac{1}{2}O_2 + 2e^- \text{ (on anode)} \tag{4.4}$$

$$2H_2O + 2e^- \rightarrow H_2 + 2OH^- \text{ (on cathode)} \tag{4.5}$$

In OER, four electrons must be transferred with multi-steps in order to evolve an O_2 molecule (figure 4.1). As a result, more energy is required to overcome the kinetic barriers of the process. Therefore, more efforts are needed in the direction of breaking down the sluggish kinetics of OER. In alkaline as well as acidic medium intermediates such as –O, –OH, –OOH are present in the reactions [18]. Therefore, the O_2 molecule develops either through the joining of two –O or through the formation of –OOH.

In alkaline electrolytes or media,

$$4OH^- \rightarrow O_2 + 2H_2O + 4e^- \quad (4.6)$$

The mechanism proposed under such conditions is proceeding with the following steps:

$$M + OH^- \rightarrow MOH \quad (4.7)$$

$$MOH + OH^- \rightarrow MO + H_2O \quad (4.8)$$

Here, M is the surface of the catalyst. Then it follows either step

$$2MO \rightarrow 2M + O_2 \quad (4.9)$$

or

$$MO + OH^- \rightarrow MOOH + e^- \quad (4.10)$$

$$MOOH + OH^- \rightarrow M + O_2 + H_2O \quad (4.11)$$

In acidic electrolytes or media,

$$2H_2O \rightarrow O_2 + 2H^+ + 4e^- \quad (4.12)$$

The mechanism proceeds with the following steps:

$$M + H_2O \rightarrow MOH + H^+ + e^- \quad (4.13)$$

$$MOH + OH^- \rightarrow MO + H_2O + e^- \quad (4.14)$$

Then it proceeds with either

$$2MO \rightarrow 2M + O_2 \quad (4.15)$$

or

$$MO + H_2O \rightarrow MOOH + H^+ + e^- \quad (4.16)$$

$$MOOH + H_2O \rightarrow M + O_2 + e^- \quad (4.17)$$

In theory, the formal potential for this process is 1.229 V versus the reversible hydrogen electrode (RHE), and the standard electrode potential for the standard hydrogen electrode (SHE) is 0.401 V [19, 20]. However, in actual operations, it involves multiple steps, resulting in a high applied overpotential.

4.3 Different engineered approaches to enhance OER performance

It is necessary to carry out further research and make improvements in order to develop a competitive, low-cost water electrolysis technology for commercially viable hydrogen production. Other than the benchmark RuO_2 and IrO_2 catalysts, a wide range of catalysts have been used in OER [21, 22]. In this section, we present various strategies chosen by researchers to achieve outstanding OER performance from 2D materials.

4.3.1 Defects

Defects can control the electrical structure of active sites and make it easier for more catalytic sites to be exposed in the electrolyte [23]. $CoSe_2$ nanosheets grafted on defective graphene ($CoSe_2$@DG) electrocatalyst deliver an overpotential of 270 mV at a current density of 10 mA cm^{-2}, a Tafel slope of 64 mV/dec, and long stability (figure 4.2(a)) [11]. The $CoSe_2$@DG composite possesses the lowest Gibbs free energy and the charge depletion on $CoSe_2$ induced by the electron transfer from $CoSe_2$ to DG is the cause of the superior OER activity (figures 4.2(b) and (c)). Insertion of Se vacancies into defective $PtSe_2$ causes electron localization around the defect, which improves OER performance [24]. The defective $PtSe_2$ shows a current density of 1.54 V @10 mA cm^{-2} and a Tafel slope of 129.3 mV/dec. Rutuparna's group studied defect-engineered 2D birnessite δ-MnO_2 for OER activity [25]. The defective catalyst reveals an overpotential of 300 mV at 20 mA cm^{-2} and a Tafel slope

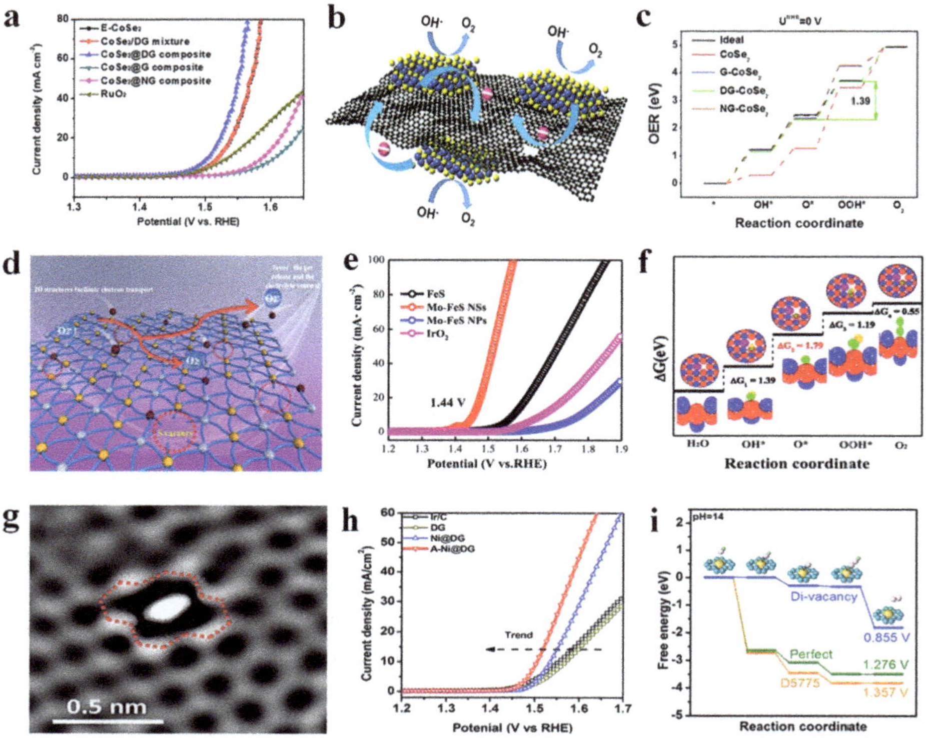

Figure 4.2. (a) LSV curves in O_2-saturated 1.0 M KOH electrolyte. (b) Schematic of the possible OER mechanism of $CoSe_2$@DG composites. (c) The free energy diagram for oxygen evolution. Reprinted from [11], Copyright (2018), with permission from Elsevier. (d) Schematic of evolution of oxygen from catalyst. (e) Polarization curves for OER. (f) Free energy diagram of Mo–FeS. Reprinted with permission from [26]. Copyright (2020) by the American Chemical Society. (g) The defective area of graphene with trapped atomic Ni. (h) LSV curves in 1.0 M KOH solution. (i) Energy profiles for OER. Reprinted from [28], Copyright (2018), with permission from Elsevier.

of 71 mV/dec in 1.0 M KOH solution. The more active surface area due to oxygen vacancies leads to quicker electron transport, which raises electrical conductivity.

As shown in figures 4.2(d)–(f), the ultrathin and amorphous Mo-FeS nanosheets with abundant sulfur defects present an overpotential of 210 mV at 10 mA cm^{-2}, a Tafel slope of 50 mV/dec, and good stability over 30 h in 1.0 M KOH solution [26]. Theoretical calculations show that the amorphous structure and sulfur-rich defects accelerate electron/mass transfer over the oxygen-containing intermediates. Also, sulfur vacancies improve the activity of its neighboring Fe-active sites. The nickel-rich nickel nitride (NR-Ni_3N) with defective graphitic carbon nitrides (GCNs) delivers an overpotential of ~290 mV at 10 mA cm^{-2}, a small Tafel slope of ~70 mV/dec and well stability until 36 h in an alkaline electrolyte [27]. The reason behind the high performance is the defective GCNs, which accelerate the generation of a Ni-rich region in Ni_3N in the nitridation process, and also its synergistic effect with robust Ni–Ni_3N heterojunction. As presented in figures 4.2(g)–(i), the defects in graphene trap atomic Ni species (A-Ni@DG) and the resultant catalyst shows an overpotential of 270 mV at 10 mA cm^{-2} [28]. The integrity of nickel species and the defects are responsible for the electrochemical reactions.

4.3.2 Alloying

It is possible to fine-tune both the electrical structure and the lattice properties more effectively by alloying various materials. Loupias *et al* prepared Ni_xFe_y nanoalloys supported on Mo_2CT_x MXene (Ni_xFe_y@Mo_2CT_x) and studied its OER activity in an alkaline medium [29]. The catalyst requires a potential of 1.50 V to reach 10 mA cm^{-2} and a Tafel slope of 34 mV/dec. According to *in situ* Raman measurements, the active phase for the OER is β-NiOOH, which is produced when the nanoalloys are oxidized under positive scan and probably contains a very small amount of Fe. Liu's group substituted Sr^{2+} for La^{3+} in $LaNiO_3$ perovskite and formed $La_{0.5}Sr_{0.5}NiO_3$ alloy showing good OER performance in 0.1 m KOH solution [12]. The $La_{0.5}Sr_{0.5}NiO_3$ catalyst required an overpotential of 0.29 V to reach the current density of 50 mA cm^{-2} (figures 4.3(a) an (b)). Alloying improves the hybridization of Ni 3d-O 2p while decreasing the charge transfer energy (figure 4.3(c)). Xie *et al* prepared molybdenum oxide decorated nickel–iron alloy nanosheets from MoO_4^{2-} intercalated layered double hydroxides (LDHs) [30]. The corresponding metal alloys made from LDHs improve conductivity. The presence of MoO_4^{2-} within the LDH layers prevents structural collapse during the high-temperature calcination to retain the layered structure of the NiFe alloy. The overpotential of NiFe–MoO_x is 276 mV at the current density of 10 mA cm^{-2} and exhibits a Tafel slope of 55 mV/dec. In addition to its high surface area and porosity, the MoO_x decorated NiFe alloy which raises the valence state of the surface Ni with the modification of MoO_x.

4.3.3 Doping

One of the most effective methods for raising catalyst OER performance has been thought to be elemental doping. During the OER process, the doping aids in optimizing the binding energy and electronic structure of the intermediates [18, 31–33].

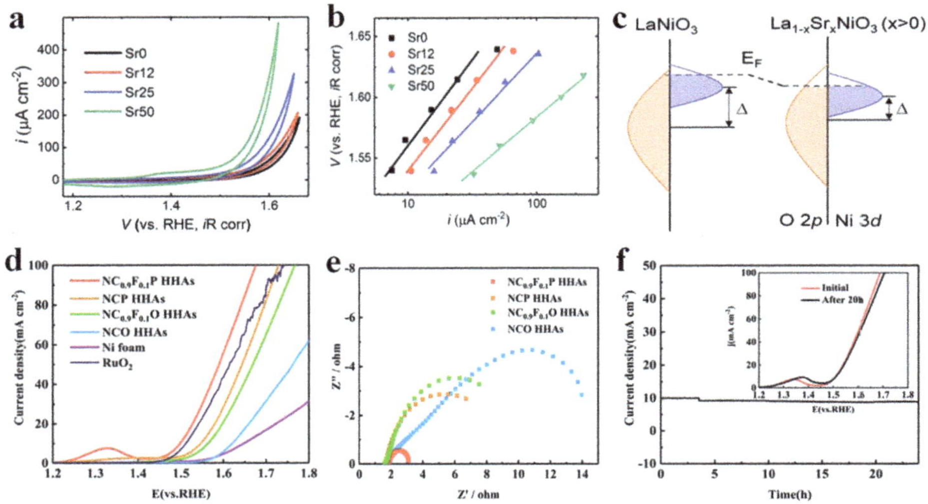

Figure 4.3. (a) Cyclic voltammetry curves in O_2-saturated 1.0 M KOH solution. (b) Tafel plot of the OER activities. (c) Schematic energy band diagram, here Δ is the charge transfer energy. Reproduced from [12] CC BY 4.0. (d) Polarization curves without iR compensation. (e) Electrochemical impedance spectroscopy for the corresponding electrocatalysts. (f) Long-time stability test at a constant current density of 10 mA cm^{-2}. Reprinted from [37], copyright (2020), with permission from Elsevier.

The catalytic properties of SnS_2 for the OER improved by doping it with cobalt in an appropriate ratio [13, 34]. The Co-doped SnS_2 nanosheet array shows an overpotential of 281 mV at 30 mA cm^{-2} and a low Tafel slope of 62 mV/dec [13]. Such performance is mainly due to high intrinsic activity and superaerophobic surface properties. In addition to improving conductivity, cobalt doping in MoS_2 also adds catalytic active sites for OER [35]. In both acidic and alkaline media, the pure MoS_2 catalyst exhibited very little OER activity, while the Co-doped MoS_2 catalysts controlled the electronic structure of MoS_2 to produce high OER. In 1.0 M KOH solution, Fe/Co/N tri-doped graphene depicts an overpotential of 288 mV at a current density of 10 mA cm^{-2}, a Tafel slope of 34.3 mV/dec, and good stability [36]. The atomically accessible iron, cobalt, and nitrogen electrocatalytic sites as well as the highly conductive graphene are attributed to the excellent OER performance. As shown in figures 4.3(d)–(f), the Fe-doped NiCoP hyperbranched hierarchical arrays grown on nickel foam ($NC_{1-x}F_xP$ HHAs) delivered an overpotential of 269 mV for OER at 10 mA cm^{-2}, under 1 M KOH solution [37]. This performance is based on more exposure to the catalytic active sites, a large electrolyte contact area, and accelerated electrolyte transport. Nitrogen-doped $Ti_3C_2T_x$ MXene exhibits an overpotential of 1.59 V to reach a current density of 10 mA cm^{-2} and a tafel slope of 76.68 mV/dec in 1 M KOH solution [38]. The energy barrier of the process is significantly lowered by the surface-adsorbed nitrogen. The Fe-doped CoP holey nanosheets show an overpotential of 220 mV at 10 mA cm^{-2} and a Tafel slope of 65.8 mV/dec [39]. The synergistic combination of Fe-doping and nanostructure fabrication is responsible for the boosted catalytic activity.

4.3.4 Strain and stress

The stability and catalytic activity of the electrocatalyst for OER have been significantly improved via strain engineering [40]. Wang *et al* explored the effect of strain on the structural and electronic properties of the $NdNiO_3$ (NNO) thin films and their OER performance (figure 4.4(a)) [40] The e_g center near the surface tends towards lower energies under compressive strain, which weakens Ni–O chemisorption and increases OER activity. The polarization curves for OER on these strained NNO films are shown in figures 4.4(b) and (c), emphasizing how compressive strain increases OER activity. Kelsey and their group moderated the tensile strain of $LaCoO_3$ and studied induced changes in the electronic structure and OER activity [14]. Tensile strain and electronic conductivity need to be balanced in order to maximize catalytic activity. Qixiang *et al* studied the OER activity with freestanding single-crystalline $SrRuO_3$ thin film [41]. The high/low-spin state transition of Ru occurs in the strain-engineered $SrRuO_3$. Through the cylindrical $SrRuO_3$, the overpotential can be ~74% reduced from 0.35 to 0.09V, while a Tafel slope from ~97 to ~31mV/dec (at 5 mA cm^{-2}) in 1.0 M KOH solution. In the $HClO_4$ electrolyte, the overpotential reduced from 0.31 to 0.07V at the current density of 5 mA cm^{-2} and a Tafel slope from ~92 to ~29mV/dec in $HClO_4$. As shown in figures 4.4(d)–(f), Benson *et al* showed that straining a thin film of n-doped rutile TiO_2 up to 3% tensile strain rises OER activities [42]. The potential essential to pass 1mA cm^{-2} shifts cathodically 89mV from 0% to 3% strain. Increased surface active

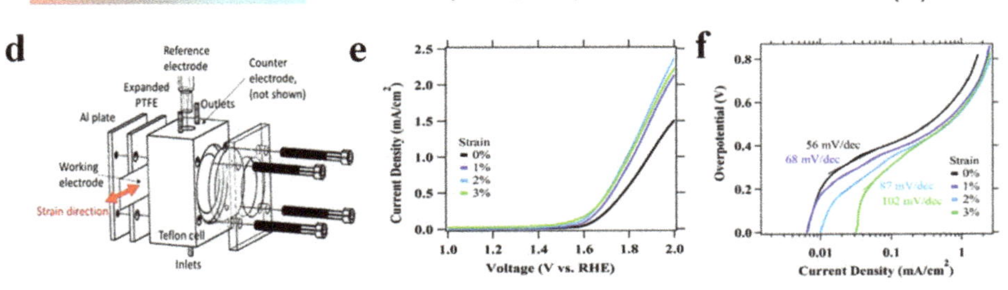

Figure 4.4. (a) Schematic of strain effect on OER performance of epitaxial $NdNiO_3$ thin films. (b) Cyclic voltammetry curves for the OER at 10 mV s^{-1} in 0.1 M KOH. (c) NNO films as a function of strain. Inset: NiO_6 octahedral rotations due to strain. Reprinted with permission from [40]. Copyright (2019) by the American Chemical Society. (d) Schematic of the electrochemical cell used for OER measurements under tensile strain. (e) LSV for TiO_2 films under tensile strain (0%–3%). (f) Corresponding Tafel plots. Reproduced from [42] CC BY 4.0.

site concentration and decreased kinetic and thermodynamic barriers along the reaction pathways result in notable increases in OER activity with tensile strain.

4.3.5 Morphology

The perceived activity of a typical electrocatalyst is significantly influenced by its morphology [43]. As a result, material morphology engineering is required, because a high specific surface area with higher active site exposure is required for significant catalytic activity. In between nanosheets, nanoflowers, nanotubes, and aggregations kind of morphologies of FeNi-MOF, nanosheets revealed a low overpotential of 246 mV at 10 mA cm^{-2} and a Tafel slope of 31.2 mV/dec [15]. The ideal morphology increases the number of active sites while also fine-tuning the electronic structure, which increases the intrinsic activity of the electrocatalysts. Cheng's group prepared structures like nanoboxes, nanocubes, nanoplates, and nanosheets of a 2D Fe–Ni MOF, and investigated their OER performance in 1.0 M KOH electrolyte [44]. The nanoboxes (FeNi–B) exhibit a low overpotential of 285 mV at 10 mA cm^{-2} and a Tafel value of 50.9 mV/dec (figures 4.5(a)–(c)). This high performance is attributed to the greater active surface area and higher intrinsic activity of the exposed crystal planes. The long-perimeter microsheets of $CuCo_2S_4$ show an overpotential of 283 mV to reach a current density of 20 mA cm^{-2} and a Tafel slope of 89 mV/dec [45]. As compared to microwires, the microsheets exhibit a long active edge perimeter, low charge-transfer resistance, and high electrochemical surface area. The $CoSe_2$ microspheres showed a small overpotential of 325 mV at 10 mA cm^{-2} and a Tafel slope of 80 mV/dec in alkaline media [46]. This high performance of spheres compared with wires and rods is due to the large specific surface area and a great number of channels for promoting mass transfer efficiency.

4.3.6 Edge

The layered materials' significant anisotropic properties, which have an impact on the basal and edge planes, provide them with new physicochemical properties [47]. In light of the fact that edge sites have an impact on catalysts' electronic structures, adding more edge sites to catalysts was therefore thought to be an extremely efficient approach to further improve OER performance. Wang *et al* performed the exfoliation of bulk CoFe LDHs into edge-rich ultrathin LDHs nanosheets with nitrogen doping and studied for OER activity in 1.0 M KOH electrolyte [16]. The catalyst delivered an overpotential of 281 mV at the current density of 10 mA cm^{-2} and a Tafel slope from 40.03 mV/dec. The N-CoFe LDHs nanosheets on Ni foam exhibited OER capability with an overpotential of 233 mV at 10 mA cm^{-2}. The improvement of the electrocatalytic activity from the edge and corner reactive sites, as well as the rearranging of the electronic configurations caused by the nitrogen dopant, worked in harmony to improve OER performance. As shown in figures 4.5(d)–(f), a layer of defective single-layer graphene is produced by the formation of oxygen bubbles, and higher scan rates result in more extended and critical structural damage [48]. At pristine graphene, the edges are where the electron transport characteristics first occur. The consecutive OER scans damage the basal

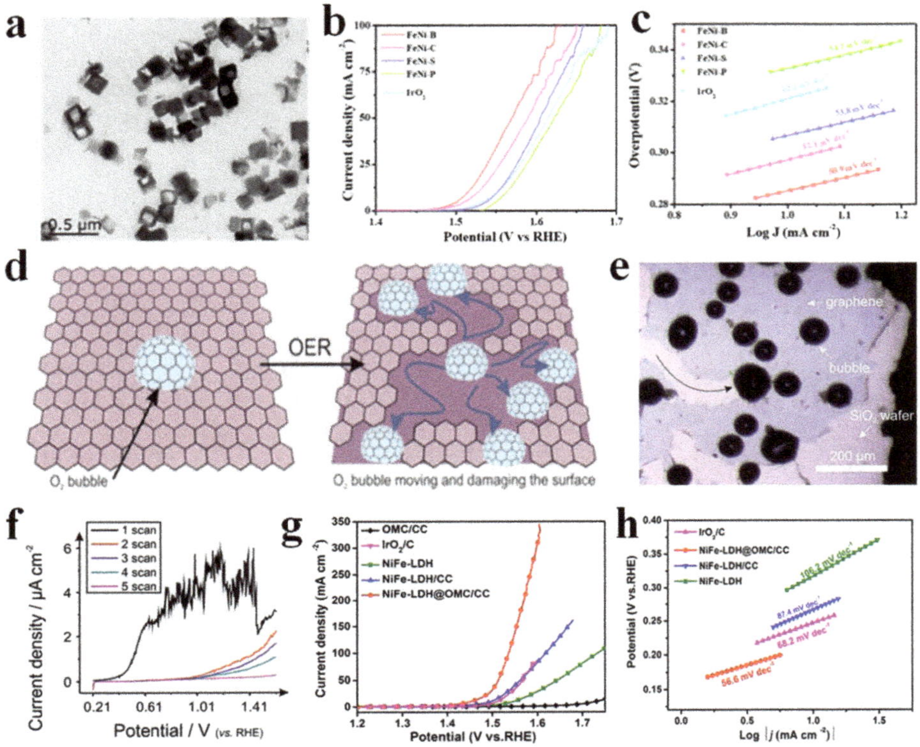

Figure 4.5. (a) Transmission electron microscopy (TEM) images of FeNi–B. (b) LSV polarization curves. (c) Tafel plots. Reprinted with permission from [44]. Copyright (2022) by the American Chemical Society. (d) Schematic overview of oxygen bubbles inducing structural damage to the surface of the mono-layer graphene. (e) Screen-shot images captured from an *in situ* experiment, which show the evolution of oxygen bubbles. (f) LSV plots in 0.1 M KOH solution. Reproduced from [48] CC BY 4.0. (g) LSV curves in 1.0 M KOH electrolyte. (h) Corresponding Tafel plots. Reprinted from [50] Copyright (2021), with permission from Elsevier.

plane of the graphene sheet and result in a weak electrochemical signal due to a loss in electrically conductive pathways. The Fe-MoS_2 reaches a current density of 50 mA cm^{-2} at 290 mV, shows a Tafel slope of 72 mV/dec, and has suitable stability in a 1.0 M KOH medium [49]. MoS_2 layers that are vertically oriented maximize the exposure of the edge sites, which is beneficial for the enhancement of electrochemical performance. The ultra-small NiFe-LDH nanoparticles with less shape anisotropy increase the number of edge active sites, which results in improved OER performance [50]. The catalyst exhibited an overpotential of 223 mV at 10 mA cm^{-2}, a Tafel slope of 56.6 mV/dec, and high mass activity (figures 4.5(g) and (h)).

4.3.7 Heterostructuring

The flexibility of 2D heterostructures allows for the application of OER through surface chemistry controls and electronic structure modulations. As shown in

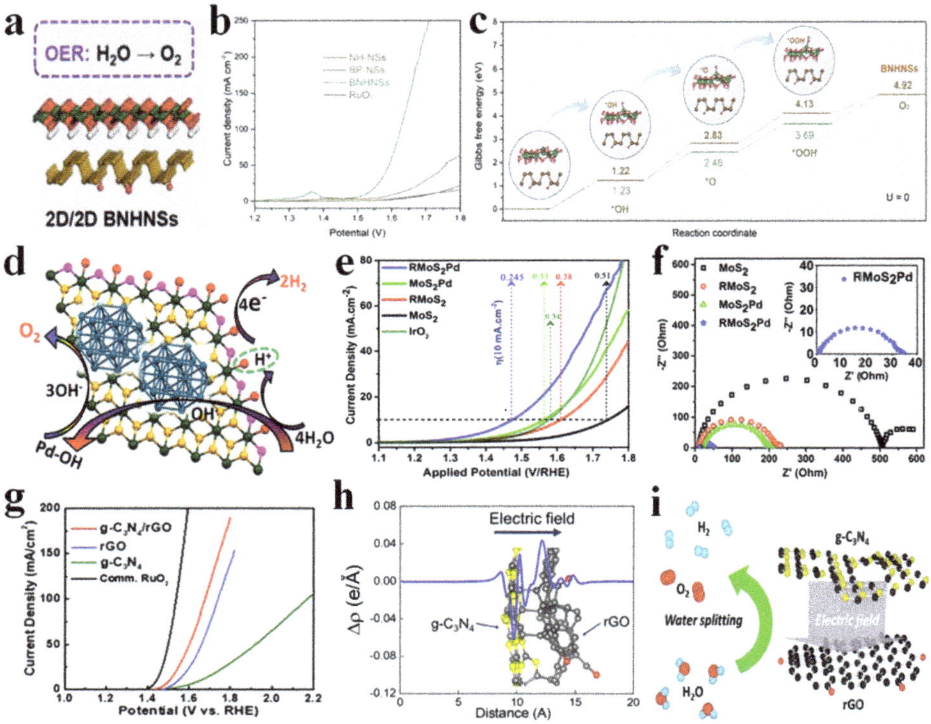

Figure 4.6. (a) Schematic of 2D/2D BNHNSs for OER. (b) LSV plots of catalyst. (c) Calculated free energy profiles for OER of 2D/2D BNHNSs (brown line) at the $U = 0$ V as compared with the ideal OER catalyst (green line). Reproduced from [51] CC BY 4.0. (d) Schematic of the proposed mechanism for OER on the $RMoS_2Pd$ catalyst. (e) Polarization curve. (f) Nyquist plot. Reprinted with permission from [52]. Copyright (2019) by the American Chemical Society. (g) LSV curves in 1.0 M KOH medium. (h) Plane-averaged charge density difference between $g\text{-}C_3N_4$ and rGO. (i) Schematic of the evolution of oxygen molecule from $g\text{-}C_3N_4$/rGO electrocatalysts. Reprinted from [17], Copyright (2022), with permission from Elsevier.

figure 4.6(a), Mei *et al*, fabricated 2D/2D heterostructured nanosheets with black phosphorus and $Ni(OH)_2$ nanosheets and studied their performance in 1.0 M KOH electrolyte for OER [51]. The heterostructured catalyst shows an overpotential of 297 mv at 10 mA cm^{-2}, which is 22% less than bare $Ni(OH)_2$ and 34% less than black phosphorus nanosheets (figure 4.6(b)). The purposeful combination of the two materials enhanced their synergistic effects and produced a heterostructure with significant interfacial interaction (figure 4.6(c)). As a result, surface adsorption and electron transport are optimized. For OER, the heterostructure interface of palladium (Pd) and reduced graphene oxide (RGO)-supported molybdenum disulfide (MoS_2) (RGO/MoS_2/Pd) shows an overpotential of 245 mV at 10 mA cm^{-2} and a Tafel slope of 42 mV/dec in an alkaline electrolyte (figures 4.6(d)–(f)) [52]. The combined synergetic effect of 1T MoS_2, sulfur vacancy, and conducting RGO sheet accelerate the OER activity. The 1,4-benzenedicarboxylate $C_8H_4O_4$ and $Ni(OH)_2$

(Ni-BDC/Ni$(OH)_2$) heterostructures show the current density of 82.5 mA cm^{-2} at 1.6 V due to high activity and favorable kinetics [53]. This is 5.5, 20.6, and 3.0 times higher than Ni-BDC, Ni$(OH)_2$, and Ir/C, respectively. The electronic structure of Ni $(OH)_2$ is well modified as a result of the strong electron interactions between Ni $(OH)_2$ and Ni-BDC, resulting in the production of Ni$(OH)_2$ with higher oxidation states. The carbon-based heterostructured electrocatalyst graphitic carbon nitride and rGO (g-C_3N_4/rGO) with 2D/2D structure exhibited lower overpotential (272 mV @10 mA cm^{-2}) and Tafel slope (97 mV/dec) for OER [17]. An intense electric field that makes it easy for electrons to move between electrodes. Theoretical calculations reveal that the electron is spontaneously transferred from g-C_3N_4 to rGO and the charge transfer induces an electric field by the interface dipole moment formation (figures 4.6(g)–(i)). The CoFe oxide/black phosphorus nanosheet heterostructure shows an overpotential of 51 mV at 10 mA cm^{-2} which is lower than that of the commercial RuO_2 catalyst [54]. The combination of these two materials offers more active sites for OER performance.

Table 4.1 shows the overpotential and Tafel slope values recorded in respective electrolytes for engineered 2D materials while evaluation of OER activity.

4.4 Conclusion

Due to its slow kinetics, the OER is frequently regarded as the main bottleneck in water splitting. The key to the widespread adoption of these energy-related technologies is the generation of high-performance and easily available OER catalysts. Therefore, this chapter presented recent advancements and progress on 2D materials as efficient and cost-effective electrocatalysts for OER. Some effective ways for increasing the catalytic activity of 2D materials include alloying, doping, defect engineering, strain/stress engineering, morphological alteration, edge engineering, and heterostructure design.

Research on 2D material alloys for OER activity is now in its early stages. Consequently, additional research into alloys between distinct periodic table groups might be advantageous for electronic structure engineering. The priority of ordered structural control in alloy synthesis must be considered. High OER performance can be attained through doping, which effectively modifies the catalyst surface. However, there is still a problem with optimized and regulated doping that needs to be resolved. In order to improve catalytic behavior, it is essential to identify the defect's basic structure and the process that led to its creation. The defects have a significant impact on the optoelectronic structure, conductivity, and intrinsic reactivity. Despite the fact that strain engineering improves the physicochemical properties of electrocatalysts, one of the biggest challenges has been how to scale up strain effects in real-world applications. In order to maximize catalytic activity, harmony within the tensile strain and electronic conductivity needs to be obtained. Edge sites have an impact on the electronic structures of catalysts, hence adding more edge sites would considerably improve OER performance. For achieving high OER performance, one intriguing strategy is to design a series of innovative catalysts with distinctive shapes and structures. The high synergistic effect between

Table 4.1. The performance of various engineered 2D materials-based electrocatalysts towards OER.

Engineered technique	Material	Electrolyte	Overpotential	Tafel slope	References
Defect	$CoSe_2$@DG	1.0 M KOH	270 mV @10 mA cm^{-2}	64 mV/dec	[11]
	δ-MnO_2	1.0 M KOH	300 mV @20 mA cm^{-2}	71 mV/dec	[25]
	Mo-FeS	1.0 M KOH	210 mV @10 mA cm^{-2}	50 mV/dec	[26]
	NR-Ni_3N/GCNs	1.0 M KOH	290 mV @10 mA cm^{-2}	70 mV/dec	[27]
	A-Ni@DG	1.0 M KOH	270 mV @10 mA cm^{-2}	47 mV/dec	[28]
Alloy	Ni_xFe_y@Mo_2CT_x	5.0 M KOH	1.50 V @10 mA cm^{-2}	34 mV/dec	[29]
	NiFe–MoO_x	1.0 M KOH	276 mV @10 mA cm^{-2}	50 mV/dec	[30]
Doping	Co-doped SnS_2	1.0 M KOH	281 mV @30 mA cm^{-2}	62 mV/dec	[13]
	Fe/Co/N tri-doped graphene	1.0 M KOH	288 mV @10 mA cm^{-2}	34.3 mV/dec	[36]
	Fe-doped NiCoP	1.0 M KOH	269 mV @10 mA cm^{-2}	72.01 mV/dec	[37]
	Fe-doped CoP	1.0 M KOH	220 mV @10 mA cm^{-2}	65.8 mV/dec	[39]
Strain and stress	$SrRuO_3$	1.0 M KOH	0.09 V @5 mA cm^{-2}	31 mV/dec	[41]
	TiO_2	1.0 M NaOH	1.80 mV @1 mA cm^{-2}	102 mV/dec	[42]
Morphology	FeNi	1.0 M KOH	285 mV @10 mA cm^{-2}	50.9 mV/dec	[44]
	$CuCo_2S_4$	1.0 M KOH	283 mV @20 mA cm^{-2}	89 mV/dec	[45]
	$CoSe_2$	1.0 M KOH	325 mV @20 mA cm^{-2}	80 mV/dec	[46]
Edge	N-CoFe	1.0 M KOH	281 mV @10 mA cm^{-2}	40.03 mV/dec	[16]
	Fe-MoS_2	1.0 M KOH	290 mV @50 mA cm^{-2}	72 mV/dec	[49]
	NiFe-LDH	1.0 M KOH	223 mV @10 mA cm^{-2}	56.6 mV/dec	[50]
Heterostructuring	Black Phosphorus/ Nickel Hydroxide	1.0 M KOH	297 mV @10 mA cm^{-2}	100 mV/dec	[51]
	RGO/MoS_2/Pd	1.0 M KOH	245 mV @10 mA cm^{-2}	42 mV/dec	[52]
	g-C_3N_4/rGO	1.0 M KOH	272 mV @10 mA cm^{-2}	97 mV/dec	[17]

two or more materials increasing the reaction kinetics makes the design and manufacture of heterostructure an effective method.

Carbon-based materials such as graphene, carbon nanotubes, activated carbon, and graphitic carbon nitride offer excellent conductivity, flexibility, and favorable durability. However, their relatively low catalytic activity can be possible to improve by doping, defects, and heterostructure formation. The incorporation of carbon-based materials into OER catalysts could significantly reduce the cost of producing water oxidation by utilizing the simple and affordable accessibility of these materials. Several kinds of 2D materials, such as MXenes, transition metal dichalcogenides, perovskites, LDH, etc, are understudied and require additional focus for OER investigations in addition to carbon materials and metal oxides. The thoughtful design of these classes will undoubtedly provide favorable outcomes in this area. There is a scope of in-depth studies of the mechanism of oxygen evolution with 2D material electrocatalysts. The combination of theoretical and experimental studies will help to better understand insights into the reaction. Overall, we believe that the easy-to-scale engineering of 2D materials can pave the way for the pursuit of cost-effectiveness for electrocatalytic OER applications.

Acknowledgments

The authors gratefully acknowledge financial assistance from the SERB Core Research Grant (Grant No. CRG/2022/000897), Department of Science and Technology (DST/NM/NT/2019/205(G)), and Minor Research Project Grant, Jain University (JU/MRP/CNMS/29/2023).

References

[1] Shinde P V, Mane P, Late D J, Chakraborty B and Rout C S 2021 Promising 2D/2D $MoTe_2$/$Ti_3C_2T_x$ hybrid materials for boosted hydrogen evolution reaction *ACS Appl. Energy Mater.* **4** 11886–97

[2] Zhao G, Rui K, Dou S X and Sun W 2018 Heterostructures for electrochemical hydrogen evolution reaction: a review *Adv. Funct. Mater.* **28** 1803291

[3] Shinde P V, Gavali D S, Thapa R, Singh M K and Rout C S 2021 Ternary VS_2/ZnS/CdS hybrids as efficient electrocatalyst for hydrogen evolution reaction: experimental and theoretical insights *AIP Adv.* **11** 105010

[4] Shinde P V, Mane P, Chakraborty B and Rout C S 2021 Spinel $NiFe_2O_4$ nanoparticles decorated 2D Ti_3C_2 MXene sheets for efficient water splitting: experiments and theories *J. Colloid Interface Sci.* **602** 232–41

[5] Patil K N, Shinde P V, Srinivasappa P M, Nabgan W, Chaudhari N K, Rout C S and Jadhav A H 2022 Rational competent electrocatalytic oxygen evolution reaction on stable tailored ternary MoO_3-NiO-activated carbon hybrid catalyst *Int. J. Energy Res.* **46** 12549–64

[6] Wang H, Zhang K H, Hofmann J P and Oropeza F E 2021 The electronic structure of transition metal oxides for oxygen evolution reaction *J. Mater. Chem.* A **9** 19465–88

[7] Cheng J and Wang D 2022 2D materials modulating layered double hydroxides for electrocatalytic water splitting *Chin. J. Catal.* **43** 1380–98

[8] Sun T, Zhang G, Xu D, Lian X, Li H, Chen W and Su C 2019 Defect chemistry in 2D materials for electrocatalysis *Mater. Today Energy* **12** 215–38

[9] Shinde P V, Samal R and Rout C S 2022 Comparative electrocatalytic oxygen evolution reaction studies of spinel $NiFe_2O_4$ and its nanocarbon hybrids *Trans. Tianjin Univ.* **28** 80–8

[10] Liu Q, Chen J, Yang P, Yu F, Liu Z and Peng B 2021 Directly application of bimetallic 2D-MOF for advanced electrocatalytic oxygen evolution *Int. J. Hydrogen Energy* **46** 416–24

[11] Wang X *et al* 2018 Grafting cobalt diselenide on defective graphene for enhanced oxygen evolution reaction *Iscience* **7** 145–53

[12] Liu J *et al* 2019 Tuning the electronic structure of $LaNiO_3$ through alloying with strontium to enhance oxygen evolution activity *Adv. Sci.* **6** 1901073

[13] Jiang M, Huang Y, Sun W and Zhang X 2019 Co-doped SnS_2 nanosheet array for efficient oxygen evolution reaction electrocatalyst *J. Mater. Sci.* **54** 13715–23

[14] Stoerzinger K A, Choi W S, Jeen H, Lee H N and Shao-Horn Y 2015 Role of strain and conductivity in oxygen electrocatalysis on $LaCoO_3$ thin films *J. Phys. Chem. Lett.* **6** 487–92

[15] Li Y, Lu M, Wu Y, Ji Q, Xu H, Gao J, Qian G and Zhang Q 2020 Morphology regulation of metal–organic framework-derived nanostructures for efficient oxygen evolution electrocatalysis *J. Mater. Chem.* A **8** 18215–9

[16] Wang Y, Xie C, Zhang Z, Liu D, Chen R and Wang S 2018 *In situ* exfoliated, N-doped, and edge-rich ultrathin layered double hydroxides nanosheets for oxygen evolution reaction *Adv. Funct. Mater.* **28** 1703363

[17] Choi H, Surendran S, Sim Y, Je M, Janani G, Choi H, Kim J K and Sim U 2022 Enhanced electrocatalytic full water-splitting reaction by interfacial electric field in 2D/2D heterojunction *Chem. Eng. J.* **450** 137789

[18] Suen N T, Hung S F, Quan Q, Zhang N, Xu Y J and Chen H M 2017 Electrocatalysis for the oxygen evolution reaction: recent development and future perspectives *Chem. Soc. Rev.* **46** 337–65

[19] Mefford J T *et al* 2021 Correlative operando microscopy of oxygen evolution electrocatalysts *Nature* **593** 67–73

[20] Nong H N *et al* 2020 Key role of chemistry versus bias in electrocatalytic oxygen evolution *Nature* **587** 408–13

[21] Shinde P V, Babu S, Mishra S K, Late D, Rout C S and Singh M K 2021 Tuning the synergistic effects of Mos_2 and spinel $NiFe_2O_4$ nanostructures for high performance energy storage and conversion applications *Sustain. Energy Fuels* **5** 3906–17

[22] Samal R, Shinde P V and Rout C S 2021 2D Vanadium diselenide supported on reduced graphene oxide for water electrolysis: a comprehensive study in alkaline media *Emergent Mater.* **4** 1047–53

[23] Yan D, Li Y, Huo J, Chen R, Dai L and Wang S 2017 Defect chemistry of nonprecious-metal electrocatalysts for oxygen reactions *Adv. Mater.* **29** 1606459

[24] Chang Y, Zhai P, Hou J, Zhao J and Gao J 2022 Excellent HER and OER catalyzing performance of Se-vacancies in defects-engineered $PtSe_2$: from simulation to experiment *Adv. Energy Mater.* **12** 2102359

[25] Samal R, Kandasamy M, Chakraborty B and Rout C S 2021 Experimental and theoretical realization of an advanced bifunctional 2D δ-MnO_2 electrode for supercapacitor and oxygen evolution reaction via defect engineering *Int. J. Hydrogen Energy* **1046** 28028–42

[26] Shao Z, Meng H, Sun J, Guo N, Xue H, Huang K, He F, Li F and Wang Q 2020 Engineering of amorphous structures and sulfur defects into ultrathin FeS nanosheets to

achieve superior electrocatalytic alkaline oxygen evolution *ACS Appl. Mater. Interfaces* **12** 51846–53
[27] Luo X, Ma H, Gao J, Yu L, Gu X and Liu J 2022 Nickel-rich Ni_3N particles stimulated by defective graphitic carbon nitrides for the effective oxygen evolution reaction *Ind. Eng. Chem. Res.* **61** 2081–90
[28] Zhang L *et al* 2018 Graphene defects trap atomic Ni species for hydrogen and oxygen evolution reactions *Chem.* **4** 285–97
[29] Loupias L *et al* 2023 Mo_2CT_x MXene supported nickel-iron alloy: an efficient and stable heterostructure to boost oxygen evolution reaction *2D Mater.* **10** 024005
[30] Xie C, Wang Y, Hu K, Tao L, Huang X, Huo J and Wang S 2017 *In situ* confined synthesis of molybdenum oxide decorated nickel–iron alloy nanosheets from MoO_4^{2-} intercalated layered double hydroxides for the oxygen evolution reaction *J. Mater. Chem.* A **5** 87–91
[31] Kanan M W and Nocera D G 2008 *In situ* formation of an oxygen-evolving catalyst in neutral water containing phosphate and Co^{2+} *Sci.* **321** 1072–5
[32] Zou X, Su J, Silva R, Goswami A, Sathe B R and Asefa T 2013 Efficient oxygen evolution reaction catalyzed by low-density Ni-doped Co_3O_4 nanomaterials derived from metal-embedded graphitic C_3N_4 *Chem. Commun.* **49** 7522–54
[33] Lu B, Cao D, Wang P, Wang G and Gao Y 2011 Oxygen evolution reaction on Ni-substituted Co_3O_4 nanowire array electrodes *Int. J. Hydrogen Energy* **36** 72–8
[34] Park J H, Ro J C and Suh S J 2022 Facile synthesis of co-doped SnS_2 as a pre-catalyst for efficient oxygen evolution reaction *Curr. Appl Phys.* **42** 50–9
[35] Xiong Q, Zhang X, Wang H, Liu G, Wang G, Zhang H and Zhao H 2018 One-step synthesis of cobalt-doped MoS_2 nanosheets as bifunctional electrocatalysts for overall water splitting under both acidic and alkaline conditions *Chem. Commun.* **54** 3859–62
[36] Wang W, Babu D D, Huang Y, Lv J, Wang Y and Wu M 2018 Atomic dispersion of Fe/Co/N on graphene by ball-milling for efficient oxygen evolution reaction *Int. J. Hydrogen Energy* **43** 10351–8
[37] Qi Y, Zhang Q, Meng S, Li D, Wei W, Jiang D and Chen M 2020 Iron-doped nickle cobalt ternary phosphide hyperbranched hierarchical arrays for efficient overall water splitting *Electrochim. Acta* **334** 135633
[38] Chen X, Zhai X, Hou J, Cao H, Yue X, Li M, Chen L, Liu Z, Ge G and Guo X 2021 Tunable nitrogen-doped delaminated 2D MXene obtained by NH_3/Ar plasma treatment as highly efficient hydrogen and oxygen evolution reaction electrocatalyst *Chem. Eng. J.* **420** 129832
[39] Xu S, Qi Y, Lu Y, Sun S, Liu Y and Jiang D 2021 Fe-doped CoP Holey nanosheets as bifunctional electrocatalysts for efficient hydrogen and oxygen evolution reactions *Int. J. Hydrogen Energy* **46** 26391–401
[40] Wang L *et al* 2019 Strain effect on oxygen evolution reaction activity of epitaxial $NdNiO_3$ thin films *ACS Appl. Mater. Interfaces* **11** 12941–7
[41] Wang Q *et al* 2022 Enhanced oxygen evolution reaction by stacking single-crystalline freestanding $SrRuO_3$ *Appl. Catal.* B **317** 121781
[42] Benson E E, Ha M A, Gregg B A, van de Lagemaat J, Neale N R and Svedruzic D 2019 Dynamic tuning of a thin film electrocatalyst by tensile strain *Sci. Rep.* **9** 15906
[43] Shinde P, Rout C S, Late D, Tyagi P K and Singh M K 2021 Optimized performance of nickel in crystal-layered arrangement of $NiFe_2O_4$/rGO hybrid for high-performance oxygen evolution reaction *Int. J. Hydrogen Energy* **46** 2617–29

[44] Cheng J, Shen X, Chen H, Zhou H, Chen P, Ji Z, Xue Y, Zhou H and Zhu G 2022 Morphology-dependent electrocatalytic performance of a two-dimensional nickel–iron MOF for oxygen evolution reaction *Inorg. Chem.* **61** 7095–102

[45] Lu M, Cui X, Song B, Ouyang H, Wang K and Wang Y 2020 Studying the effect of $CuCo_2S_4$ morphology on the oxygen evolution reaction using a flexible carbon cloth substrate *ChemElectroChem.* **7** 1080–3

[46] Lan K, Li J, Zhu Y, Gong L, Li F, Jiang P, Niu F and Li R 2019 Morphology engineering of $CoSe_2$ as efficient electrocatalyst for water splitting *J. Colloid Interface Sci.* **539** 646–53

[47] Lu Z *et al* 2017 Identifying the active surfaces of electrochemically tuned $LiCoO_2$ for oxygen evolution reaction *J. Am. Chem. Soc.* **139** 6270–6

[48] García-Miranda Ferrari A, Brownson D A and Banks C E 2019 Investigating the integrity of graphene towards the electrochemical oxygen evolution reaction *ChemElectroChem.* **6** 5446–53

[49] Tang B, Yu Z G, Seng H L, Zhang N, Liu X, Zhang Y W, Yang W and Gong H 2018 Simultaneous edge and electronic control of MoS_2 nanosheets through Fe doping for an efficient oxygen evolution reaction *Nanoscale* **10** 20113–9

[50] Huang G, Zhang C, Liu Z, Yuan S, Yang G and Li N 2021 Ultra-small NiFe-layered double hydroxide nanoparticles confined in ordered mesoporous carbon as efficient electrocatalyst for oxygen evolution reaction *Appl. Surf. Sci.* **565** 150533

[51] Mei J, Shang J, He T, Qi D, Kou L, Liao T, Du A and Sun Z 2022 2D/2D black phosphorus/nickel hydroxide heterostructures for promoting oxygen evolution via electronic structure modulation and surface reconstruction *Adv. Energy Mater.* **12** 2201141

[52] Pandey A, Mukherjee A, Chakrabarty S, Chanda D and Basu S 2019 Interface engineering of an RGO/MoS_2/Pd 2D heterostructure for electrocatalytic overall water splitting in alkaline medium *ACS Appl. Mater. Interfaces* **11** 42094–103

[53] Zhu D, Liu J, Wang L, Du Y, Zheng Y, Davey K and Qiao S Z 2019 A 2D metal–organic framework/$Ni(OH)_2$ heterostructure for an enhanced oxygen evolution reaction *Nanoscale* **11** 3599–605

[54] Zhao M, Cheng X, Xiao H, Gao J, Xue S, Wang X, Wu H, Jia J and Yang N 2023 Cobalt-iron oxide/black phosphorus nanosheet heterostructure: electrosynthesis and performance of (photo-) electrocatalytic oxygen evolution *Nano Res.* **16** 6057–66

Chapter 5

Engineered 2D materials for oxygen reduction reaction (ORR)

Abhinandan Patra and Chandra Sekhar Rout

As the pressing need for the progression of energy conversion and storage technologies looms large, research endeavours have undergone a conspicuous intensification within the sphere of oxygen reduction reactions (ORRs). ORR is pivotal in various applications, including fuel cells, metal–air batteries, and environmental remediation. However, the quest for efficient and cost-effective ORR catalysts is fraught with challenges, encompassing issues of catalytic activity, durability, and scalability. Two-dimensional (2D) materials, such as graphene, transition metal oxides (TMOs), transition metal dichalcogenides (TMDs), transition metal trichalcogenides (TMTCs), and MXenes, have emerged as frontrunners in the pursuit of superior ORR catalysts. This chapter explores their potential by delving into the fundamental mechanisms, strategies for catalytic enhancement, and considerations for material stability. Techniques including doping, alloying, defect engineering, edge activation, and heterostructuring are elucidated as effective means of elevating 2D materials to the forefront of ORR catalysts. The imperative for interdisciplinary collaboration, encompassing theoretical and experimental approaches, is underscored. In conclusion, 2D materials hold substantial promise in reshaping the landscape of ORR catalysis, offering pathways toward sustainable and efficient energy conversion technologies. Nevertheless, challenges persist at the crossroads of materials science, electrochemistry, and catalysis, necessitating concerted happenings to unlock their full potential.

5.1 Introduction

The oxygen reduction reaction (ORR) holds a pivotal role in scientific research across diverse disciplines. It underpins the advancement of energy conversion and storage technologies, informs environmental studies by relating to microbial respiration and pollution control, and drives materials science, catalysis research,

doi:10.1088/978-0-7503-5719-7ch5

© IOP Publishing Ltd 2024

and nanotechnology through the quest for improved ORR catalysts and nanomaterials [1, 2]. In the realm of electrochemistry, ORR serves as a fundamental process in technologies such as hydrogen fuel cells and metal–air batteries, where it underpins electricity generation and energy storage [3–5]. The high cost, limited availability, and susceptibility to degradation of noble metal catalysts, like platinum, palladium etc have prompted the rise of 2D materials as promising alternatives for ORR in energy applications [5–8]. 2D materials such as graphene, TMOs, TMDs, TMTCs and MXenes etc offer cost-effectiveness, improved durability, high surface area, high conductivity, ample active sites/edges and environmental sustainability, while their tunable physicochemical properties enable precise control over ORR kinetics, making them a compelling choice to address the challenges associated with noble metal catalysts [9, 10]. The basic mechanism of the ORR on 2D electrocatalysts involves the adsorption of oxygen molecules, followed by electron transfer and the formation of hydroxide ions. The efficiency of these steps is influenced by the 2D material's properties, making it crucial to engineer and optimize these materials for enhanced ORR performance in energy applications [11, 12].

However, challenges in developing 2D materials as ORR catalysts include achieving competitive catalytic activity with noble metals, ensuring long-term durability, and optimizing mass transport. Enhancing 2D materials for the ORR in electrocatalysis can be achieved through numerous strategies like doping, alloying, heterostructuring, morphology control, edges, and defect engineering which has been the prime focus in this book chapter [13–16]. Doping with foreign elements can modify electronic properties, improving ORR catalytic activity. Alloying, often with transition metals, can enhance activity and reduce the use of expensive noble metals. Heterostructures formed by stacking different 2D materials create synergistic effects, benefiting ORR performance. Morphology control and defect engineering can expose more active sites and improve mass transport, ultimately optimizing 2D materials for efficient ORR in applications such as fuel cells and metal–air batteries [13–17]. In a nutshell, this chapter on the electrocatalytic ORR of 2D materials has provided insights into both the fundamental mechanisms of ORR on these materials and innovative approaches for enhancing catalytic activity through defects, edges, morphology, doping, and heterostructures. These discoveries bear the potential to propel the significance of 2D materials in the realm of sustainable and efficient ORR electrocatalysis, thereby making notable contributions to the development of cleaner and more environmentally friendly energy technologies.

5.2 ORR-mechanism and insights

ORR is of great importance in the development of energy conversion devices, such as fuel cells, where it serves as the cathodic half-reaction, converting oxygen into water while releasing energy that can be harnessed as electrical power through fuel cells and batteries [5–8]. Researchers in the scientific field study ORR to develop more efficient catalysts and better understand the underlying mechanisms for applications in clean energy production and environmental protection. The presence of a substantial amount of oxygen in the atmosphere makes it the preferred oxidizing agent for cathodes in fuel cells. However, the O_2 molecule's high bond energy, which is 498 kJ mol^{-1}, imparts significant stability, resulting in ORR being notably slow [18–20]. Typically, ORR occurring on cathodes necessitates high overpotentials, exceeding 0.3 V (depending on the electrode material), when electrocatalysts are not employed owing to the sluggish reaction kinetics and ORR rate hysteresis [18–21]. The electrochemical reduction of oxygen involves a complex mechanism characterized by multi-electron irreversible transfer processes spanning numerous elementary steps and involving various intermediate species. This mechanism is influenced by factors such as reaction temperature, oxygen partial pressure, and potential. Without considering specific reaction process details, ORR on the electrode can be categorized as either 'partial' 2-electron reduction or 'direct' four-electron reduction, with the key distinction being the presence or absence of peroxide intermediates in the liquid phase, whether the environment is alkaline or acidic [17–20].

The $2e^-$ reduction of O_2 produces intermediate oxygen compounds that may revert to O_2, diminishing battery efficiency and posing electrode damage risks. In contrast, four-electron reduction yields H_2O (in acid) or OH^- (in alkaline conditions) directly from O_2 [19, 20]. Most electrodes favour $2e^-$ or combined $2e^-/4e^-$ pathways due to O_2's high bond energy, making direct $4e^-$ reduction challenging. The two-electron pathway generates H_2O_2 or HO_2 with lower activation energy (146 kJ mol^{-1}), promoting ORR likelihood. In fuel cells (FCs), the essential $4e^-$ process requires selective catalysts and high potential to boost FC output voltage and efficiency. The cathode ORR catalyst indirectly governs FC energy conversion rates, impacting FC application. Mostly, for 2D materials, the $2e^-$ pathway is likely favourable and widely referred to, which can be ascribed to to its high surface area, edge sites, tunable electronic structure, and intrinsic catalytic activity, promoting the formation of two-electron reduction products. Oxygen molecules that adsorb onto the catalyst surface, undergo dissociation into oxygen atoms (O^*), and subsequently desorb as water or hydroxide ions, involving crucial steps of adsorption, dissociation, and desorption. For the clear understanding of the reactions and of molecules participating in ORR, a schematic diagram is given figure 5.1(a). Mechanisms and potentials (versus reversible hydrogen electrode, RHE) of ORR in neutral and acidic media as follows in figures 5.1(a) and (b). A typical ORR polarization curve (figure 5.1(c)) exhibits three regions: a slow rate of O_2 reduction with decreasing potential (kinetically controlled), a marked increase in current density with a drop in potential (mixed kinetic- and diffusion-controlled), and a region where current density depends on reactant diffusion rate (diffusion-controlled). Catalyst activity is quantitatively assessed by onset potential (E_{onset}) and half-wave potential ($E_{1/2}$), with higher potentials indicating greater ORR activity, while J_L represents the diffusion-limited current density [20–22]. The other figure of merits like Tafel slope, kinetic current density (J_K), electron transfer number (n) and measured current density (J) to access electrocatalytic activity of the electrocatalyst through studying the electrocatalytic kinetics. The measured current density and electron transferred

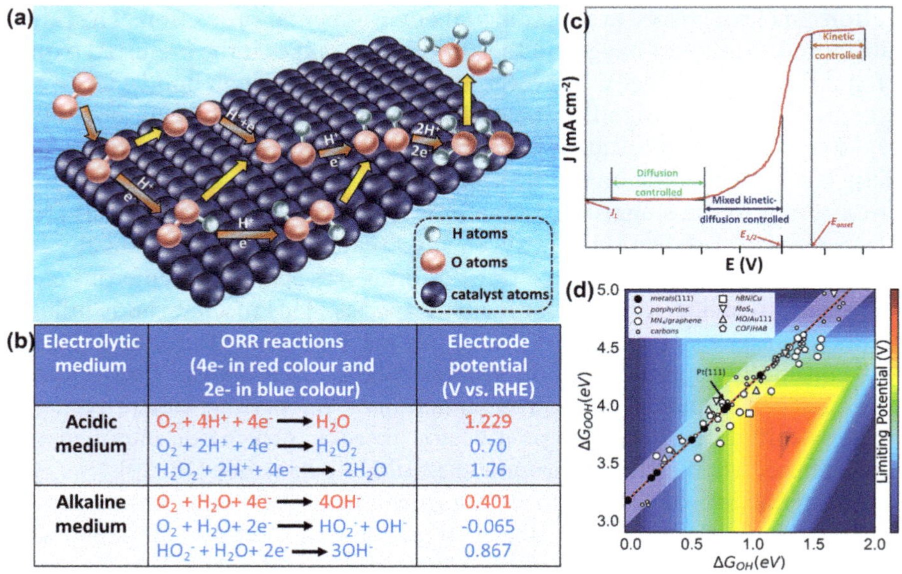

Figure 5.1. (a) Schematic illustration of the reaction mechanism of ORR, (b) mechanism and potentials of ORR in tabular form, (c) graphical representation of a typical ORR polarisation curve, and (d) 2D volcano plot. Reprinted with permission from [3]. Copyright (2018) American Chemical Society.

number can be evaluated by Koutecký–Levich (K–L) method and the rotating ring-disk electrode (RRDE) method; [22–24]

$$\frac{1}{J} = \frac{1}{J_K} + \frac{1}{J_L}$$

$$n = 4 \times \left[\frac{I_d}{I_d + \frac{I_r}{N}} \right]$$

herein, N, I_d and I_r are the collection co-efficient (percentage of current drawn by ring from disk), disk current and the ring current.

A lot of materials are being investigated for ORR and thereby further utilized in FC and metal–air batteries. However, 2D materials have garnered a lot of consideration in the past decade owing its tunable properties. The performance of 2D ORR catalyst can be retrieved through 2D volcano plot which is a graphical representation used to assess and compare the catalytic activity of 2D materials or catalysts, plotting a performance metric against a relevant parameter, typically displaying an optimal range for efficient catalysis in the form of a volcano-shaped curve (figure 5.1(d)) [3]. Additionally, to boost the catalytic activity of 2D ORR electrocatalysts, a plethora of innovative tactics such as doping, alloying, defects, edges, heterostructuring, and surface engineering can be adapted.

5.3 Different engineered approaches to enhance ORR performance

5.3.1 Defects

Defect engineering in ORR involves intentionally creating structural imperfections in catalyst materials to enhance their catalytic activity, increase active sites, and modulate electronic properties for improved oxygen reduction [25]. In view of that, an effective and stable electrocatalyst for the ORR in alkaline electrolytes is created using a simple wet-chemistry approach, resulting in a defect-rich ultrathin porous Pd metallene [26]. This metallene offers an abundance of highly active sites and vacancy defects, leading to superior ORR activity (0.892 A mgPd−1 at 0.9 V versus RHE). Compared to commercial Pt/C and Pd/C, it exhibits 5.1- and 16.8-times higher mass activity, respectively, while maintaining performance after 5000 cycles. The outstanding ORR performance is attributed to the strain effect and tunable electronic structure. Likewise, the remarkable electrocatalytic performance of ultrathin nitrogen-doped graphene nanomesh in acidic electrolytes was attributed to its extremely thin 2D structure, which features a hierarchical pore arrangement and a structure enriched with defects [27]. Another electrocatalyst, 2D graphene material possessing carbon defects exhibited excellent catalytic performance while reducing O_2 (E_{onset} = 0.91 V versus RHE, $E_{1/2}$ = 0.76 V and n = 3.87) owing to the local modulated electronic environment associated with the defects [28]. Similarly, the abundant topological defects and edge defects aided the catalytic activity of nitrogen-doped graphene mesh and 2D porous turbostratic carbon nanomesh [29, 30]. In this regard Zhu and their group produced an *in situ* galvanic replacement technique which was employed to create a Pt nanotubes/N-doped graphene (Pt NTs/NG) catalyst [31]. The performance of this Pt NTs/NG catalyst was thoroughly examined, revealing significantly improved activity and long-term stability in ORR under acidic conditions, highlighting its promising potential for fuel cell applications. Specifically, the Pt NTs/NG catalyst exhibited with a 2.7-fold enhancement in mass activity at 0.9 V compared to the state-of-the-art. Moreover, accelerated durability tests demonstrated that the Pt NTs/NG catalyst retained 88.9% of its initial activity after the testing period (figure 5.2).

5.3.2 Alloying

Alloying in the context of the ORR involves mixing different metals to enhance catalytic activity, durability, and cost-effectiveness of ORR catalysts in applications like fuel cells. The 2D porous bimetallic PtPd alloy demonstrated outstanding performance in the ORR, achieving a notable positive shift of approximately 43 mV in the half-wave potential compared to commercial Pt/C (50 wt%) in a 0.1 M KOH solution [32]. Additionally, this catalyst exhibited a 2.4-fold increase in mass activity and a 3.5-fold improvement in specific activity at 0.80 V relative to Pt/C. Theoretically, Back *et al* made over 50 combinations of 2D materials with doping of single atoms which displayed amazing catalytic activity, specifically, Rh embedded in N-doped graphene outstood of all for fuel cell applications [33]. Lyu and their co-workers developed an excellent alloyed catalyst, PtPd bimetallic

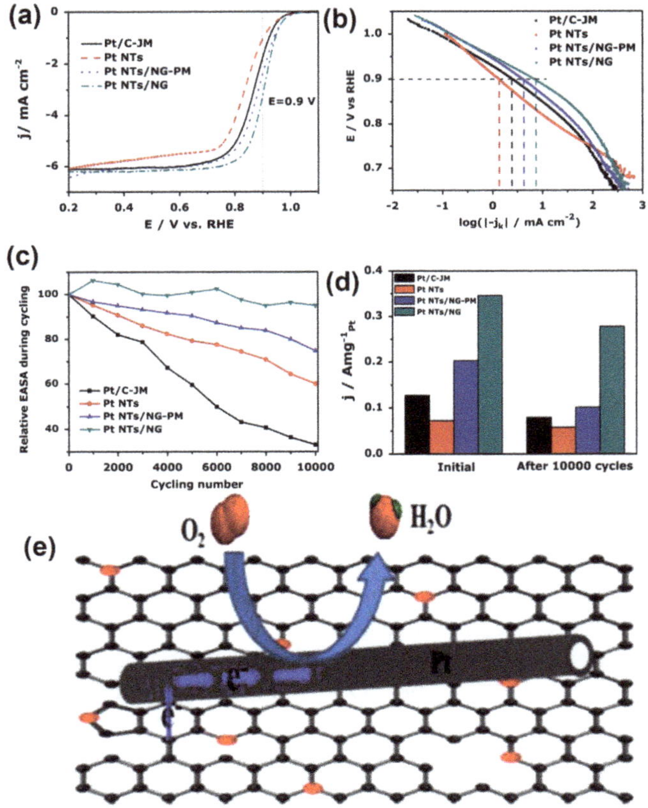

Figure 5.2. (a) Polarization curves of the catalysts, (b) extracted Tafel plot, (c) measured electrochemical active surface area (ECSA) of the catalysts, (d) data showing the comparison of ECSA before and after stability and (e) mechanism of ORR through schematic representation. Reprinted from [31], Copyright (2015), with permission from Elsevier.

nanostructures with varying Pt and Pd ratios synthesized using organized SiO_2 spheres as templates (figure 5.3(a)) [34]. The most effective catalysts exhibited an onset potential and half-wave potential of 0.967 and 0.889 V, respectively, along with mass and specific activity enhancements of 2.1 and 2.6 times, respectively, compared to Pt/C (figures 5.3(b) and (c)). This improved catalytic performance in ORR can be primarily attributed to the porous structure providing ample active sites and the synergistic effect of Pt alloyed with Pd effectively reducing OH adsorption, thereby mitigating poisoning of the ORR active site.

Likewise, Salmon *et al* synthesized an excellent catalyst by subjecting Pt-decorated $Ni(OH)_2$ nanosheets to controlled thermal treatments and hierarchical 2D nanoframes were fabricated, featuring a catalytically active Pt–Ni alloy phase [35]. The detailed synthesis protocol is provided in figure 5.3(d). These nanoframes exhibited an ORR specific activity 10 times higher (5.8 mA cm^{-2}) than Pt/C, along with remarkable stability under accelerated testing up to 1.3 V versus RHE

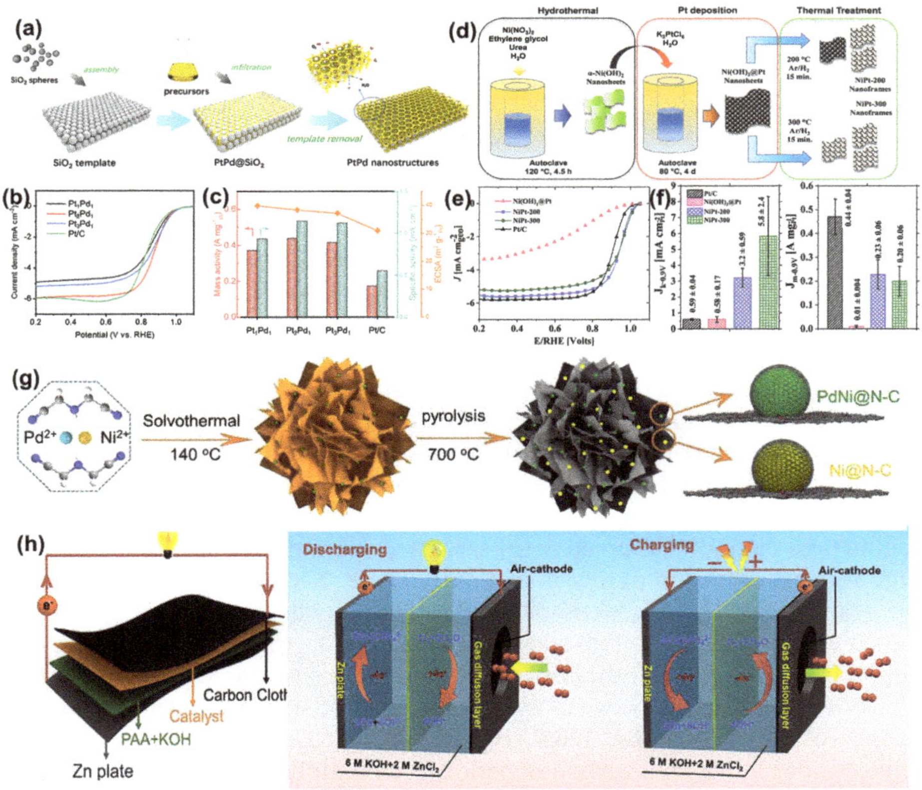

Figure 5.3. (a) Schematic illustration of the synthesis of PtPd alloy, (b) ORR polarisation curve the catalysts, (c) mass activity and ESCA of the catalysts. Reprinted with permission from [34]. Copyright (2020) American Chemical Society. (d) Schematic diagram of the synthesis of $Ni(OH)_2$@Pd, (e) ORR polarisation curve for the catalysts, (f) mass activity and specific activity calculations in a graphical way. Reprinted with permission from [35]. Copyright (2017) American Chemical Society. (g) Schematic representation of the synthesis procedure of PdNi/Ni@N-C and (h) device architecture and the charge storage mechanism. Reprinted from [36], Copyright (2021), with permission from Elsevier.

(figures 5.3(e) and (f)). The unique nanoarchitecture and local structure of these metallic 2D nanoframes deliver both high specific activity and elevated potential stability, facilitating a four-electron reduction process for ORR with the first electron transfer as the rate-determining step. In another interesting study, a novel and straightforward OCP-polymer approach was employed to derive hierarchical porous N-doped carbon nanosheets anchored with PdNi and Ni nanoparticles (PdNi/Ni@N-C) using an organic ligand (figure 5.3(g) [36]. Excellent electrocatalytic activity was demonstrated, with a 60 mV positive shift in the half-wave potential for the ORR compared to commercial Pd/C. Density functional theory (DFT) calculations revealed that alloying Pd and Ni caused significant changes in reaction site geometry and electronic structure, resulting in distinct catalytic behaviour led to showcase promising performance as an air-cathode in rechargeable

Zn–air batteries and flexible solid-state Zn–air batteries, offering high-power density and superior cycling life compared to batteries driven by Pd/C+RuO_2 (figure 5.3(h)).

5.3.3 Doping

Doping pertaining to ORR involves deliberately introducing foreign atoms or species into catalyst materials to enhance their catalytic activity by modifying electronic and chemical properties through defect creation which is confirmed through theoretically as well [37–39]. Among the 2D materials, the pioneer, graphene has been widely investigated regarding to ORR and a lot of single atoms like boron, nitrogen, phosphorus etc have been used as dopants to enhance its catalytic activity [40]. In view of that, the reaction mechanism changes to $4e^-$ pathway instead of 2+2 and $4e^-$ pathway which can be attributed towards the nitrogen doping and solvent treatment which not only affects the activity but also changes the morphology of the catalyst [41]. The catalyst exhibited an onset potential of 1.1 V and a half-wave potential of 0.84 V versus RHE. Additionally, during extended potential cycling, there was minimal degradation in the half-wave potential (less than 3%) observed in O_2-saturated solutions. In the same way, using a straightforward two-step pyrolysis method, 2D Fe–N-doped carbon sheets resembling graphene (FeNGC) were produced, and it was found that FeNGC-950 (sintered @950 °C) exhibited superior performance to many non-noble metal catalysts [42]. Tan *et al* synthesized a 2D structure consisting of reduced graphene oxide (rGO) sandwiched between two monolayers of small nitrogen-doped carbon nanospheres (N-HCNS) for ORR (figure 5.4(a)). They demonstrated the best catalyst among others showcased superior electrocatalytic oxygen reduction compared to physically mixed rGO and N-HCNS, surpassing a 20 wt % Pt/C catalyst in alkaline conditions with a 0.872 V half-wave potential, 12.0 mV higher than Pt/C (figures 5.4(b) and (c)) [43]. In contrast, a distinct lamellar microstructure, Nd-doped $Bi_4Ti_3O_{12}$ nanosheet was created through a modified sol–gel hydrothermal method, showing exceptional selectivity for the ORR to generate hydrogen peroxide in an alkaline medium [44]. This microstructure outperformed pure nanosheets (E_{onset} of 0.7 V versus RHE) in both current density and selectivity, maintaining a hydrogen peroxide yield of 93%–95% in the 0–0.50 V versus RHE potential range, compared to pure nanosheets' 83%–89% which can be attributed towards surface adsorption enhancement through doping. Yue *et al* described a straightforward method for creating porous nitrogen-doped Ti_3C_2 nanosheets (figure 5.4(d)) [45]. The designed porous structure offered a large surface area and abundant channels, enhancing oxygen adsorption and transport, thus improving ORR kinetics owing to the doping effect. Compared to pristine material, the modified nanosheets exhibited a higher half-wave potential (0.72 V), greater number of electron transfer and minimal potential decay (2 mV after 5000 CV cycles) (figures 5.4(e) and (f)).

5.3.4 Strain and stress

Strain engineering in the realm of ORR within 2D materials entails the deliberate imposition of mechanical or structural strain to modulate their electronic structure

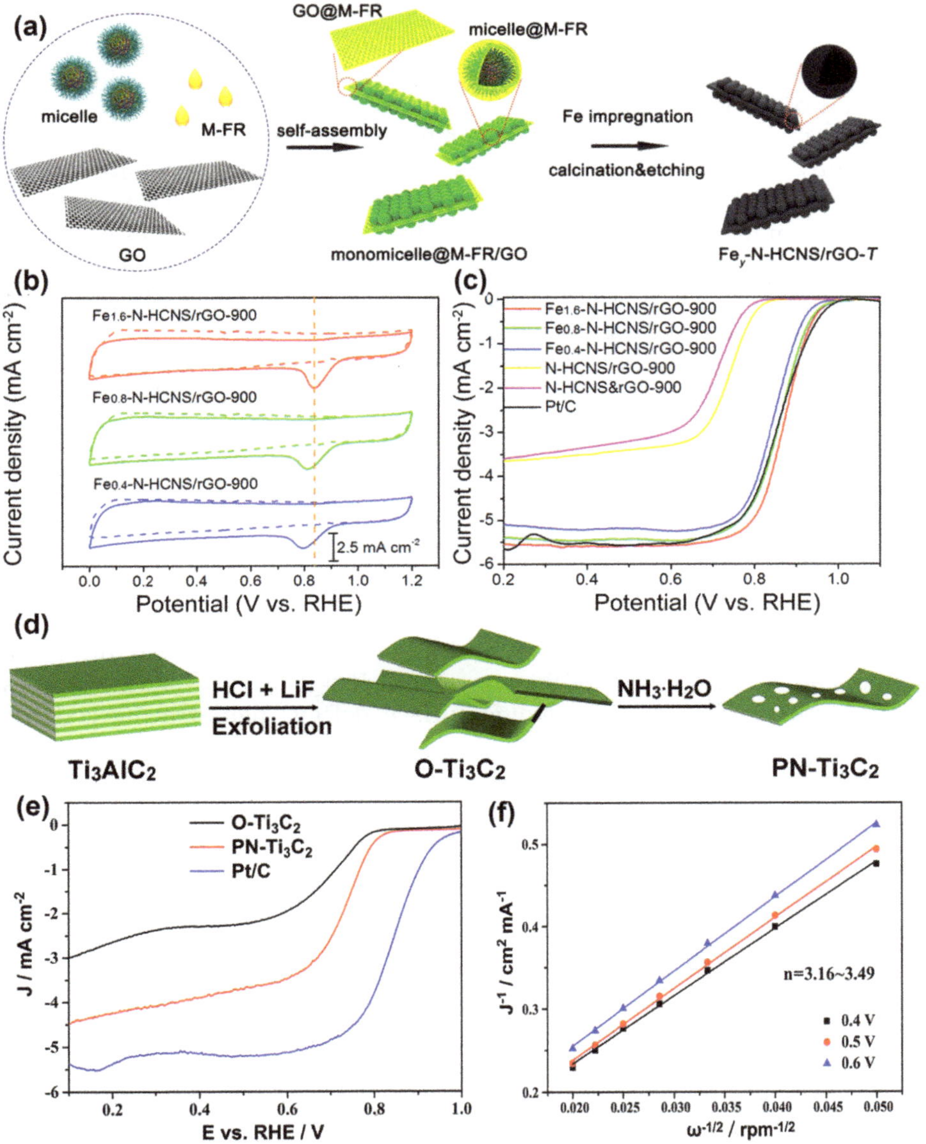

Figure 5.4. (a) Schematic representation showing the synthesis procedure of N-HCNS, (b) CV plots of the catalysts, (c) LSV polarisation curves of the catalysts. Reprinted with permission from [43]. Copyright (2018) American Chemical Society. (d) Diagram showing synthesis process of N-doped MXene, (e) polarisation curves of the catalyst and (e) K–L plot. Reprinted with permission from [45]. Copyright (2022) American Chemical Society.

and catalytic properties. This method is employed to fine-tune ORR efficiency and longevity in applications like fuel cells and metal–air batteries, leveraging strain-induced alterations in surface interactions, active sites, and electronic band structures to enhance catalytic performance. A lot of theoretical studies suggested the above thing which include strain engineering of Pt-doped Ti_2CO_2, Ta_9Se_{12}, $GaPS_4$, MXene and carbon nitride etc and exhibited amazing catalytic properties in the context of ORR [46–50].

Keeping this in mind, Yin *et al* came up with a solution to address the situation. Nanoporous gold (NPG) films possess excellent electrical conductivity, high corrosion resistance, a large specific surface area, and a unique concave/convex curvature. Pt monolayers were fabricated by their group on NPG films with varying lattice strains, resulting in a ~1.8% reduction in Pt lattice constant and a threefold increase in intrinsic ORR activity (figure 5.5(a)) [51]. The NPG-1.8%-Pt sample exhibited about 5 and 16 times higher intrinsic and mass ORR activity, respectively, compared to commercial Pt/C. Furthermore, after 10 000 durability cycles, NPG-Pt showed only a modest 1.8% decrease in activity, contrasting with the 35% decrease observed for Pt/C. This highlights the potential of constructing compressive-strained Pt in nanoporous materials for highly efficient and durable Pt-based ORR catalysts (figures 5.5(b)–(d)). This can be encouraging for the scientific community to put more efforts towards developing strategic strain engineered ORR catalysts [51].

5.3.5 Morphology

In the realm of ORR within 2D materials, morphology refers to the physical structure and surface features of these materials. It significantly influences catalytic activity, with considerations including surface defects, nanostructures, dopants, layer thickness, supporting substrates, and the application of mechanical strain to engineer optimal ORR performance [52]. Huang *et al* synthesized a 2D carbon morphology via hydrothermal carbonization using guanine and carbohydrates. These materials exhibited exceptional performance in the ORR with $E_{1/2}$ of 0.87 V (versus RHE) [53]. The morphology of a layer double hydroxide using Ni and Co were studied for ORR and exhibited outstanding performance where it achieved 3.7 times of electrons with respect to the multilayer catalysts [54]. In another report, a puff-like reduced graphene oxide (PG) support was synthesized via spray drying and thermal shock treatments, followed by impregnation with iron phthalocyanine (FePc/PG) [55]. PG's unique nanostructure significantly increased both Brunauer–Emmett–Teller (BET) surface area (29-fold) and electrochemically active surface area (ECSA) (73-fold), mitigating restacking issues. FePc dispersion on PG created more anchoring sites, improving active site exposure and mass transport. The ORR activity revealed with a $E_{1/2}$ of 0.909 V (versus RHE) and J_K of 89.2 mA cm^{-2} at 0.80 V, surpassing Pt/C. Additionally, in a gas diffusion electrode (GDE) setup, FePc/PG displayed high current densities, low mass-transport overpotential, and achieved a peak power density of 228 mW cm^{-2} in a zinc–air battery (ZAB). Another morphology dependent study, CeO_2 nanostructures, varying in shape and size such as hollow/solid spheres, triangular flakes, nanotubes, and flower-like structures, exhibit distinct ORR activities [56]. The order of activity, from highest to lowest, is hollow spheres > triangular flakes > flowers > nanotubes > bowling balls, influenced by factors like oxygen vacancies, porosity, and surface

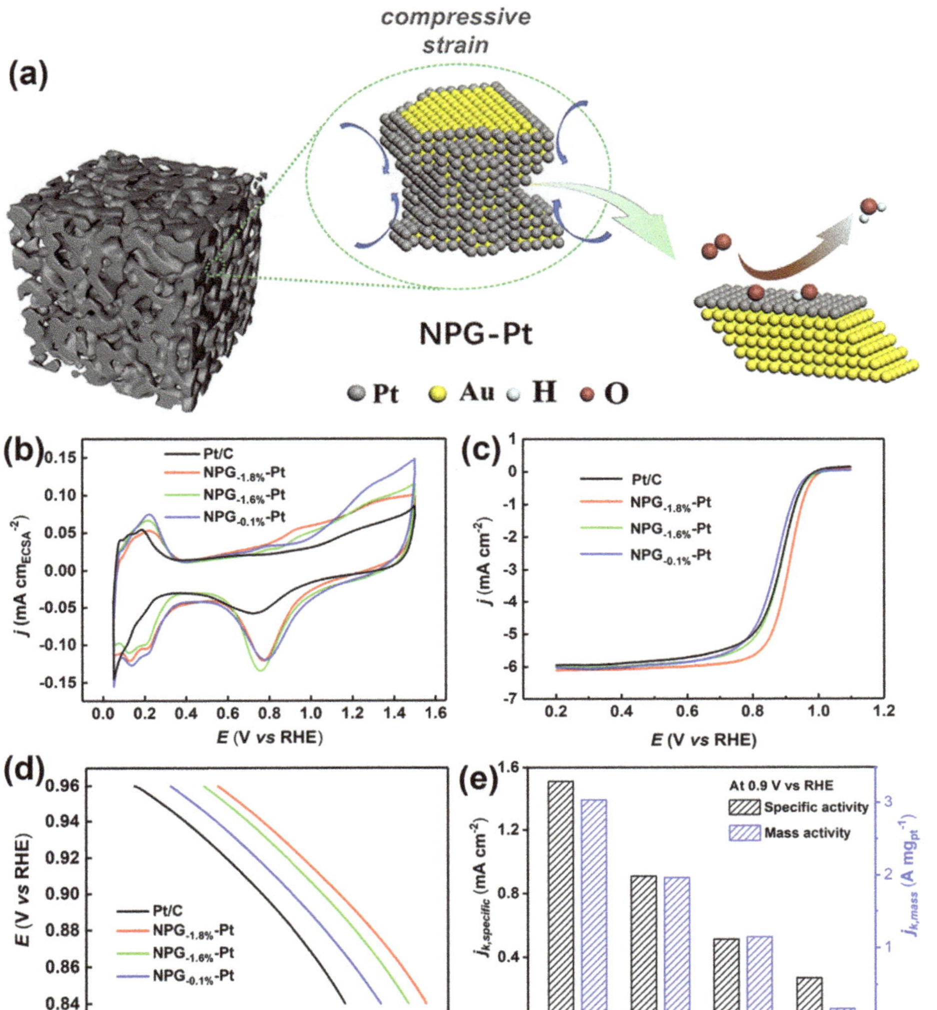

Figure 5.5. (a) Schematic illustration of the synthesis protocol, (b) CV profiles of the catalysts 0.1 M $HClO_4$, (c) polarized LSV curves of the catalysts, (d) Tafel plots and (e) mass and specific activity calculations in bar graph form. Reprinted with permission from [51]. Copyright (2020) American Chemical Society.

area. These CeO_2 nanostructures primarily follow a four-electron ORR pathway, generating H_2O as a by-product. Notably, CeO_2 hollow spheres surpass Pt/C in terms of durability and methanol tolerance. Likewise, different Cu_2O crystal shapes were examined for their electrocatalytic performance in ORR in alkaline conditions [57]. It was observed that the truncated octahedral Cu_2O exhibited higher surface-specific activity for ORR compared to spherical or octahedral morphologies which was attributed to the preferential exposure of (100) crystal planes on the truncated

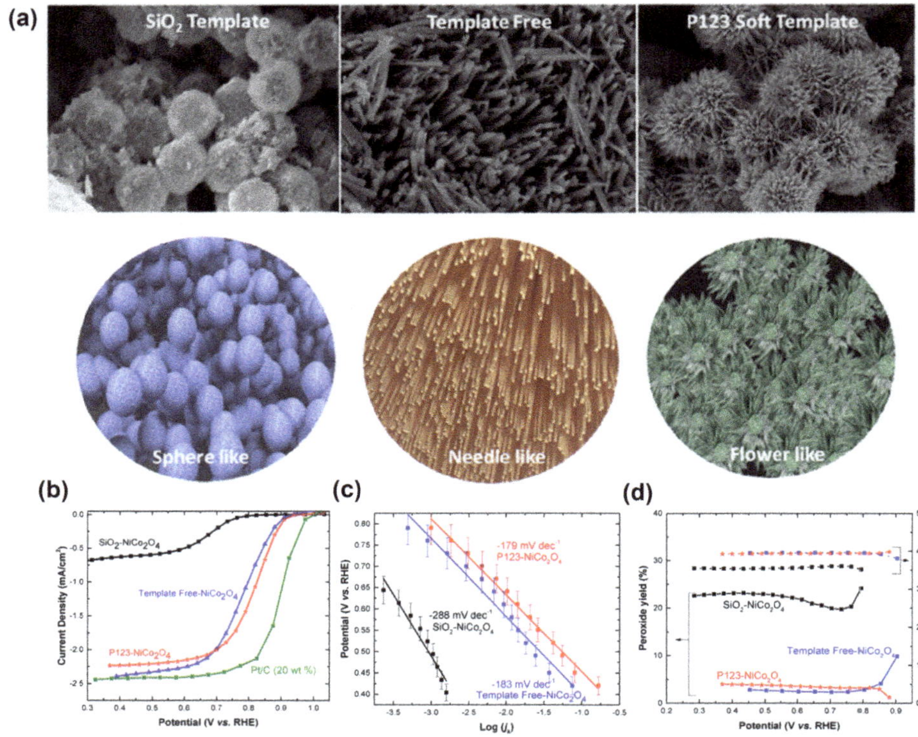

Figure 5.6. (a) Schematic illustration of the three distinct morphologies of Ni–Co hydroxides, (b) polarisation corves of the catalysts, (c) Tafel plots and (d) yield of peroxide and the number of electrons transferred during ORR. Reprinted with permission from [58]. Copyright (2017) American Chemical Society.

octahedron, facilitating stronger O_2 adsorption and easier activation of adsorbed O_2 on the Cu_2O surface, as supported by both electrocatalysis experiments and periodic spin DFT calculations. Moreover, Wu *et al* supposed of three distinct $NiCo_2O_4$ catalysts using different templates: temple-free, Pluronic-123 (P-123) soft, and SiO_2 hard templates, followed by hydrothermal methods and calcination [58]. The soft-template method resulted in a unique nanoneedle cluster assembly with meso- and macropores, while the hard-template produced dense spherelike structures (figure 5.6(a)). The flower-like nanoneedle assembly $NiCo_2O_4$ catalyst obtained via the soft-template approach exhibited the highest catalytic activity and stability for the ORR in alkaline media, with onset and half-wave potentials of 0.94 and 0.82 V versus the RHE, respectively.

5.3.6 Edge

The scientific exploration of edge sites in 2D materials for ORR is a prominent research area within materials science and electrochemistry. These atomically thin materials possess unique edge site properties, such as altered electronic structures,

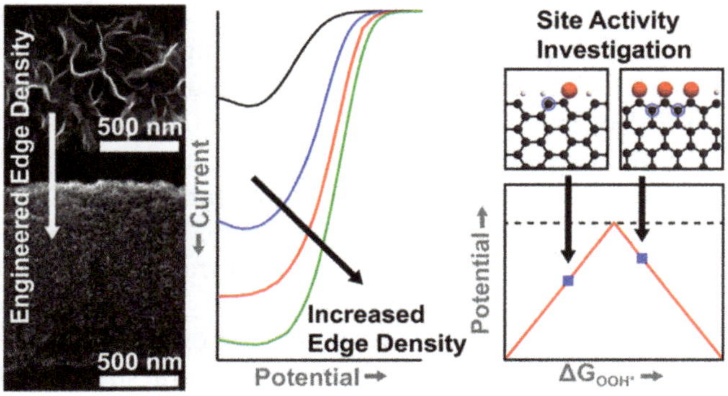

Figure 5.7. Schematical representation of the graphene nanosheets and the edge site investigation. Reprinted with permission from [63]. Copyright (2020) American Chemical Society.

which significantly influence their catalytic performance in ORR and hold promise for advancing clean energy technologies. The advantageous edges were noticed by Wang *et al* where they inspected a leaf-like 2D zeolitic imidazolate framework, ZIF-L, with exposed edge active sites. Heteroepitaxial growth of ZIF-L-Co onto ZIF-L-Zn allowed control over exposed facets, resulting in high electrocatalytic activity for the ORR [59]. Similarly, theoretically the edge sites for Ni_2SbTe_2 could provide ample activity for ORR which was investigated by Li *et al* through first-principles calculations [60]. Further, a highly efficient ORR electrocatalysts was created by utilizing edge-rich and dopant-free graphene, carbon nanotubes, and graphite, demonstrating the pivotal role of edge carbon in enhancing ORR activity via a one-step, four-electron pathway [61]. Similarly, the edge planes of 2D-MoS_2 were explored for ORR catalyst and displayed amazing electrocatalytic activity owing to the edge cut planes [62]. Karni *et al* fabricated a graphene-based electrocatalyst, nanowire-templated three-dimensional fuzzy graphene (NT-3DFG) for efficient $2e^-$ ORR. NT-3DFG demonstrated a high efficiency (onset potential of 0.79 ± 0.01 V versus RHE), selectivity ($94 \pm 2\%$ H_2O_2 production), and tunable ORR activity based on graphene edge site density. Functionalization of NT-3DFG edge sites with carbonyl (C=O) and hydroxyl (C–OH) groups under alkaline ORR conditions enabled selective $2e^-$ ORR (figure 5.7). A geometric descriptor predicts site activity within ~0.1 V of computed values [63].

5.3.7 Heterostructuring

Heterostructures and composites incorporating 2D materials, like graphene and transition metal dichalcogenides, hold promise for enhancing the ORR through improved kinetics, band structure engineering, and synergistic effects with other catalysts. However, practical challenges in scaling up production and cost-effectiveness need to be addressed for their broader application in clean energy technologies. In view of that, a strategy was employed to convert a limitation into an advantage by coupling two slow ORR materials, hexagonal boron nitride (hBN) and MoS_2 [64]. Within the heterostructure, boron vacancies were preferentially formed in the presence of MoS_2, serving as active sites for oxygen adsorption. This induced B-vacancies facilitated rapid electron transfer, overcoming kinetic limitations observed in pure hBN nanosheets during ORR kinetics. The resulting catalyst exhibited a low Tafel slope (66 mV/dec) and a high onset potential (0.80 V versus RHE) with an unaltered ESCA after extended cycling. Similarly, a new hierarchical structure comprising tiny α-Fe_2O_3 nanoparticles enclosed within a MoS_2/N-doped graphene nanosheets (NGNS) heterostructure was developed [65]. This catalyst demonstrated remarkable catalytic efficacy, with a high electron-transfer number ranging from 3.91 to 3.96, comparable to Pt/C. Additionally, it exhibited superior stability, retaining 96.1% of its performance after 30 000 s, and excellent resistance to alcohol, surpassing the performance of Pt/C. Gou *et al* synthesized a 2D sandwiched heterostructure by combining N-doped mesoporous defective carbon and nitrogen-modified titanium carbide (Ti_3C_2), resulting in excellent electrochemical performance not only experimentally but also through DFT calculations and withstood with a high onset potential of 0.90 V and a low current density of 5.50 mA cm^{-2} [66]. This structure also reduced the energy required for O_2^* to OOH^* conversion from 0.73 to 0.61 eV by enhancing oxygen adsorption through its electric bandgap.

Likewise, other heterostructures like $Ni(OH)_2/ZrO_2$, Fe–N–C@$Ti_3C_2T_x$, N-doped C/Ti_3C_2, FePc/$Ti_3C_2T_x$, Co-CNT/$Ti_3C_2T_x$, g-C_3N_4/Ti_3C_2, $NiCo_2O_4$/MXene, Mn_3O_4/$Ti_3C_2T_x$ and CoS_2@MXene etc unveiled excellent ORR activity which can be accounted for the synergistic effect taking place between the individual materials [66–70]. Another interesting report regarding a novel design featured ultrasmall α-Fe_2O_3 nanoparticles encased within MoS_2/NGNS heterostructures was came to light and the synthesis procedure is provided in figure 5.8(a) [65]. This arrangement anchored α-Fe_2O_3 nanoparticles was confirmed through the SEM images shown in figure 5.8(b) and promoted interactions with MoS_2/NGNS shells, and improved charge and mass transport. The resulting synergistic effects among α-Fe_2O_3 nanoparticles, MoS_2 layers, and NGNS led to excellent catalytic performance for ORR. The catalyst exhibited remarkable ORR activity, with a high electron-transfer number of 3.91–3.96, comparable to Pt/C. Additionally, it displayed exceptional stability, retaining 96.1% of its performance after 30 000 s, and superior alcohol tolerance, surpassing Pt/C performance. The polarization curve along with Tafel plots and Nyquist plots are given in figures 5.8(c)–(e). Similarly, Fe–N–C nanosheets (with a Zeta potential of +30.4 mV) were electrostatically assembled with anionic MXene (with a Zeta potential of −39.7 mV), resulting in the formation of a superlattice-like heterostructure (figure 5.8(f)) characterized by lateral dimensions in the tens of nanometres range, a surface area of 30 m^2 g^{-1}, featuring repeating dimensions of 0.4 and 2.1 nm [71]. The SEM and TEM images displayed successful segregation of the heterostructure through the surface (figures 5.8(g) and (h)). The potential application of this synthesized Fe–N–C/MXene heterostructure was demonstrated for electrocatalytic ORR. It exhibited a favourable onset potential of 0.92 V, facilitated a four-electron transfer pathway, and demonstrated remarkable durability over 20 h in an alkaline electrolyte (figures 5.8(i) and (j)) [71].

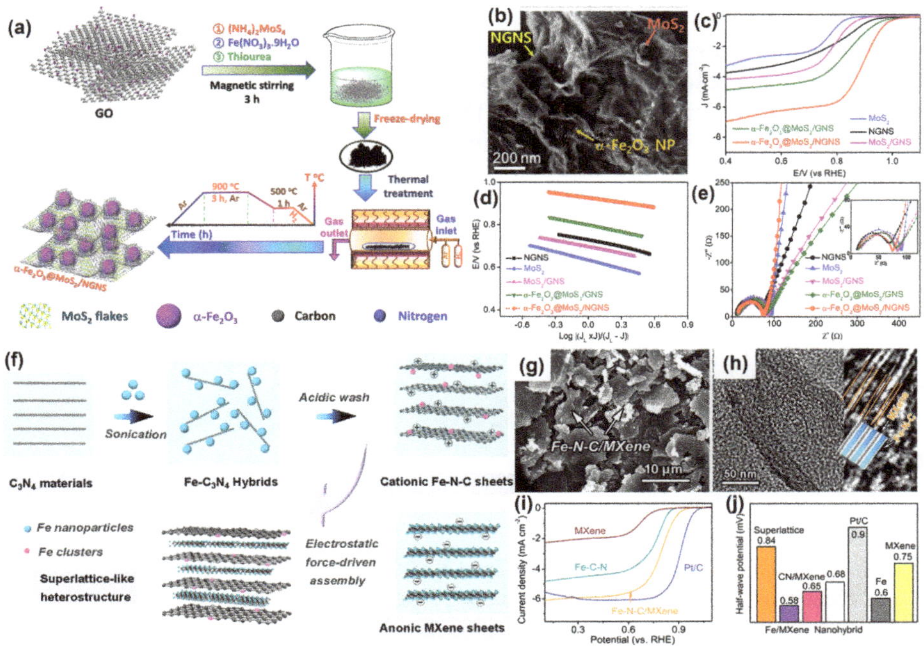

Figure 5.8. (a) Schematic representation of the synthesis procedure of MoS_2/NGNS, (b) SEM image of MoS_2/NGNS, (c) polarisation curves of the catalysts, (d) Tafel plots, (e) Nyquist plots. Reprinted with permission from [65]. Copyright (2018) American Chemical Society. (f) Schematic illustration of the synthesis protocol for Fe–N–C/MXene heterostructure, (g) SEM image of the catalyst, (h) HRTEM images of the catalyst, (i) LSV curves of the catalysts and (j) half-wave potentials of the catalysts. Reprinted with permission from [71]. Copyright (2020) American Chemical Society.

5.4 Conclusion

In summary, this chapter provided a comprehensive exploration of the electrocatalytic behaviour of 2D materials in the context of ORR. This critical discussion elucidated the fundamental mechanisms underlying ORR on 2D materials and strategies for enhancing their catalytic activity. The intrinsic characteristics of 2D materials, including their high surface area and tunable electronic properties, render them promising catalysts for ORR. Notably, the chapter emphasizes the pivotal roles played by defects, edges, alloying, doping, heterostructuring, and morphology in tailoring ORR kinetics. Prospects include the precise design of 2D materials with customized properties, a deeper atomistic understanding of ORR mechanisms through advanced techniques, and scalable synthesis methods for industrial applications. Ensuring the stability and durability of 2D material-based catalysts over extended periods, as well as addressing environmental sustainability concerns, are essential objectives. The integration of 2D material catalysts into energy devices and the exploration of their multifunctional capabilities are areas of significant interest. Commercialization hurdles, bridging the gap between research and practical applications, require concerted efforts. Overall, this field holds promise for innovations but necessitates interdisciplinary collaboration to overcome the multifaceted challenges lying ahead. This chapter collectively has underscored the remarkable potential of 2D materials in advancing the field of ORR electrocatalysis, providing valuable insights for future research and the development of sustainable energy conversion technologies.

Acknowledgments

The authors gratefully acknowledge financial assistance from the SERB Core Research Grant (Grant No. CRG/2022/000897), Department of Science and Technology (DST/NM/NT/2019/205(G)), and Minor Research Project Grant, Jain University (JU/MRP/CNMS/29/2023).

References

[1] Klebe R J 1975 Cell attachment to cllagen: the requirement for energy *J. Cell. Physiol.* **86** 231–6

[2] Wang J, Kim J, Choi S, Wang H and Lim J 2020 A review of carbon-supported nonprecious metals as energy-related electrocatalysts *Small Methods* **4** 2000621

[3] Kulkarni A, Siahrostami S, Patel A and Nørskov J K 2018 Understanding catalytic activity trends in the oxygen reduction reaction *Chem. Rev.* **118** 2302–12

[4] Sui S, Wang X, Zhou X, Su Y, Riffat S and Liu C 2017 A comprehensive review of Pt electrocatalysts for the oxygen reduction reaction: nanostructure, activity, mechanism and carbon support in PEM fuel cells *J. Mater. Chem.* A **5** 1808–25

[5] Stacy J, Regmi Y N, Leonard B and Fan M 2017 The recent progress and future of oxygen reduction reaction catalysis: a review *Renew. Sustain. Energy Rev.* **69** 401–14

[6] Patra A and Sekhar Rout C 2020 Anisotropic quasi-one-dimensional layered transition-metal trichalcogenides: synthesis, properties and applications *RSC Adv.* **10** 36413–38

[7] Thomas S A, Patra A, Al-Shehri B M, Selvaraj M, Aravind A and Rout C S 2022 MXene based hybrid materials for supercapacitors: recent developments and future perspectives *J. Energy Storage* **55** 105765

[8] Niu W J, He J Z, Gu B N, Liu M C and Chueh Y L 2021 Opportunities and challenges in precise synthesis of transition metal single-atom supported by 2D materials as catalysts toward oxygen reduction reaction *Adv. Funct. Mater.* **31** 2103558–8

[9] Perivoliotis D K and Tagmatarchis N 2017 Recent advancements in metal-based hybrid electrocatalysts supported on graphene and related 2D materials for the oxygen reduction reaction *Carbon* **118** 493–510

[10] Sun T, Zhang G Q, Xu D, Lian X, Li H and Chen W *et al* 2019 Defect chemistry in 2D materials for electrocatalysis *Mater. Today Energy* **12** 215–38

[11] Singh S K, Takeyasu K and Nakamura J 2018 Active sites and mechanism of oxygen reduction reaction electrocatalysis on nitrogen-doped carbon materials *Adv. Mater.* **31** 1804297

[12] Singh H, Zhuang S, Ingis B, Nunna B B and Lee E S 2019 Carbon-based catalysts for oxygen reduction reaction: a review on degradation mechanisms *Carbon* **151** 160–74

[13] Tang L, Meng X, Deng D and Bao X 2019 Confinement catalysis with 2D materials for energy conversion *Adv. Mater.* **31** 1901996–6

[14] Chia X and Pumera M 2018 Characteristics and performance of two-dimensional materials for electrocatalysis *Nat. Catal.* **1** 909–21

[15] Khan K, Tareen A K, Aslam M, Zhang Y, Wang R and Ouyang Z *et al* 2019 Recent advances in two-dimensional materials and their nanocomposites in sustainable energy conversion applications *Nanoscale* **11** 21622–78

[16] Tong X, Zhan X, Rawach D, Chen Z, Zhang G and Sun S 2020 Low-dimensional catalysts for oxygen reduction reaction *Prog. Nat. Sci.: Mater. Int.* **30** 787–95

[17] Kiran G K, Sreekanth T V M, Yoo K and Kim J 2023 Bifunctional electrocatalytic activity of two-dimensional multilayered vanadium carbide (MXene) for ORR and OER *Mater. Chem. Phys.* **296** 127272–2

[18] Li Y, Tong Y and Peng F 2020 Metal-free carbocatalysis for electrochemical oxygen reduction reaction: activity origin and mechanism *J. Energy Chem.* **48** 308–21

[19] Gómez–Marín A M and Ticianelli E A 2018 A reviewed vision of the oxygen reduction reaction mechanism on Pt-based catalysts *Curr. Opin. Electrochem.* **9** 129–36

[20] Yao-Lin A, Du Z Y, Huajie Z, Wang X, Zhang Y and Zhang H *et al* 2023 Understanding the molecular mechanism of oxygen reduction reaction using *in situ* raman spectroscopy *Curr. Opin. Electrochem.* **42** 101381–1

[21] Yi S, Jiang H, Bao X, Zou S, Liao J and Zhang Z 2019 Recent progress of Pt-based catalysts for oxygen reduction reaction in preparation strategies and catalytic mechanism *J. Electroanal. Chem.* **848** 113279–9

[22] Du C, Sun Y, Shen T, Yin G and Zhang J 2014 *Applications of RDE and RRDE Methods in Oxygen Reduction Reaction* (Elsevier eBooks) pp 231–77

[23] Ge X, Sumboja A, Wuu D, An T, Li B and Goh F W T *et al* 2015 Oxygen reduction in alkaline media: from mechanisms to recent advances of catalysts *ACS Catal.* **5** 4643–67

[24] Chlistunoff J 2011 RRDE and voltammetric study of ORR on pyrolyzed Fe/polyaniline catalyst. On the origins of variable tafel slopes *J. Phys. Chem.* C **115** 6496–507

[25] Goswami C, Hazarika K K and Bharali P 2018 Transition metal oxide nanocatalysts for oxygen reduction reaction *Mater. Sci. Energy Technol.* **1** 117–28

[26] Yu H, Zhou T, Wang Z, Xu Y, Li X and Wang L *et al* 2021 Defect-rich porous palladium metallene for enhanced alkaline oxygen reduction electrocatalysis *Angew. Chem.* **133** 12134–8

[27] Xia W, Tang J, Li J, Zhang S, Wu K C, W and He J *et al* 2019 Defect-rich graphene nanomesh produced by thermal exfoliation of metal–organic frameworks for the oxygen reduction reaction *Angew. Chem.* **131** 13488–93

[28] Jia Y, Zhang L, Du A, Gao G, Chen J and Yan X *et al* 2016 Defect graphene as a trifunctional catalyst for electrochemical reactions *Adv. Mater.* **28** 9532–8

[29] Tang C, Wang H F, Chen X, Li B Q, Hou T Z and Zhang B *et al* 2016 Topological defects in metal-free nanocarbon for oxygen electrocatalysis *Adv. Mater.* **28** 6845–51

[30] Lai Q, Zheng J, Tang Z, Bi D, Zhao J and Liang Y 2020 Optimal configuration of n-doped carbon defects in 2D turbostratic carbon nanomesh for advanced oxygen reduction electrocatalysis *Angew. Chem.* **59** 11999–2006

[31] Zhu J, Xiao M, Zhao X, Liu C, Ge J and Wang X 2015 Strongly coupled Pt nanotubes/N-doped graphene as highly active and durable electrocatalysts for oxygen reduction reaction *Nano Energy* **13** 318–26

[32] Chen L, Jin M X, Zhang L, Wang A, Yuan J and Zhang Q *et al* 2019 One-pot aqueous synthesis of two-dimensional porous bimetallic PtPd alloyed nanosheets as highly active and

durable electrocatalyst for boosting oxygen reduction and hydrogen evolution *J. Colloid Interface Sci.* **543** 1–8

[33] Back S, Kulkarni A R and Siahrostami S 2018 Single metal atoms anchored in two-dimensional materials: bifunctional catalysts for fuel cell applications *Chem. Cat. Chem.* **10** 3034–9

[34] Lyu X, Zhang W, Li G, Shi B, Zhang Y and Chen H *et al* 2020 Two-dimensional porous PtPd nanostructure electrocatalysts for oxygen reduction reaction *ACS Appl. Nano Mater.* **3** 8586–91

[35] Godinez-Salomon F, Mendoza-Cruz R, Arellano-Jiménez M J, Jose-Yacaman M and Rhodes C J 2017 Metallic two-dimensional nanoframes: unsupported hierarchical nickel–platinum alloy nanoarchitectures with enhanced electrochemical oxygen reduction activity and stability *ACS Appl. Mater. Interfaces* **9** 18660–74

[36] Li Z, Li H, Li M, Hu J, Liu Y and Sun D *et al* 2021 Iminodiacetonitrile induce-synthesis of two-dimensional PdNi/Ni@carbon nanosheets with uniform dispersion and strong interface bonding as an effective bifunctional eletrocatalyst in air-cathode *Energy Storage Mater.* **42** 118–28

[37] Singh A and Pakhira S 2022 Unraveling the electrocatalytic activity of platinum doped zirconium disulfide toward the oxygen reduction reaction *Energy Fuels* **37** 567–79

[38] Upadhyay S N and Pakhira S 2021 Mechanism of electrochemical oxygen reduction reaction at two-dimensional Pt-doped $MoSe_2$ material: an efficient electrocatalyst *J. Mater. Chem.* C **9** 11331–42

[39] Shao Y, Jiang Z, Zhang Q and Guan J 2019 Progress in nonmetal-doped graphene electrocatalysts for the oxygen reduction reaction *Chem. Sus. Chem.* **12** 2133–46

[40] Sheng Z H, Gao H L, Bao W J, Wang F B and Xia X H 2012 Synthesis of boron doped graphene for oxygen reduction reaction in fuel cells *J. Mater. Chem.* **22** 390–5

[41] Dumont J H, Martinez U, Artyushkova K, Purdy G M, Dattelbaum A M and Zelenay P *et al* 2019 Nitrogen-doped graphene oxide electrocatalysts for the oxygen reduction reaction *ACS Appl. Nano Mater.* **2** 1675–82

[42] Zhou T, Ma R, Zhang T, Li Z, Yang M and Liu Q *et al* 2019 Increased activity of nitrogen-doped graphene-like carbon sheets modified by iron doping for oxygen reduction *J. Colloid Interface Sci.* **536** 42–52

[43] Tan H, Tang J, Henzie J, Li Y, Xu X and Chen T *et al* 2018 Assembly of hollow carbon nanospheres on graphene nanosheets and creation of iron–nitrogen-doped porous carbon for oxygen reduction *ACS Nano.* **12** 5674–83

[44] Zhang Z, Dong Q, Li P, Shemsu L F, Guo J and Fang Z *et al* 2021 Highly catalytic selectivity for hydrogen peroxide generation from oxygen reduction on Nd-doped $Bi_4Ti_3O_{12}$ nanosheets *J. Phys. Chem.* C **125** 24814–22

[45] Zhang J, Zhang X and Yue W 2022 Nanoporous nitrogen-doped Ti_3C_2 nanosheets as efficient electrocatalysts for oxygen reduction *ACS Appl. Nano Mater.* **5** 11241–8

[46] Ma N, Li N, Zhang Y, Wang T, Zhao J and Fan J 2022 Strain adjustment Pt-doped Ti2CO2 as an efficient bifunctional catalyst for oxygen reduction reactions and oxygen evolution reactions by first-principles calculations *Appl. Surf. Sci.* **590** 153149

[47] Huang H, Wang T, Li J, Chen J, Bu Y and Cheng S 2022 A strain-engineered self-intercalation Ta9Se12 based bifunctional single atom catalyst for oxygen evolution and reduction reactions *Appl. Surf. Sci.* **602** 154378 8

[48] Liu X, Liu T, Xiao W, Wang W, Zhang Y and Wang G *et al* 2022 Strain engineering in single-atom catalysts: $GaPS_4$ for bifunctional oxygen reduction and evolution *Inorg. Chem. Front.* **9** 4272–80

[49] Wei B, Fu Z, Legut D, Germann T C, Du S and Zhang H *et al* 2021 Rational design of highly stable and active MXene-based bifunctional ORR/OER double-atom catalysts *Adv. Mater.* **33** 2102595

[50] Li F, Ai H, Shen S, Geng J, Ho Lo K and Pan H 2022 Two-dimensional dirac nodal line carbon nitride to anchor single-atom catalyst for oxygen reduction reaction *Chem. Sus. Chem.* **15** e202102537

[51] Zhang J, Yin S and Yin H 2020 Strain engineering to enhance the oxidation reduction reaction performance of atomic-layer Pt on nanoporous gold *ACS Appl. Energy Mater.* **3** 11956–63

[52] Liang Z, Zheng H and Cao R 2019 Importance of electrocatalyst morphology for the oxygen reduction reaction *Chem. Electro. Chem.* **6** 2600–14

[53] Huang B, Liu Y and Xie Z 2017 Biomass derived 2D carbonsviaa hydrothermal carbonization method as efficient bifunctional ORR/HER electrocatalysts *J. Mater. Chem.* A **5** 23481–8

[54] Wang L, Lin C, Huang D, Zhang F X, Wang M and Jin J 2014 A comparative study of composition and morphology effect of $Ni_xCo_{1-x}(OH)_2$ on oxygen evolution/reduction reaction *ACS Appl. Mater. Interfaces* **6** 10172–80

[55] Gao M, Liu J, Ye G, Zhao Z, Liu J and He G *et al* 2023 Molecular iron phthalocyanine catalysts on morphology-engineered graphene towards the oxygen reduction reaction *Sci. China Mater.* **66** 3865–74

[56] Ghosh D, Parwaiz S, Mohanty P and Pradhan D 2020 Tuning the morphology of CeO_2 nanostructures using a template-free solvothermal process and their oxygen reduction reaction activity *Dalton Trans.* **49** 17594–604

[57] Zhang X, Zhang Y, Huang H, Cai J, Ding K and Lin S 2018 Electrochemical fabrication of shape-controlled Cu_2O with spheres, octahedrons and truncated octahedrons and their electrocatalysis for ORR *New J. Chem.* **42** 458–64

[58] Devaguptapu S V, Hwang S, Stavros K, Zhao S, Gupta S and Su D *et al* 2017 Morphology control of carbon-free spinel $NiCo_2O_4$ catalysts for enhanced bifunctional oxygen reduction and evolution in alkaline media *ACS Appl. Mater. Interfaces* **9** 44567–78

[59] Wang Y, Sun T, Bagherzadeh H, Goncalves T J, Liang Z and Zhou Y *et al* 2022 Two-dimensional metal–organic frameworks with unique oriented layers for oxygen reduction reaction: tailoring the activity through exposed crystal facets *Ccschemistry* **4** 1633–42

[60] Li L H, Yuan J, Xue K, Xu M, Xu M and Wang J *et al* 2020 Synergic effect in a new electrocatalyst Ni_2SbTe_2 for oxygen reduction reaction *J. Phys. Chem.* C **124** 3671–80

[61] Shahzeb A S, Fang Z, Shi P, Zhu J, Lu C and Su Y *et al* 2023 Porous carbon nanosheets for oxygen reduction reaction and Zn–Air batteries *2D Mater.* **10** 022001

[62] Rowley-Neale S J, Fearn J M, Brownson D A C, Smith G C, Ji X and Banks C E 2016 2D molybdenum disulphide (2D-MoS_2) modified electrodes explored towards the oxygen reduction reaction *Nanoscale* **8** 14767–77

[63] Daniel S R, Krishnamurthy D, Garg R, Hafiz H, Lamparski M and Nuhfer N T *et al* 2020 Engineering three-dimensional (3D) out-of-plane graphene edge sites for highly selective two-electron oxygen reduction electrocatalysis *ACS Catal.* **10** 1993–2008

[64] Roy D, Karamjyoti P, Bikram Kumar D, Uday Kumar G, Bhattacharjee S and Samanta M *et al* 2021 Boron vacancy: a strategy to boost the oxygen reduction reaction of hexagonal boron nitride nanosheet in hBN–MoS_2 heterostructure *Nanoscale Adv.* **3** 4739–49
[65] Nguyen M C, Tran D T, Kim N and Joong H L 2018 Hierarchical heterostructures of ultrasmall Fe_2O_3-encapsulated MoS_2/N-graphene as an effective catalyst for oxygen reduction reaction *ACS Appl. Mater. Interfaces* **10** 24523–32
[66] Gou Z, Qu H, Liu H, Ma Y, Zong L and Li B *et al* 2022 Coupling of N-doped mesoporous carbon and N-Ti_3C_2 in 2D sandwiched heterostructure for enhanced oxygen electroreduction *Small* **18** 2106581
[67] Li R, Zhang R, Qiao Y, Zhang D, Cui Z and Wang Q 2022 Heterostructure $Ni(OH)_2/ZrO_2$ catalyst can achieve efficient oxygen reduction reaction *Chem. Eng. Sci.* **250** 117398 8
[68] Wang W, Batool N, Zhang T, Liu J, Han X F and Tian J *et al* 2021 When MOFs meet MXenes: superior ORR performance in both alkaline and acidic solutions *J. Mater. Chem. A, Mater. Energy Sustain* **9** 3952–60
[69] Chen J, Yuan X, Lyu F, Zhong Q, Hu H and Pan Q *et al* 2019 Integrating MXene nanosheets with cobalt-tipped carbon nanotubes for an efficient oxygen reduction reaction *J. Mater. Chem.* A **7** 1281–6
[70] Li Z, Zhuang Z, Lv F, Zhu H, Zhou L and Luo M *et al* 2018 The marriage of the FeN_4 moiety and MXene boosts oxygen reduction catalysis: Fe 3d electron delocalization matters *Adv. Mater.* **30** 1803220–0
[71] Jiang L, Duan J, Zhu J, Chen S and Antonietti M 2020 Iron-cluster-directed synthesis of 2D/2D Fe–N–C/MXene superlattice-like heterostructure with enhanced oxygen reduction electrocatalysis *ACS Nano.* **14** 2436–44

Chapter 6

Engineered 2D materials for CO_2 reduction reaction (CO_2 RR)

Abhinandan Patra and Chandra Sekhar Rout

In response to the urgent need to combat climate change and reduce carbon dioxide (CO_2) emissions, extensive research has been directed toward sustainable CO_2 conversion technologies. Among these strategies, electrochemical CO_2 reduction has emerged as a promising avenue for transforming CO_2 into valuable chemicals and fuels. However, this endeavour has been stained by inherent limitations in existing materials, including issues with selectivity, stability, cost, and mass transport. Two-dimensional (2D) materials, such as graphene, transition metal oxides (TMOs), transition metal dichalcogenides (TMDs), transition metal trichalcogenides (TMTCs), and MXenes, which have emerged as the frontiers in the field of electroreduction of CO_2 reduction. 2D materials offer unique advantages for catalysis due to their ultrathin structure, large surface area, and versatile chemistry. This chapter explores their potential in CO_2 electroreduction, focusing on reaction mechanisms, catalytic enhancement, and material stability. Techniques like doping, alloying, defect engineering, edge activation, and heterostructuring are discussed as effective approaches to upgrade them as the prime and sole candidates for the job. Collaboration across multiple disciplines, integrating both theoretical and experimental approaches, is essential. In the end, 2D materials have the potential to be pivotal in achieving carbon-neutral technologies, yet challenges persist at the crossroads of materials science, electrochemistry, and catalysis.

6.1 Introduction

The global challenge of mitigating climate change and reducing carbon dioxide (CO_2) emissions has spurred intense research into sustainable and efficient technologies for CO_2 conversion and utilization. Among the myriad strategies explored, the electrochemical reduction of CO_2 has emerged as a promising avenue, holding the potential to transform this greenhouse gas into valuable chemicals and fuels [1–3].

doi:10.1088/978-0-7503-5719-7ch6 © IOP Publishing Ltd 2024

While various materials have been explored for this purpose, inefficient selectivity, stability issues, cost limitations, and mass transport constraints are the key bottlenecks which limits their practice [4, 5]. In view of that, the ascent of 2D materials has significantly reshaped the landscape of CO_2 reduction [6–8]. These atomically thin sheets, such as graphene, TMOs, TMDs, TMTCs, and MXenes etc offer a remarkable platform for catalytic processes owing to their high surface area, ample electrochemical active sites/edges, tunable electronic properties, exceptional catalytic potential, and versatile surface chemistry [8–10]. This chapter embarks on an insightful journey, exploring the scientific inquiries and challenges that underpin the electroreduction of CO_2 into high energy density products such as carbon monoxide (CO), formic acid (HCOOH), methane (CH_4), or even more complex hydrocarbon fuels with 2D material catalysts [11–13]. It delves into the intricacies of reaction mechanisms, the optimization of catalytic activity and selectivity, and the quest for enhanced material stability under the demanding innovative strategies such as doping, alloying, defects, edges, and heterostructuring etc [14, 15]. Through 2D material embraces promising characteristics for electroreduction of CO_2, stability under electrochemical conditions, the precise control of selectivity for target reduction products, and the development of scalable synthesis techniques for practical large-scale applications are the scientific stumblings that contain them, which can be overcome by the aforementioned strategies [12–16]. Doping introduces specific atoms or ions, altering the material's electronic structure and reactivity for improved catalytic activity [16]. Alloying combines different elements, creating tailored alloys optimized for CO_2 electroreduction [17]. Defects provide active sites for catalytic reactions, while material edges increase surface area for electrochemical reactions [18]. Heterostructuring combines 2D materials, creating interfaces for efficient charge transfer and catalytic reactions [19]. These strategies represent the vanguard of scientific inquiry, offering the potential to fine-tune the properties of materials, thus unlocking unprecedented catalytic performance. Furthermore, the integration of theoretical insights and computational modelling with experimental investigations plays a pivotal role in unravelling the mysteries of CO_2 electroreduction. This interdisciplinary synergy between scientific theory and empirical validation offers a holistic approach to understanding the catalytic behaviour of 2D materials and guides the rational design of catalysts [20].

In the grander perspective, this chapter foresees the wider consequences of employing 2D materials for electrochemically reducing CO_2. It acknowledges the capacity of these materials to have a revolutionary impact in achieving carbon-neutral technologies and propelling the worldwide shift toward a sustainable, low-carbon future. The chapter begins by elucidating the fundamental mechanisms of CO_2 electroreduction through 2D materials, proceeds with innovative strategies to enhance the electrochemical performance of these materials, and culminates in contemporary perspectives and challenges that underpin the field. The application of 2D materials as catalysts in the electroreduction of CO_2 stands at the intersection of various fields, encompassing materials science, electrochemistry, and catalysis, providing exceptional prospects and posing distinct scientific hurdles.

6.2 CO_2 RR-mechanism and insights

Carbon dioxide molecules exhibit notable inertness and stability due to the carbon atoms within CO_2 existing in their highest oxidation state. Consequently, the enhancement of electrochemical processes aimed at reducing CO_2, which is inherently slow, necessitates the advancement of effective electrocatalysts. Viable avenues for achieving electrochemical reduction of CO_2 involve facilitating multi-electron transfers in aqueous solutions using appropriately designed electrocatalytic materials [21]. By and large, CO_2 is subjected to electrochemical reactions on the surface of a catalyst, typically made of metal nanoparticles or other conductive 2D materials. These catalysts facilitate the conversion of CO_2 into various products through a series of reduction steps. The specific products formed depend on factors such as the choice of catalyst, reaction conditions, and applied voltage [22, 23]. The general reduction pathway for CO_2 can involve multiple stages, with intermediates formed along the way. The process generally involves the transfer of electrons from an external power source to CO_2 molecules adsorbed on the catalyst surface. This leads to the formation of intermediate species, which subsequently undergo further reduction steps to yield final products. As CO_2 electroreduction is a solid–liquid-gas reaction, CO_2 molecules can be reduced to form high energy density products such as carbon monoxide (CO), formic acid (HCOOH), methane (CH_4), or even more complex hydrocarbon fuels, depending on the reaction conditions and catalyst properties [23, 24]. The chance of formation of mixture of gases and liquids inside the electrochemical chamber is very likely persuasive. The overall reaction is driven by the energy supplied through the applied voltage and the catalytic activity of the chosen materials, which is explained in figure 6.1.

The primary stride in CO_2 reduction involves the chemical transformation of CO_2 into less oxidized carbon species. This undertaking is intricate due to the sluggish kinetics witnessed during CO_2 electroreduction. In a conventional single-electron CO_2 reduction protocol, distinct anodic and cathodic chambers are utilized. The anode orchestrates water oxidation, leading to molecular oxygen generation, while the cathode oversees CO_2 reduction, resulting in the formation of reduced carbon species. The thermodynamic potential requisite for catalysing the one-electron CO_2 reduction to form CO_2^- is situated at −1.90 V versus SHE in aqueous solutions at pH 7, indicative of a highly energy-intensive and unfavourable course (figure 6.2) [24, 25]. Foremost among these stages is the generation of CO_2^-, a pivotal juncture acting as the rate-determining phase, subsequently determining the identity of the two-electron reduction product, be it CO or formate. Conversely, multi-electron proton-assisted electron transfer mechanisms exhibit heightened favourability within the potential interval of −0.2 to −0.6 V versus SHE [22–25].

This divergence results in a diverse spectrum of CO_2^- derived products, contingent on the chosen catalyst and electrolyte. Recent inquiries have embraced aqueous and non-aqueous electrolytes for CO_2 reduction, with the prevalent selection being a 0.5 M $NaHCO_3$ (or $KHCO_3$) solution at pH 7, serving as a pH buffer to uphold electrode surface pH and suppress undesired HER (hydrogen evolution reaction) [25, 26]. Notably, the ensuing reduction steps transpire nearly

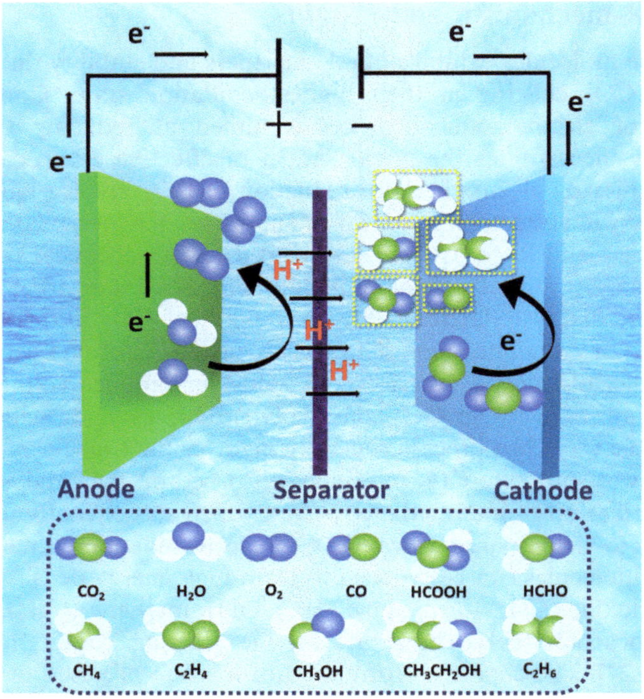

Figure 6.1. Schematic illustration of the general mechanism of electroreduction of CO_2.

instantaneously in relation to the inaugural step. Therefore, the stabilization of this high-energy intermediate stands as a critical determinant in enabling a prompt and energy-efficient CO_2 reduction process concerning 2D materials [24–26].

In the realm of electrocatalytic CO_2 reduction reactions of 2D electrocatalysts, a plethora of pivotal figures of merit assumes paramount importance as evaluative parameters for catalyst efficacy, unravelling insights into their capability to effectuate the conversion of carbon dioxide into valuable derivatives. These quantitative metrics serve as invaluable indicators of efficiency, selectivity, and overall operational. Herein, a compendium of salient figures of merit of interest to electrocatalytic CO_2 reduction reactions is explicated [13–18]:

- **Faradic efficiency (FE):** This metric encapsulates the proportion of electrons judiciously harnessed for the conformation of the targeted end product during the electrocatalytic cascade. Mathematically, it ensues as the quotient of electrons participatory in the synthesis of the desired compound concerning the total quantum of electrons transiting through the system which can be denoted by;

$$\varepsilon_{\text{Faradic}} = \frac{\alpha n F}{Q} \tag{6.1}$$

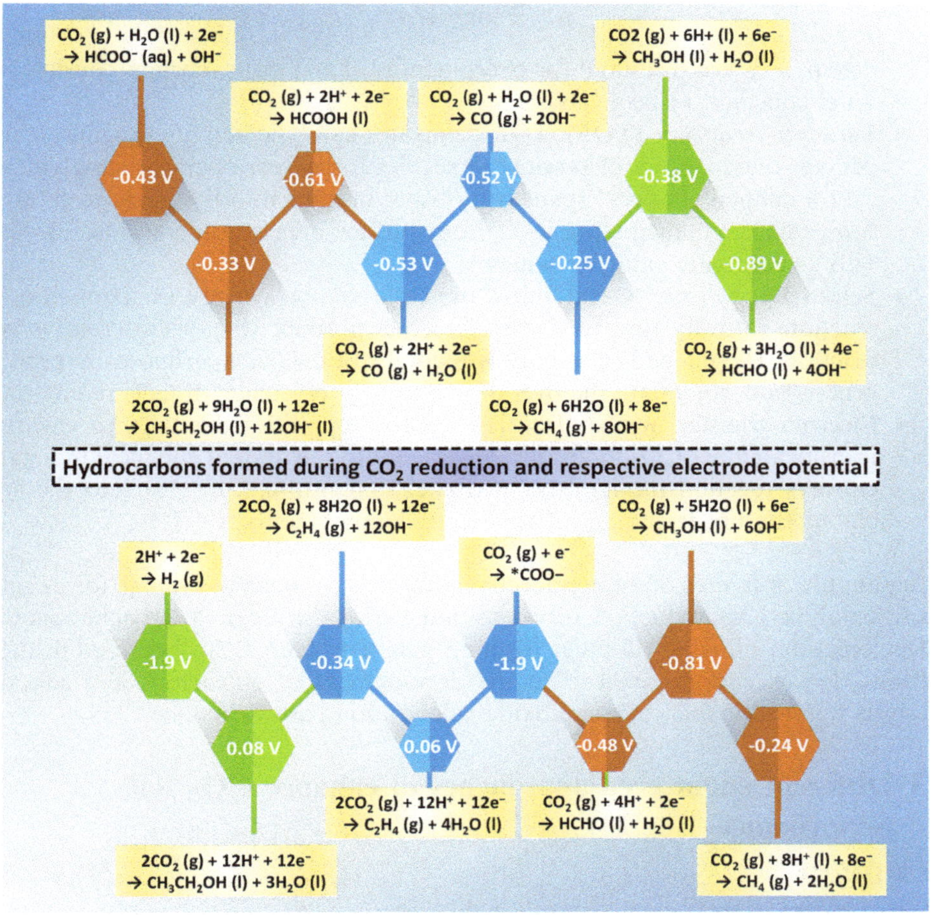

Figure 6.2. Schematic illustration of possible by products formed in course of electroreduction of CO_2 and respective electrode potentials (V versus SHE) in water at pH 7.

here, α, n, F and Q represent the number of electrons transferred, the number of moles for a given product, F is Faraday's constant [96 485 C mol^{-1}], and the charge passed throughout the electrolysis process respectively.

- **Overpotential (η)**: Overpotential constitutes the adjunctive electric potential requisite for propelling the electrocatalytic reaction beyond its thermodynamic equilibrium threshold. This parameter reflects the activation energy indispensable for reaction initiation, thus offering a quantitative index for assessing energy efficiency.
- **Tafel slope:** The Tafel slope embodies the rate of exponential change in current density concerning variations in overpotential. It furnishes insights into the reaction kinetics and the dominant reaction step's mechanistic attributes. The Tafel equation is given as;

$$\eta = b \log j + a \tag{6.2}$$

herein, η, b, j and a stand for overpotential, Tafel slope, current density, and Tafel constant, respectively.

- **Turnover frequency (TOF):** TOF delineates catalytic activity magnitude and denotes the quantum of product molecules forged per operative catalytic site over a defined temporal span. It furnishes insights into the intrinsic catalytic competence, emancipating itself from surface area considerations. Elevated TOFs accentuate catalytic potency.
- **Selectivity:** Selectivity epitomizes the catalyst's propensity to channelize the reaction pathway toward a specific target product vis-à-vis alternative side reaction. Heightened selectivity is pivotal to obviate superfluous by-product genesis and potentiate the overall operational efficiency of CO_2 reduction.
- **Electron transfer number (*n*):** This metric quantifies the tally of electrons translocated per CO_2 molecule converted into the desired chemical moiety. It conveys reaction mechanistic attributes and proffers insights into the stoichiometric underpinnings of the electrocatalytic journey.

This panoply of figures of merit collectively furnishes a holistic vantage for gauging electrocatalytic CO_2 reduction reactions and various strategic approaches such as defect creation, alloying, doping, forming heterostructures, surface modification, cutting edge execution can lead to the development of efficacious and selective catalysts for sustainable carbon dioxide valorization pathways.

6.3 Different engineered approaches to enhance CO_2 RR performance

6.3.1 Defects

Diverse functional aspects emerge in defects and interfaces due to varying contact styles and scenarios. Defects at boundaries stabilize interfaces, while nearby interface environments foster defect formation. Defects possess the ability to alter the chemical surroundings such as strain in bonds and generate electrocatalyst active sites, whereas interfaces prove advantageous in enhancing the capacity for CO_2 adsorption. These defects can be point, line, volume or surface defects in the case of manoeuvring the structure of 2D materials for electroreduction of CO_2. Pan *et al* synthesized Cu nanosheets for the electroreduction of CO_2 to CO and demonstrated a remarkable FE of 74.1% along with a record-breaking partial current density of 23.0 mA cm^{-2} and a turnover frequency of 0.092 s^{-1} for CO production [26]. Furthermore, at a potential of −1.0 V versus the reversible hydrogen electrode (RHE), these electrodes also exhibit an FE of 24.8% for hydrogen (H_2) generation which can be attributed to the vacancy defects taking place in Cu nanosheets (figures 6.3(a)–(d)). Additionally, theoretical computations elucidated that the abundant lattice vacancies present on the surface of extremely thin copper nanosheets enhance the generation of carbon monoxide during the reduction of CO_2. These enhancements can be accredited to the swift movement of mass and electrons,

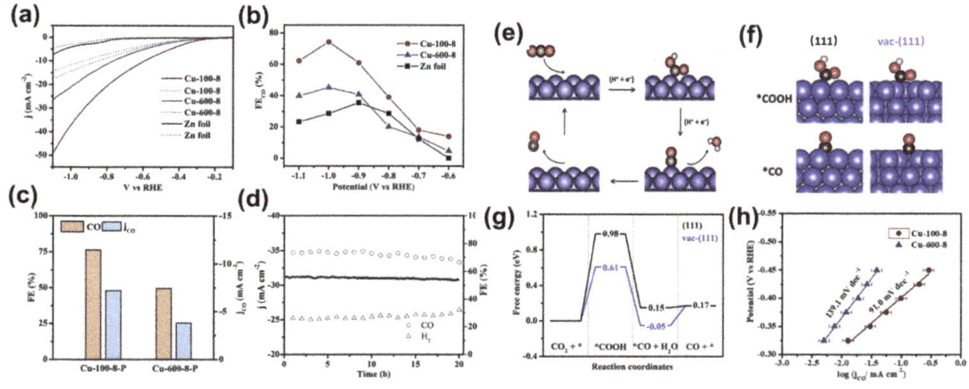

Figure 6.3. (a) Graphs of electrode polarization in Ar (dashed line) and CO_2-saturated 0.5 M $KHCO_3$ electrolyte (solid line), (b) CO Faradic efficiencies (FEs) at various potentials for these electrodes, (c) examination of FEs and CO current density at −1.0 V for Cu-100-8-P and Cu-600-8-P traditional electrodes, (d) extended −1.0 V electrolysis study with Cu-100-8 electrode, (e) diagram depicting the process of converting CO_2 to CO through a two-electron transfer process, (f) profile view of the binding arrangements of *COOH and *CO (* denotes the active sites) on both the Cu (111) and vac-(111) surfaces, (g) reaction pathways for CO_2 reduction on the Cu (111) and vac-(111) surfaces and (h) Tafel plots displaying the CO production rates for Cu-100-8 nanosheets and Cu-600-8 nanoparticles electrodes. Reprinted from [26], Copyright (2019), with permission from Elsevier.

as well as the numerous vacancy defects within the active sites. These intrinsic activities are a result of the distinctive hierarchical structure of the copper nanosheet electrode, which is free of binders. The mechanism, reaction pathway and Tafel plots are given in figures 6.3(e)–(h) [26].

Similarly, Gao and their group synthesized an oxygen-deficient cobalt oxide (Co_3O_4) for electroreduction of CO_2 into formate and according to their study, the existence of oxygen (II) vacancies reduces the activation barrier, which is the rate-limiting step, from 0.51 to 0.40 eV [27]. This reduction occurs by stabilizing the formate anion radical intermediate, as confirmed by the decreased onset potential from 0.81 to 0.78 V and a lower Tafel slope, which decreases from 48 to 37 mV/dec. Consequently, cobalt oxide single-unit-cell layers rich in vacancies display current densities of 2.7 mA cm^{-2} with approximately 85% selectivity for formate during 40 h testing (figures 6.4(a)–(f)) [27].

Density functional theory (DFT) computations indicate that the primary defect is the presence of oxygen (II) vacancies, and this is corroborated by x-ray absorption fine structure (XAFS) spectroscopy, which provides information on the specific concentrations of these oxygen vacancies. In the absence of defects in the Co_3O_4 single-unit-cell layers, the existence of an O(II) vacancy played a crucial role in enhancing the stability of the $HCOO^-$ intermediate, thereby promoting the hydrogenation process. The mechanism and step-by-step reduction of CO_2 into formate on defect-rich cobalt oxide is given in figure 6.4(g) along with the free energy diagram [27]. Another oxide of tin has been synthesized by Amal *et al* for converting CO_2 into formate, achieving an impressive formate FE of 85% and a current density of −23.7 mA cm^{-2} at an applied potential of −1.1 V versus E_{RHE} [28].

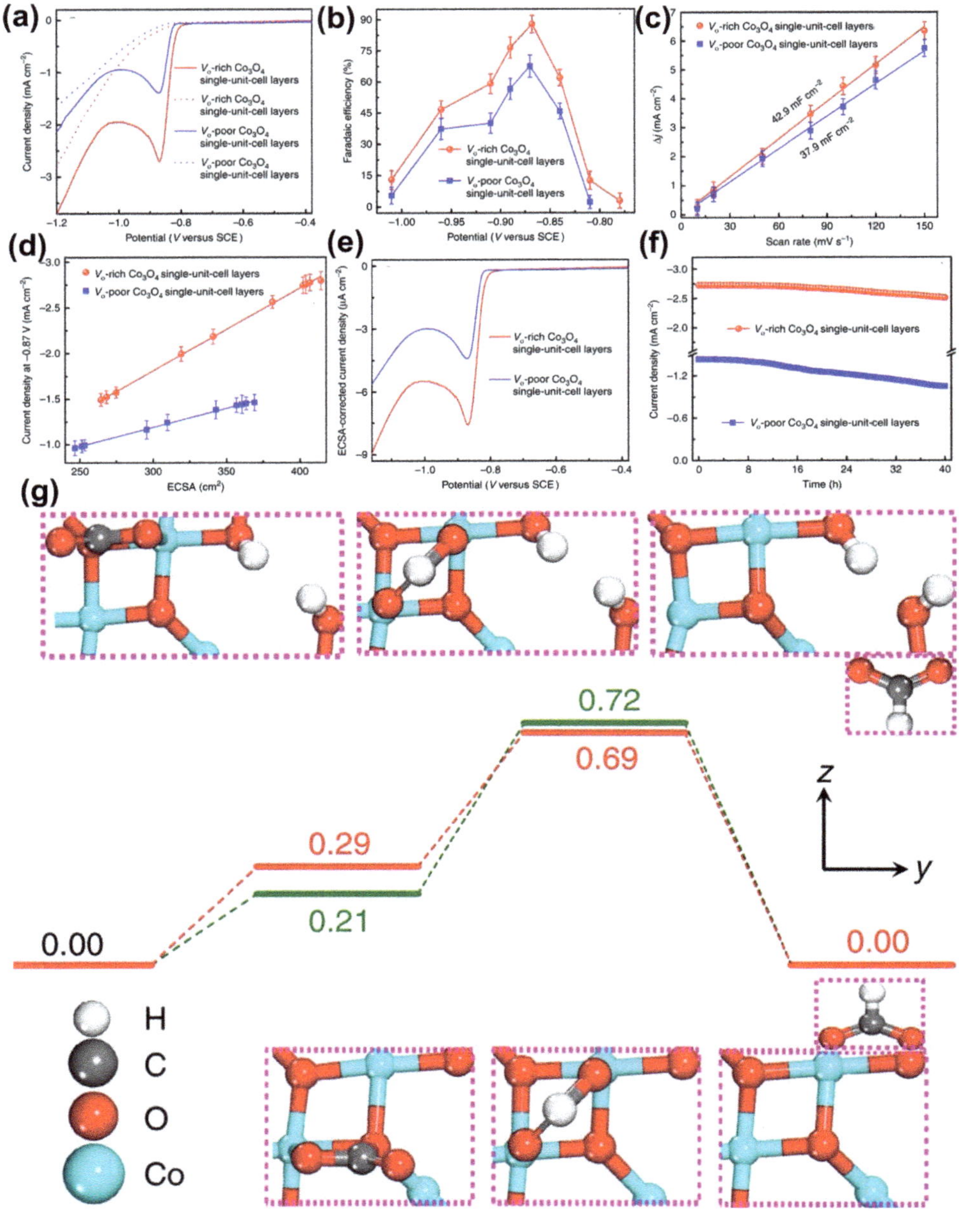

Figure 6.4. (a) Linear sweep voltammetry in CO_2-saturated and N_2-saturated 0.1M $KHCO_3$, (b) FE for formate at different potentials, (c) charging current density differences versus scan rates, (d) current density at −0.87 V versus SCE plotted against ECSA for various materials and loadings, (e) ECSA-corrected current densities versus applied potentials and (f) chronoamperometry at −0.87 V versus SCE with error bars showing standard deviations from five independent measurements and (g) step-by-step reaction mechanism illustration by a schematic diagram. Reprinted by permission from Macmillan Publishers Ltd [*Nature Communications*] [27] CC BY 4.0.

This achievement was made possible by intentionally adjusting the flame synthesis conditions to control the formation of oxygen hole centres which played a pivotal role in the activation of CO_2, thus dictating the remarkable activity demonstrated by the FSP-SnO_2 catalysts in formate production. This controlled generation of defects through a straightforward and scalable fabrication method represents an ideal approach for the deliberate design of effective catalysts for CO_2 reduction reactions [28]. According to theoretical predictions based on DFT calculations, MoS_2 and $MoSe_2$ exhibited potential as catalysts for the electroreduction of CO_2. This is attributed to their ability to selectively bind COOH and CHO to the bridging S or Se atoms and CO to the metal atom, which should, in principle, facilitate the conversion of CO_2 into CO. Additionally, when doped with Ni, MoS_2 and $MoSe_2$ have the potential to further reduce CO into hydrocarbons and alcohols [29]. Another study by Wu *et al* involved the synthesis of 3D graphene foam with nitrogen (N) doping, introducing a novel form of N-defect structure through chemical vapor deposition (CVD). This work revealed that pyridinic-N defects could lower the free energy barrier for the formation of adsorbed $COOH^*$, thus promoting CO production, in agreement with DFT calculations [30].

6.3.2 Alloying

The electroreduction of CO_2 through alloying 2D materials is a burgeoning field aiming to use ultrathin substances to create efficient catalysts for converting CO_2 into valuable chemicals, addressing environmental and energy challenges. A lot of transition metal dichalcogenides such as MoS_2, VS_2, $MoSe_2$ etc have shown a tremendous effect on enhancing the electrocatalytic activity regarding CO_2 reduction; moreover, the strategy of alloying of these TMDs can be an efficacious way to boost the electroreduction of CO_2 to a greater extent [31–33]. In view of that, Xie *et al* successfully synthesized monolayers of MoSeS alloy. x-ray absorption fine structure spectroscopy analysis revealed that these monolayers exhibited shortened Mo-S bonds and lengthened Mo–Se bonds [34]. This structural modification had the effect of tailoring the d-band electronic structure of Mo atoms. Notably, the MoSeS alloy monolayers demonstrated a significant improvement in electrocatalytic performance compared to MoS_2 and $MoSe_2$ monolayers. Specifically, at −1.15 V versus RHE, the MoSeS alloy monolayers achieved a current density of 43 mA cm^{-2}, which was approximately 2.7 and 1.3 times higher than that of MoS_2 and $MoSe_2$ monolayers, respectively. Furthermore, the MoSeS alloy monolayers exhibited an impressive FE of 45.2% for CO production under the same conditions. This efficiency surpassed that of MoS_2 monolayers (16.6%) and $MoSe_2$ monolayers (30.5%) at −1.15 V versus RHE [34]. Theoretical calculations additionally demonstrated that the presence of off-centre charge surrounding Mo atoms had a dual effect. Firstly, it enhanced the stability of the $COOH^-$ intermediate, and secondly, it facilitated the rate-limiting step of CO desorption. Likewise, another study unravelled the electroreduction of CO_2 through 2D mixed alloy of palladium and copper nanodendrites (nds) [35]. 2D PdCu nanodendrites (nd-PdCu) were synthesized via a low-temperature solution method, with precise control of a set of experimental parameters.

The modulation of electronic structures, known as the electronic effect, and the subsequent alteration of the interaction between the catalyst and key reaction intermediates have been observed both experimentally and computationally in Pt-based or Pd-based electrocatalysis. These alloyed products demonstrate significantly improved CO tolerance, enabling selective and exceptionally stable CO_2 reduction to formate even at relatively high overpotentials which can be accredited towards the combined influence of the electronic effect and nanostructuring. Specifically, in a 0.1 M $KHCO_3$ solution, the most effective catalyst, nd-PdCu-1, displayed the ability to selectively and consistently reduce CO_2 to formate, even at a cathodic working potential as low as −0.4 V [35]. Analysis of the products indicated that formate was the primary reduction product in the catholyte, with negligible amounts of H_2 or CO observed over the examined potential range. Based on chronoamperometric tests, the FE for formate production by nd-PdCu-1 was measured to be approximately 82% at 0 V (figures 6.5(a)–(e)) [35]. DFT calculations unveiled that alloying Pd with Cu enhanced the resistance of our product to CO surface poisoning and also improved the adsorption of the $^{-}$OCHO intermediate, which is crucial for achieving high reaction selectivity (figures 6.5(f)–(i)). Apart from this, some other studies revealed the improved catalytic activity regarding electroreduction of CO_2 by alloying of some other metals such as Pd–Sn, Sn–Pb, Cu–Pd, Cu–Cd, Cu–In etc [36–40]. As a result, the introduction of highly nanoporous materials can give rise to the formation of extensive grain boundaries, while a confined geometry allows for the attainment of quantum confinement. Additionally, the electronic properties can be directly altered through alloying and doping processes involving various metals.

6.3.3 Doping

To achieve superior catalytic performance in the reduction of CO_2, it is imperative to break away from the linear scaling relations that typically govern the binding energies of reduction intermediates, and doping offers a promising avenue for achieving this. Doping, involving the introduction of heteroatoms (comprising both nonmetal and metal atoms), not only induces changes in the electronic structure of neighbouring skeleton atoms but also furnishes new active sites. This, in turn, optimizes the binding interactions between intermediates and active sites within 2D materials. According to predictions from DFT calculations, the augmentation of Density of States (DOS) at the Fermi level resulting from doping can significantly enhance the catalytic activity of the catalyst. Accordingly, Zeng and colleagues had conducted an investigation of atomically thin SnS_2 nanosheets through the incorporation of Ni doping [41]. This modification aimed to enhance their performance in the electroreduction of CO_2. The outcome of this modification was that the Ni-doped SnS_2 nanosheets had demonstrated an increase in both current density and FE for the production of carbonaceous products compared to the pristine SnS_2 nanosheets. Specifically, when the Ni content had reached 5 atomic weight percentages, the Ni-doped SnS_2 nanosheets had achieved a remarkable FE of 93% for carbonaceous products, accompanied by a current density of 19.6 mA cm^{-2} at −0.9 V versus RHE [41]. Throughout the potentiostatic test, the Ni-doped SnS_2

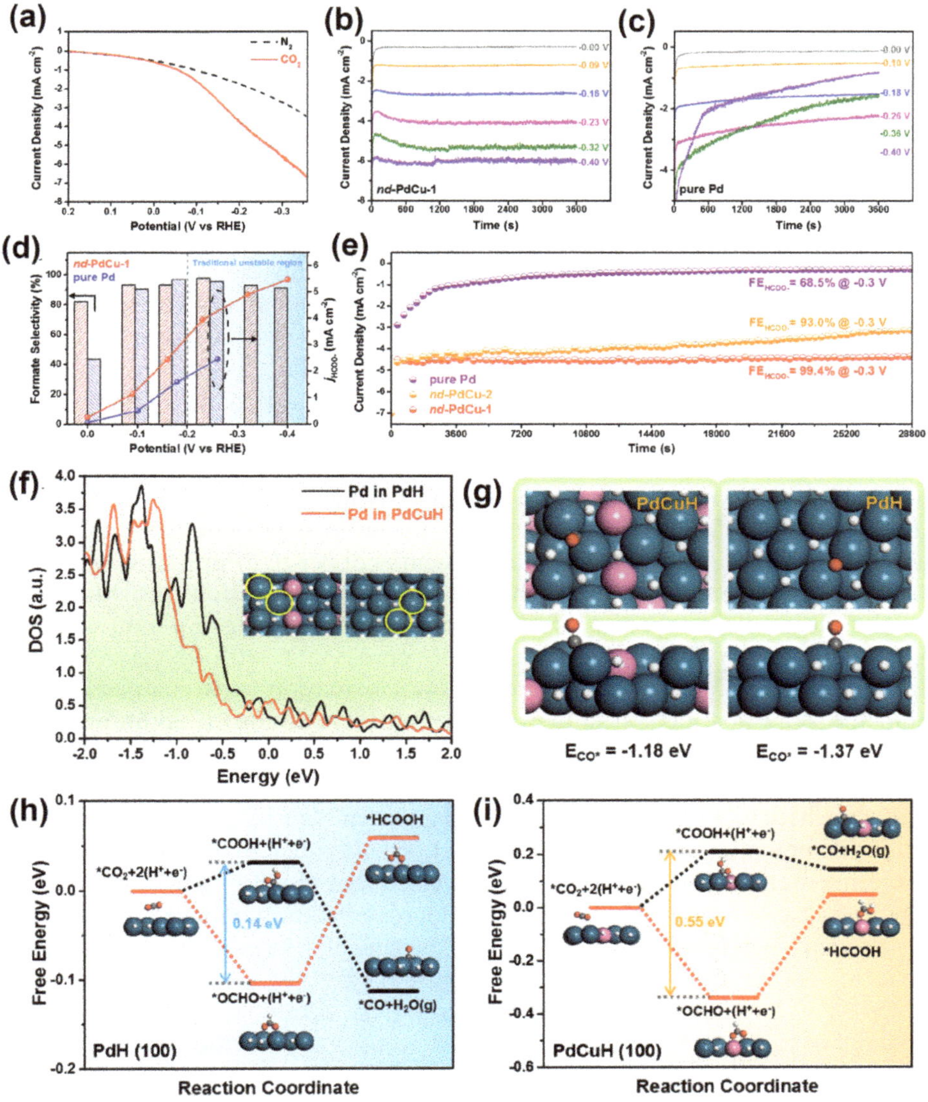

Figure 6.5. (a) Polarization curves generated for nd-PdCu-1 in electrolytes saturated with either CO_2 or N_2, chronoamperometric curves recorded for (b) nd-PdCu-1 and (c) pure Pd at various working potentials, (d) formate FE and partial current density at different working potentials, (e) long-term chronoamperometric stability of pure Pd, nd-PdCu-1, and nd-PdCu-2 for a duration of 8 h, DFT simulations; (f) projected density of states (PDOS) of Pd atoms, highlighted by yellow circles, were analysed on the (100) surfaces of both PdH and PdCuH, (g) adsorption configuration and binding energy of CO on the (100) surfaces of PdH and PdCuH and free energy diagrams illustrating the CO_2 reduction reaction pathway constructed for (h) PdH (100) and (i) PdCuH (100). Reprinted with permission from [35]. Copyright (2021) American Chemical Society.

nanosheets had maintained a consistent FE for carbonaceous products without any observable decline in current density. An examination of the mechanism had revealed that Ni doping had generated a defect level at the conduction band edge and had led to a reduction in the work function of the SnS_2 nanosheets. These changes had proven advantageous for the activation of CO_2, consequently resulting in an improvement in the catalytic performance of CO_2 electroreduction. In a separate investigation, exfoliated graphitic carbon nitride (GCN) underwent a modification by incorporating single copper atoms. This modification aimed to harness the catalytic capabilities of the Cu–GCN hybrid system for electrochemical CO_2 reduction [42]. The study focused on examining the influence of visible light on the catalytic process. Cu–GCN was employed as an electrocatalyst for the CO_2 reduction reaction in two distinct electrolytes, with the objective of assessing its impact on product selectivity. When utilized in a bicarbonate solution, the sole product observed in the liquid phase was formate, simplifying its separation. Conversely, in a phosphate solution, hydrogen was the exclusive product detected. Another report of copper doing came to light by Ding and their group and according to them, a high-performance electrocatalyst for the conversion of CO_2 to CO in an aqueous solution was developed by introducing copper (Cu) into ZnO nanosheets, resulting in Cu-doped and defect-rich ZnO [43]. DFT simulations provided insights into the electronic properties of ZnO catalysts containing Cu-induced oxygen vacancies, which differed from those of pure ZnO with oxygen vacancies. Furthermore, the Cu dopant exhibited a lower oxidation state compared to Cu^{2+} within the ZnO host crystal lattice, promoting the creation of additional oxygen vacancies. Experimental investigations conducted in a flow cell demonstrated that the as-prepared Cu–ZnO catalysts outperformed in the CO_2 reduction reaction, achieving a high CO FE exceeding 80% and a current density exceeding 45 mA cm^{-2} over a wide potential range from −0.76 to −1.06 V (figure 6.6) [43]. The combination of experimental and theoretical findings revealed that modifying the local electronic structure of oxygen vacancies through Cu doping effectively reduced the adsorption of hydrogen species while preserving the beneficial role of oxygen vacancies in facilitating the electroreduction of CO_2 (figure 6.6).

Similarly, CuO nanosheets as an electrocatalyst, doped with Sn and containing abundant oxygen vacancy defects (referred to as Vo–CuO(Sn), were synthesized using a straightforward one-pot method [44]. The Vo–CuO(Sn) electrode demonstrated remarkable electrocatalytic performance, with FE exceeding 95% over a wide potential range (−0.48 to −0.93 V versus RHE). At a competitive overpotential of 420 mV, the efficiency even reached a maximum of 99.9%. Impressively, after an extended reaction period of 180 h, the efficiency of Vo–CuO(Sn) remained at approximately 96%, surpassing the stability of most previously reported CO_2 reduction electrocatalysts. DFT calculations revealed that Vo–CuO(Sn) effectively enhanced the adsorption and activation of CO_2 molecules, thus reducing the energy barrier for the formation of $COOH^-$ intermediates and the desorption of $^*CO$ [44]. This led to outstanding performance in the reduction of CO_2 to CO. Consequently, this study expands the possibilities for designing environmentally friendly CO_2 reduction catalysts by incorporating metal doping and engineering defects to achieve

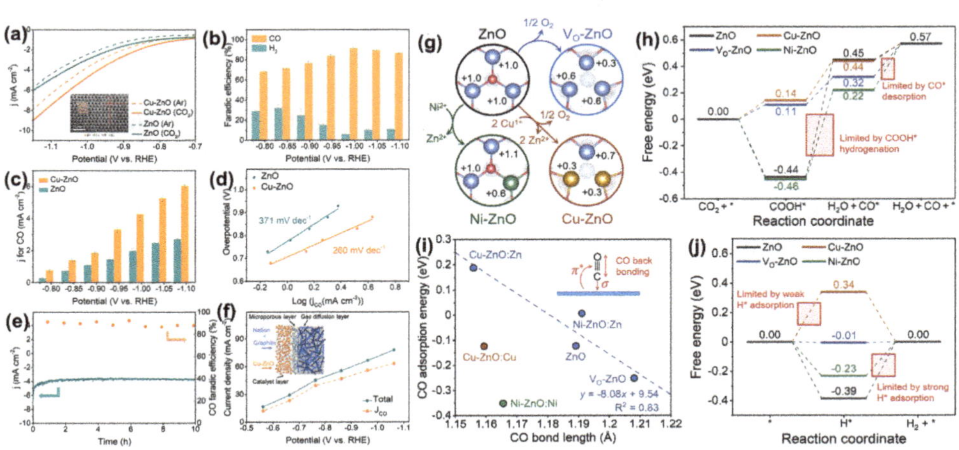

Figure 6.6. (a) Polarization curves generated for Cu–ZnO and ZnO catalysts in a 0.1 M $KHCO_3$ solution saturated with CO_2 and inset showing the doping effect through position of atoms, (b) FEs for various reduction products determined over Cu–ZnO at different applied potentials, (c) CO partial current densities measured for both Cu–ZnO and ZnO catalysts, (d) Tafel plots, (e) the long-term potentiostatic stability of Cu–ZnO, (f) total current densities and partial current densities for the production of CO on Cu–ZnO in a flow cell and the inset depicts the gas diffusion electrodes, (g) top view illustration of different surface models of catalysts, (h) diagram depicting the free energy profile for CO production, (i) the correlation between the bond length of CO and its adsorption energy on different sites and (j) free energy diagram illustrating the hydrogen production process in these models. Reprinted from [43], Copyright (2022), with permission from Elsevier.

efficient CO_2 reduction. DFT calculations indicate that N doping reduces the energy barrier for $COOH^-$ adsorption and lowers the energy barrier for the direct protonation of $COOH^-$ to produce CO gas and H_2O, which requires less energy compared to the formation of H_2 via the Heyrovsky step. Therefore, a lot of studies have been carried out keeping the aforementioned fact. Among the plethora of studies, recently, $ZnIn_2S_4$ (ZIS) was synthesized as an electrocatalyst for the CO_2 reduction reaction, primarily producing C1 species, and to enhance its performance, ultrathin ZIS nanosheet arrays were successfully grown on nitrogen-doped carbon cloth through a straightforward hydrothermal process and achieve a remarkable 42% FE for ethanol production during electrocatalytic reduction of CO_2 at an applied potential of −0.7 V versus RHE in a CO_2-saturated 0.5 M $KHCO_3$ aqueous solution [45]. Likewise, Ni-doped MoS_2, gold doping in CeO_x nanosheets, In-doped $Cu@Cu_2O$ etc exhibited enhancement in CO_2 electrode-reduction to CO (FE of 90.2%, 90.1% and 2.2%, respectively) through doping [46–48]. Additionally, the modification of electronic structure and properties to improve CO_2 reduction performance can not only be tuned through doping but also through surface stress/strain.

6.3.4 Strain and stress

Surface strain is the stress that arises due to lattice mismatch when introducing one metal into a host composed of other metal constituents, and it exerts a substantial impact on the catalytic efficiency. This lattice-induced strain can boost the effectiveness and specificity of electrochemical reactions by disrupting the linear scaling relationship. However, it is worth noting that the explicit utilization of strain to manipulate the process of the CO_2 reduction reaction is seldom documented in the literature, though a lot of strain engineering reports in the field of the core–shell synthesis are out there which relate the challenge in complexity of intentionally engineering and quantifying surface strain in materials, especially in 2D materials for CO_2 reduction. The modification of electronic structure in heterogeneous catalysts is influenced by interfacial strain between different materials, enhancing charge transfer in CO_2 reduction reactions. For instance, 2D Bi@Sn, where a Bi nanoparticle coated with Sn generated 8.5% compressive strain in the Sn shell [49]. This reduced the Gibbs free energy of $^{-}$HCOO by approximately 0.21 eV compared to pristine Sn, leading to high selectivity for HCOOH. Additionally, introducing Li atoms onto the Sn nanoparticle surface, creating surface-lithium-doped tin, induced 5% tensile strain in the Sn lattice, lowering the Gibbs free energy from CO_2 to HCOOH by about 0.06 eV compared to pristine Sn catalyst. The LSV profiles and Nyquist plots were corroborated and aligned with the experimental data (figures 6.7(a)–(c)) [49]. The combination of nanoporous morphology and reduced thermodynamic reaction barriers resulted in excellent electrochemical properties and high HCOOH selectivity (FEHCOOH = 95.9% at −0.8 V RHE) (figures 6.7(d) and (e)). The stability assessment exhibited the robustness of the electrocatalysts for over 30 h which can be seen in figure 6.7(f). These findings underscored the role of interfacial strain in promoting charge transfer at heterostructure interfaces, leading

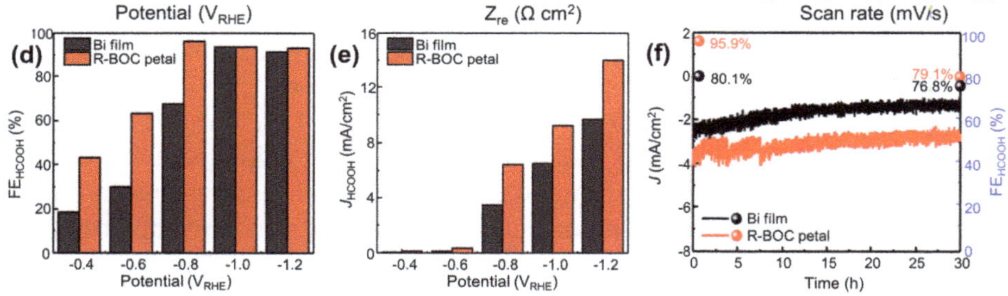

Figure 6.7. (a) Linear sweep voltammetry (LSV) curves for both a Bi film and R-BOC petals recorded in 0.1 M $KHCO_3$ electrolytes containing at a scan rate of 50 mV s^{-1}, (b) Nyquist plots, (c) current–density profiles at various scan rates were derived from double-layer charge/discharge curves obtained at 0.33 V RHE, (d) FE, (e) the current density for HCOOH production were evaluated for both the Bi film and R-BOC petals and (f) stability assessments over a period of 30 h at −0.8 V RHE. Reproduced from [49] CC-BY 4.0.

to electron redistribution and optimization of CO_2 binding energies to intermediates, ultimately controlling the catalytic pathway.

6.3.5 Morphology

The catalyst's performance relies heavily on its surface morphology and thereby textured electrochemical surface area, especially evident in catalysts with unique porous or hollow structures; altering metal oxide morphology enhances CO_2 adsorption, active site exposure, and durability, ultimately improving product selectivity and CO_2 reduction rates. Li *et al* observed during CV, an *in situ* morphology transformation from Bi_2O_3 to Bi nanosheets (BiNSs), with the presence of oxygen and the initial Bi_2O_3 morphology playing crucial roles in this phase change. SEM analysis revealed that this transformation was particularly facile in the case of precipitated Bi_2O_3 particles compared to Bi nanoparticles (BiNPs) and Oxi-BiNPs, underscoring the importance of oxygen and the initial Bi precursor morphology in the exfoliation of BiNSs [50]. OD-BiNSs had demonstrated superior catalytic performance in comparison to BiNPs and Oxi-BiNPs. In an H-type cell, it achieved a remarkable 93% FE with a current density of 62 mA cm^{-2} at −0.95 V versus RHE. The improved performance of OD-BiNSs could be attributed to its higher intrinsic activity and increased electrochemical surface area (ECSA). In a flow cell, it achieved FE of 94% with a high current density of 200 mA cm^{-2}, indicating the potential of OD-BiNSs as a promising catalyst for practical applications in electrode-reduction of CO_2 [50]. However, Luo *et al* provided a different insight regarding the morphology affecting the CO_2 reduction performance, typically, three different oxidized-Zn (OD-Zn) catalysts were prepared through the electrochemical reduction of ZnO precursors with varying morphologies (nanowires, nanoflowers, and nanoparticles-figure 6.8(a)) [51]. Their catalytic performance was assessed in both H-cell and flow cell setups, demonstrating the significant potential of OD-Zn catalysts for practical applications owing to their high activity, CO selectivity, and stability. It was observed that, in contrast to other OD-metal catalysts, ZnO underwent substantial reconstruction during electrochemical reduction (figures 6.8(b) and (c)), resulting in the formation of hexagonal Zn crystals regardless of their initial properties. Furthermore, it was revealed that neither the oxidation method nor the morphology of ZnO influenced the intrinsic CO_2 reduction performance but rather impacted the electrochemical surface area of the reduced Zn electrodes, which played a decisive role in determining geometric activity and CO selectivity [51].

Wang *et al* introduced a novel group of 2D Cu(II) oxide electrocatalysts for CO_2 electroreduction under neutral pH conditions [52]. These Cu(II)O nanosheets, with a strong preference for the (001) facet orientation due to their 2D structure, exhibited exceptional catalytic performance when integrated into commercial electrolyzer gas diffusion electrodes (GDEs) operating under industrially relevant neutral pH conditions. Their stable dendritic structure, high catalytic reactivity, and sustained performance make them a promising alternative to conventional cubic Cu_2O catalysts for converting CO_2 into C_2H_4 [52]. Under an applied bias, the

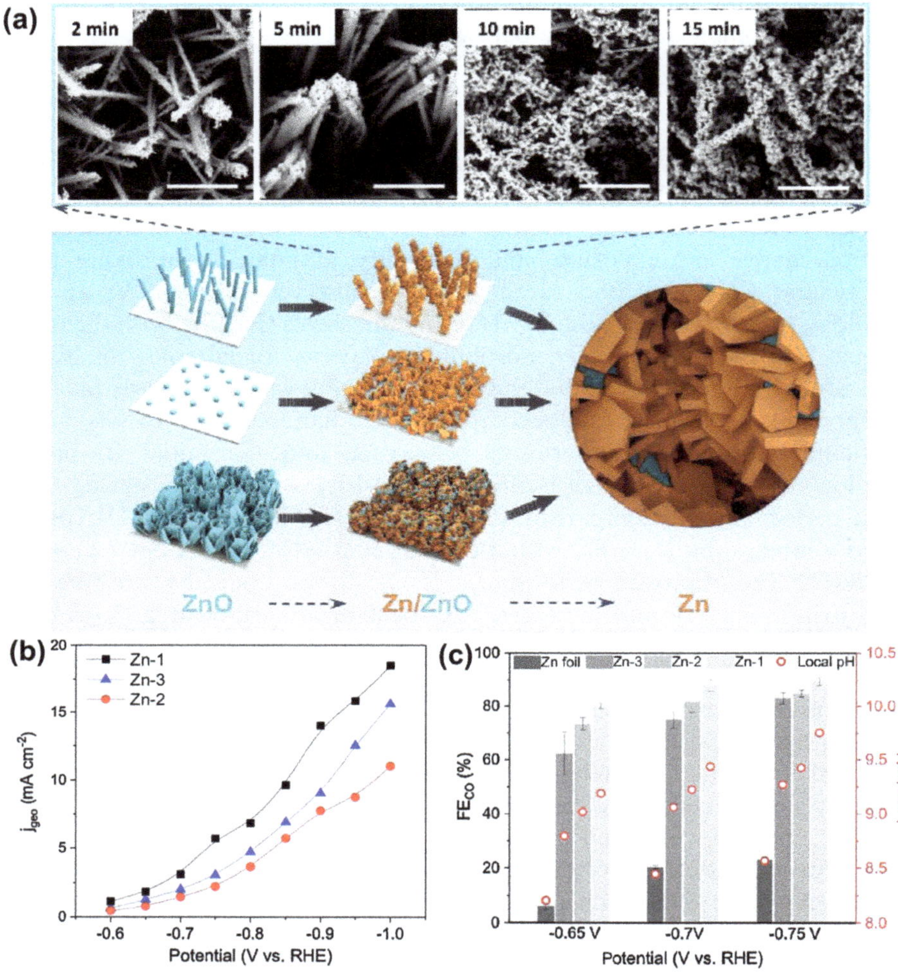

Figure 6.8. (a) Three different morphologies of Zn oxide, namely Zn-1, Zn-2 and Zn-3, (b) current densities of three samples in 0.1 M $KHCO_3$, (c) FE and local pH of the three samples at various potentials. Reprinted from [51], Copyright (2020), with permission from Elsevier.

(001)-oriented CuO nanosheets gradually transformed into highly branched metallic CuO dendrites, a common morphology under electrolyte flow conditions. Another interesting study explained; Bi_2O_3 electrocatalysts with varying morphologies, including nanoparticles and thin nanorods, were synthesized for the electroreduction of CO_2 into formate [53]. Notably, Bi_2O_3 nanoparticles exhibited superior catalytic activity compared to Bi_2O_3 thin nanorods achieving FE of up to 91% for formate production at a moderate potential of −1.2 V versus RHE in an aqueous solution. Furthermore, the maximum current density for formate production reached 22 mA cm^{-2}, nearly three times higher than that of Bi_2O_3 thin nanorods. Importantly, the catalytic activity of Bi_2O_3 nanoparticles remained relatively stable throughout a 23 h electrolysis period, underscoring their potential in the electroreduction of CO_2 to formate [53]. Similarly, a controllable reduction-melting-crystallization (RMC) protocol was employed to synthesize free-standing bismuth nanocrystals with tunable dimensions, morphologies, and surface structures, all in a surfactant-free manner [54]. Notably, ultrathin bismuth nanosheets with flat or jagged surfaces/edges were selectively prepared. Specifically, during the crystallization step, monodispersed round bismuth nanoparticles (r-BiNPs) were obtained by rapidly freezing nanosized bismuth droplets. In contrast, regular flat bismuth nanoflakes (f-BiNFs) were prepared through a much slower crystallization process at higher temperatures. The jagged bismuth nanosheets, characterized by abundant surface steps and defects, exhibited enhanced electrocatalytic CO_2 reduction performance in acidic, neutral, and alkaline aqueous solutions, achieving near unity selectivity at a current density of 210 mA cm^{-2} for formate evolution under ambient conditions [54]. A morphology detailed investigation was carried out and large quantities of CuO nanoparticles with five distinct morphologies (samples a–e, where sample a exhibited a 3D spherical structure with uniform nanorods, sample b displayed a 3D spherical structure with nonuniform nanoparticles, sample c showcased a 2D structure assembled from nanoparticles, sample d featured a nanosheet structure, and sample e presented a uniform nanorod structure) were synthesized through a hydrothermal method [55]. The potentials resulting from CO_2 electroreduction at a constant current density of 10 mA cm^2 were utilized, yielding values of 1.49, 1.54, 1.60, 1.62, and 1.64 V for the five samples, respectively. Under identical reaction conditions, sample a demonstrated a clear advantage over the others, with total FEs 1.9, 2.4, 2.8, and 3.9 times higher than those of samples b, c, d, and e, respectively [55]. Furthermore, it was empirically verified through multiple techniques that CuO nanoparticles with varying morphologies exhibited distinct electrocatalytic activities.

6.3.6 Edge

Active edges play a crucial role in the electroreduction of CO_2 as they provide sites with unique electronic properties and adsorption characteristics. These active edge sites facilitate the activation of CO_2 molecules, leading to the formation of reaction intermediates and products with enhanced selectivity and efficiency. The predominant catalytic activity in CO_2 reduction within TMDs is primarily attributed to their metal-terminated edges, characterized by a high density of d-electrons and metallic properties, as indicated by DFT calculations. This was investigated by Nørskov and co-workers and another group through the active edge sites of metal tellurides and MoS_2, $MoSe_2$, and Ni-doped MoS_2 simulated by DFT method [56–58]. Likewise, Tang *et al* also through DFT simulation provided the idea of the importance of active edges by examining the electrocatalytic performance of carbon-doped and line-defect-embedded boron-nitride nanoribbons in the context of the CO_2 electroreduction reaction. It was found that these defective BNs, using the neighbouring bare edge B atoms and C_2 dimer as active sites, demonstrated high catalytic activity and selectivity [59].

Keeping this line of thought, a liquid-exfoliation approach was employed to produce ultrathin 2D bismuth (Bi) nanosheets for efficient electrocatalytic CO_2 conversion [60]. Compared to bulk Bi, the increased presence of edge sites on these ultrathin Bi nanosheets played a pivotal role in enhancing CO_2 adsorption and reaction kinetics, notably facilitating CO_2-to-formate ($HCOOH/HCOO^-$) conversion. Capitalizing on their high conductivity and abundant edge sites, the Bi nanosheets achieved FE of 86.0% for formate production and a substantial current density of 16.5 mA cm^{-2} at −1.1 V (versus RHE), surpassing the performance of bulk Bi (figures 6.9(a)–(d)) [60]. In figure 6.9(e), these Bi nanosheets exhibited excellent long-term catalytic stability, maintaining their activity over consecutive 10 h of testing. DFT calculations revealed that the formation of ^-OCOH intermediates tended to occur predominantly on the edge sites, supported by their lower Gibbs free energies (figures 6.9(f)–(h)). Likewise, Wu and Zhang and their co-workers synthesized a Ni–graphene oxide and antimony nanosheets catalyst performed amazingly owing to the advantageous Ni-edges and exfoliated nanosheets edges, respectively, the electrocatalytic activity with more than 90% of FE [61, 62]. On the other hand, tailoring the active edges of MoS_2, Abbasi *et al* revealed that 5% niobium doped vertically aligned MoS_2 displays a one-order-of-magnitude higher turnover frequency (TOF) for CO formation compared to pristine MoS_2 within an overpotential range of 50–150 mV, and this TOF is also two orders of magnitude greater than that of Ag nanoparticles across the entire studied overpotential range of 100–650 mV [63]. In another report, the ZnO sheet array/Zn foil achieved outstanding stable CO_2 reduction performance, with an 85% FE for CO at −2.0 V (versus Ag/AgCl) and a current density of 11.5 mA cm^{-2}, surpassing free-standing ZnO sheets and particles owing to the vertical growth of the ZnO sheet array with exposed $(1\bar{1}00)$a edge facets improved electron transfer and active site availability. DFT simulations confirmed its lower Gibbs free energy for CO_2 activation due to increased $(1\bar{1}00)$ edge surface exposure. A schematic representation of the reaction mechanism and active edges are being displayed in figure 6.9(i) [20].

6.3.7 Heterostructuring

Heterostructures are designed to harness the strengths of different materials to optimize electrochemical CO_2 reduction, making them a promising avenue for improving the efficiency and selectivity of this environmentally significant reaction which can be ascribed to the synergistic effect taking place between the materials involved. Though a lot of heterostructures have been investigated so far, herein, the innovative strategies are only elaborated for fruitful enhancement of CO_2 reduction performance [64]. For instance, an effective heterostructure composed of crystalline–amorphous In_2O_3 and CeO_x demonstrated remarkable catalytic performance in the conversion of CO_2 to formate, achieving a maximum FE of 94.8% and maintaining above 90% efficiency over a wide potential range from −0.8 to −1.2 V versus RHE [65]. In this heterostructure, In_2O_3 served as the active site for CO_2 activation and formate formation, while amorphous CeO_x facilitated electron transfer, leading to the reconfiguration of In_2O_3's electronic structure. Consequently, the In_2O_3–CeO_x

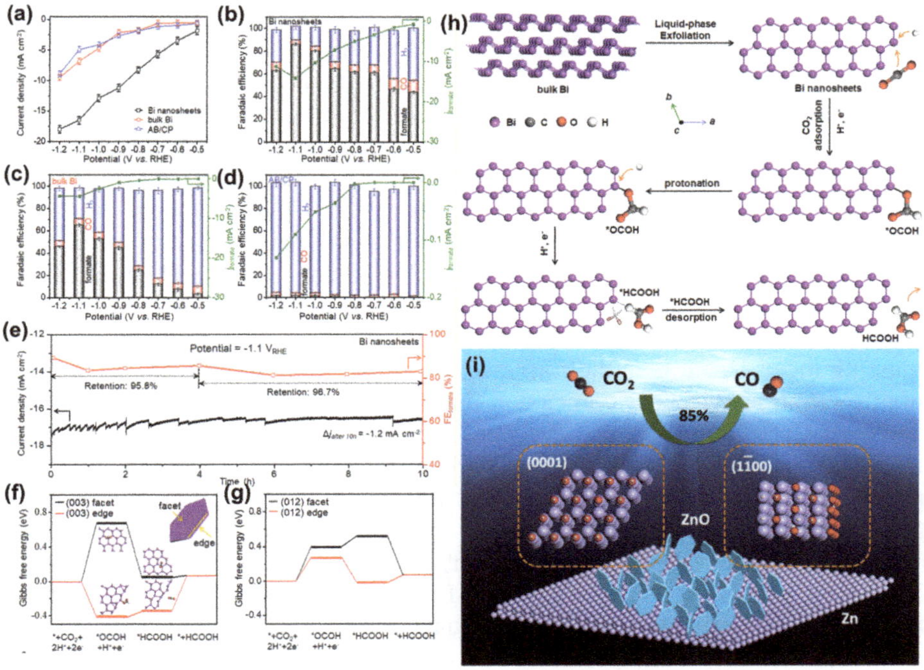

Figure 6.9. (a) Current densities of Bi nanosheets, in 0.1 M $KHCO_3$ aqueous solution, (b–d) FEs and partial current densities of the formate product for (b) Bi nanosheets, (c) bulk Bi, and (d) AB/CP at various applied potentials over, (e) current density and FE formate of Bi nanosheets during an extended 10 h CO_2 electroreduction, (f) DFT-computed ΔG values for the reaction pathways involved in the conversion of CO_2 into formate, considering both facet and edge sites, on the (003) plane of Bi. (g) DFT-calculated ΔG values for the same reaction pathways but on the (012) plane of Bi, and (h) schematic representation of the mechanism underlying the selective formation of formate on Bi nanosheets. Reprinted from [60], Copyright (2018), with permission from Elsevier. (i) Schematic illustration of the reaction mechanism and the active sites along with FE. Reprinted with permission from [20]. Copyright (2021) American Chemical Society.

heterostructure improved the adsorption of $^{-}OCHO$ intermediates and reduced the energy barrier for $^{-}HCOOH$ formation from $^{-}OCHO$. Likewise, the introduction of Sn atoms through deposition contributes electrons to the Bi_2O_3 material, enhancing its electrical conductivity. In the SnM–Bi_2O_3 catalyst with the optimized Sn content, a notable FE of 95.8% was achieved at −1.0 V for formate production, accompanied by a substantial partial current density of 41.8 mA cm^{-2} [66]. Furthermore, this SnM–Bi_2O_3 heterostructure demonstrates remarkable long-term stability during electrolysis. The incorporation of Sn species not only stabilizes reaction intermediates but also suppresses the hydrogen evolution reaction (HER) pathway, leading to a synergistic enhancement of catalytic activity. Fu *et al* yielded an *in situ* electrochemical synthesis of Cr_2O_3–Ag heterostructure electrocatalyst that displayed exceptional efficiency in the electrocatalytic reduction of CO_2 to CO [67]. Cr_2O_3@Ag achieved an impressive FECO of 99.6% at −0.8 V (versus RHE) while maintaining a high J_{CO} of 19.0 mA cm^{-2}, and demonstrated notable operational stability. Notably, operando Raman spectroscopy investigations revealed that the significantly improved performance can be attributed to the stabilizing influence of the Cr_2O_3–Ag heterostructure on $CO_2^{\bullet-}$/$^{-}$COOH intermediates. DFT calculations further revealed that the metallic Ag catalyses the CO_2 reduction to CO with an energy input of 1.45 eV, which is 0.93 eV. In another study, an ultrathin metalloporphyrin-based covalent organic framework was epitaxially grown on graphene, forming a 2D van der Waals heterostructure for CO_2 reduction [68]. Strong interlayer coupling enhanced electron-deficient metal centres, accelerating electrocatalysis. It achieved a 97% CO FE at 8.2 mA cm^{-2} in an H-cell, remaining stable for 30 h. In a liquid flow cell, CO selectivity neared 99%, with a partial current density of 191 mA cm^{-2}, and a TOF of 50 400 h^{-1} at −1.15 V versus RHE, surpassing most organometallic frameworks, highlighting interlayer van der Waals coupling's vital role in CO_2 conversion dynamics.

Further, CuSe/g-C_3N_4 electrodes were synthesized, combining CuSe nanoplates with g-C_3N_4 nanosheets for use in electrochemical CO_2 reduction reactions [69]. Figures 6.10(a) and (b) display the optimized structure of the heterostructure catalyst along with the SEM images showing the successful heterostructure formation. The electrode containing 50 wt% CuSe on g-C_3N_4 exhibited the highest FE, reaching 85.28% at −1.2 V versus RHE, surpassing pure CuSe nanoparticles by a factor of 1.47 (figure 6.10(c)). This enhancement is attributed to the increased specific surface area facilitated by planar-structured g-C_3N_4 nanosheets and improved electron transfer rates at the interface between CuSe nanoparticles and g-C_3N_4 nanosheets. DFT calculations further supported the presence of electronic

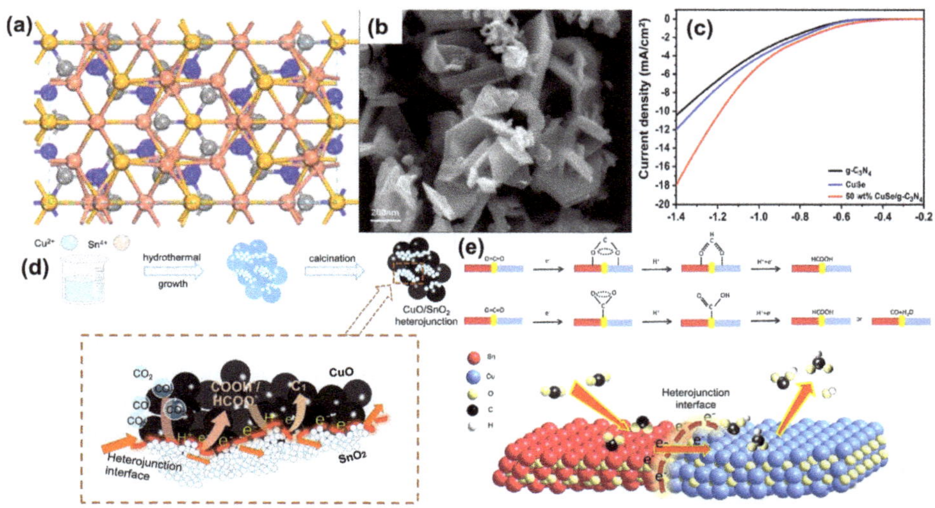

Figure 6.10. (a) Optimised structure of CuSe/g-C_3N_4, (b) SEM image of CuSe/g-C_3N_4, (c) polarization curve for CO_2 reduction. Reprinted from [69], Copyright (2021), with permission from Elsevier. (d) Schematic illustration of the synthesis of CuO/SnO_2 heterostructure and (e) reaction mechanism involving CO_2 reduction through the interface of the heterostructure. Reprinted from [72], Copyright (2023), with permission from Elsevier.

coupling at the electrode interface, leading to the formation of an internal electric field and electron flow from g-C_3N_4 nanosheets to CuSe nanoparticles [69]. Likewise, a high-quality heterostructure was designed with band alignment to facilitate interfacial charge transfer from Zn_2SnO_4 to SnO_2, effectively modifying the electronic properties of Zn_2SnO4/SnO_2 to reduce the kinetic barriers of CO_2 reduction [70]. DFT calculations further revealed that, in comparison to pure Zn_2SnO_4 or SnO_2, Zn_2SnO_4/SnO_2 promoted the favourable stabilization of the $HCOO^-$ intermediate on its surface, primarily due to an enhanced hydrogen coverage effect. Consequently, this hybrid catalyst demonstrated robust CO_2 reduction capabilities, maintaining a stable HCOOH selectivity of 77% over a 24-h period at an applied potential of −1.08 V versus RHE. In another study, SnO_2/Sn electrode showed a high 93% FE and 28.7 mA cm^{-2} current density for formate production in an H-type cell, stable for 9 h, and 174.86 mA cm^{-2} in a flow cell at −1.18 V versus RHE. *In situ*-formed SnO_2/Sn heterostructures reduced the energy barrier for formate production, enhancing activity and selectivity [71]. Shen *et al* designed CuO/SnO_2 heterostructure to enhance electroreduction of CO_2 performance and product selectivity [72]. The catalyst was prepared via hydrothermal reaction and high-temperature annealing, offering a scalable production method (figure 6.10(d)) [72]. Characterizations revealed abundant composite interfaces, facilitating electron transfer. The heterojunction catalyst maintained over 80% FE for C_1 products in a wide potential window (−0.85 to −1.06 V versus RHE) with a maximum partial current density of 24 mA cm^{-2}, surpassing individual CuO or SnO_2. It effectively suppressed HER, achieving higher C_1 product selectivity in electroreduction of CO_2 and the reaction mechanism can be seen in figure 6.10(e).

6.4 Conclusion

The electrochemical reduction of CO_2 through 2D materials represents a promising avenue in the ongoing efforts to mitigate the escalating challenges posed by climate change and greenhouse gas emissions. At the heart of this transformative endeavour lies the catalytic prowess of 2D materials, an extraordinary class of nanomaterials with exceptional properties that enable precise control over electrochemical reactions. The tunable physicochemical properties, unique electronic structure, high surface area, ample electrochemical active sites, and excellent catalytic properties of these materials offer a fertile ground for designing efficient CO_2 electroreduction catalysts. Furthermore, the mechanistic insights into the electrochemical processes occurring at the interface of these materials have shed light on the intricate pathways involved in CO_2 reduction, enabling us to fine-tune and optimize these catalysts for enhanced selectivity and activity through various strategies like doping, alloying, defects, edges, heterostructuring etc. However, understanding the underlying principles governing catalytic selectivity remains a fundamental scientific enigma. Ensuring 2D materials' long-term stability under harsh electrochemical conditions remains a critical challenge. Investigating degradation mechanisms and developing strategies to enhance material durability is imperative. High-resolution techniques, *in situ* spectroscopy and theoretical viewpoints, offer avenues to uncover the intricacies of reaction pathways. Moreover, the integration of computational modelling and experimental validation has proven to be indispensable in unravelling the underlying mechanisms and guiding the rational design of these catalysts. This interdisciplinary approach will undoubtedly propel us closer to realizing sustainable and scalable technologies for CO_2 conversion and utilization. As we forge ahead, the fusion of fundamental research with innovative technologies offers a promising path toward addressing the global challenge of CO_2 emissions and steering us toward a sustainable and greener future.

Acknowledgments

The authors gratefully acknowledge financial assistance from the SERB Core Research Grant (Grant No. CRG/2022/000897), Department of Science and Technology (DST/NM/NT/2019/205(G)), and Minor Research Project Grant, Jain University (JU/MRP/CNMS/29/2023).

References

[1] Sullivan I, Goryachev A, Digdaya I A, Li X, Atwater H A and Vermaas D A *et al* 2021 Coupling electrochemical CO_2 conversion with CO_2 capture *Nat. Catal.* **4** 952–8

[2] Xia R, Overa S and Jiao F 2022 Emerging electrochemical processes to decarbonize the chemical industry *JACS Au* **2** 1054–70

[3] Rahimi M, Khurram A, Hatton T A and Gallant B 2022 Electrochemical carbon capture processes for mitigation of CO_2 emissions *Chem. Soc. Rev.* **51** 8676–95

[4] Mi Woo L, Park K, Lee W, Lim H, Kwon Y and Kang S 2020 Current achievements and the future direction of electrochemical CO_2 reduction: a short review *Crit. Rev. Environ. Sci. Technol.* **50** 769–815

[5] Zhao G, Huang X, Wang X and Wang X 2017 Progress in catalyst exploration for heterogeneous CO_2 reduction and utilization: a critical review *J. Mater. Chem.* A **5** 21625–49

[6] Liu J, Guo C, Vasileff A and Qiao S 2016 Nanostructured 2D materials: prospective catalysts for electrochemical CO_2 reduction *Small Methods* **1** 1600006

[7] Kwon K C, Suh J M, Varma R S, Shokouhimehr M and Jang H W 2019 Electrocatalytic water splitting and CO_2 reduction: sustainable solutions via single-atom catalysts supported on 2D materials *Small* **3** 1800492–2

[8] Sun Z, Ma T, Tao H, Fan Q and Han B 2017 Fundamentals and challenges of electrochemical CO_2 reduction using two-dimensional materials *Chem.* **3** 560–87

[9] Thomas S A, Patra A, Al-Shehri B M, Selvaraj M, Aravind A and Rout C S 2022 MXene based hybrid materials for supercapacitors: recent developments and future perspectives *J. Energy Storage* **55** 105765

[10] Patra A and Sekhar Rout C 2020 Anisotropic quasi-one-dimensional layered transition-metal trichalcogenides: synthesis, properties and applications *RSC Adv.* **10** 36413–38

[11] Lu Q and Jiao F 2016 Electrochemical CO_2 reduction: electrocatalyst, reaction mechanism, and process engineering *Nano Energy* **29** 439–56

[12] Garza A J, Bell A T and Head-Gordon M 2018 Mechanism of CO_2 reduction at copper surfaces: pathways to C2 products *ACS Catal.* **8** 1490–9

[13] Zhao C, Yi B, Wang G and Jiang Q 2017 CO_2 reduction mechanism on the Pb(111) surface: effect of solvent and cations *J. Phys. Chem.* C **121** 19767–73
[14] Ding M, Chen Z, Liu C, Wang Y, Li C and Li X *et al* 2023 Electrochemical CO_2 reduction: progress and opportunity with alloying copper *Mater. Rep.: Energy* **3** 100175
[15] Yuan M, Kummer M J and Minteer S D 2019 Strategies for bioelectrochemical CO_2 reduction *Chem. Eur. J.* **25** 14258–66
[16] Dou S, Song J, Xi S, Du Y, Wang J and Huang Z *et al* 2019 boosting electrochemical CO_2 reduction on metal–organic frameworks via ligand doping *Angew. Chem. Int. Ed.* **58** 4041–5
[17] He J, Johnson N R, Huang A and Berlinguette C P 2018 Electrocatalytic alloys for CO_2 reduction *Chem. Sus. Chem.* **11** 48–57
[18] Wang Y, Han P, Lv X, Zhang L and Shao Z 2018 Defect and interface engineering for aqueous electrocatalytic CO_2 reduction *Joule* **2** 2551–82
[19] Wang S, Wang Y, Zang S Q and David W 2019 Hierarchical hollow heterostructures for photocatalytic CO_2 reduction and water splitting *Small Methods* **4** 1900586–6
[20] Xiang Q, Li F, Wang J, Chen W, Miao Q and Zhang Q *et al* 2021 Heterostructure of ZnO nanosheets/Zn with a highly enhanced edge surface for efficient CO_2 electrochemical reduction to CO *ACS Appl. Mater. Interfaces* **13** 10837–44
[21] Keith J A and Carter E A 2013 Theoretical insights into electrochemical CO_2 reduction mechanisms catalyzed by surface-bound nitrogen heterocycles *J. Phys. Chem. Lett.* **4** 4058–63
[22] Fernandez S G, Franco F, Martínez Belmonte M, Friães S, Royo B and Luis J M *et al* 2023 Decoding the CO_2 reduction mechanism of a highly active organometallic manganese electrocatalyst: direct observation of a hydride intermediate and its implications *ACS Catal.* **13** 10375–85
[23] Keith J A and Carter E A 2013 Electrochemical reactivities of pyridinium in solution: consequences for CO_2 reduction mechanisms *Chem. Sci.* **4** 1490
[24] Popović S, Smiljanić M, Jovanovič P, Vavra J, Buonsanti R and Hodnik N 2020 Stability and degradation mechanisms of copper-based catalysts for electrochemical CO_2 reduction *Angew. Chem.* **132** 14844–54
[25] Lim H K and Kim H 2017 The mechanism of room-temperature ionic-liquid-based electrochemical CO_2 reduction: a review *Molecules* **22** 536
[26] Pan J, Sun Y, Deng P, Yang F, Chen S and Zhou Q *et al* 2019 Hierarchical and ultrathin copper nanosheets synthesized via galvanic replacement for selective electrocatalytic carbon dioxide conversion to carbon monoxide *Appl. Catal.* B **255** 117736
[27] Gao S, Sun Z, Liu W, Jiao X, Zu X and Hu Q *et al* 2017 Atomic layer confined vacancies for atomic-level insights into carbon dioxide electroreduction *Nat. Commun.* **8** 14503
[28] Daiyan R, Lovell E C, Bedford N M, Saputera W H, Wu K and Lim S *et al* 2019 Modulating activity through defect engineering of tin oxides for electrochemical CO_2 reduction *Adv. Sci.* **6** 1900678
[29] Chan K, Tsai C, Hansen H A and Nørskov J K 2014 Molybdenum sulfides and selenides as possible electrocatalysts for CO_2 reduction *Chem. Cat. Chem.* **6** 1899–905
[30] Wu J, Liu M, Sharma P, Yadav R N, Ma L and Lou J *et al* 2015 Incorporation of nitrogen defects for efficient reduction of CO_2 via two-electron pathway on three-dimensional graphene foam *Nano Lett.* **16** 466–70
[31] Hussain N, Ali Abdelkareem M, Alawadhi H, Elsaid K and Olabi A 2022 Synthesis of Cu-g-C_3N_4/MoS_2 composite as a catalyst for electrochemical CO_2 reduction to alcohols *Chem. Eng. Sci.* **258** 117757

[32] Zhang C, Yang K, Li L, Liu W, Yang Z and Cao Q *et al* 2023 Computational study of nitrogen-doped vanadium disulfide-loaded single-atom catalysts for the electrocatalytic reduction of CO_2 *Appl. Surf. Sci.* **640** 158279

[33] Najafi L, Oropesa-Nuñez R, Bellani S, Martín-García B, Pasquale L and Serri M *et al* 2021 Topochemical transformation of two-dimensional VSe_2 into metallic nonlayered VO_2 for water splitting reactions in acidic and alkaline media *ACS Nano.* **16** 351–67

[34] Sun Y, Xu J, Li X, Liu W, Ju Z and Yao T *et al* 2017 Carbon dioxide electroreduction into syngas boosted by a partially delocalized charge in molybdenum sulfide selenide alloy monolayers *Angew. Chem.* **129** 9249–53

[35] Zhou R, Fan X, Ke X, Xu J, Zhao X and Lin J *et al* 2021 Two-dimensional palladium–copper alloy nanodendrites for highly stable and selective electrochemical formate production *Nano Lett.* **21** 4092–8

[36] Bai X, Chen W, Zhao C, Li S, Song Y and Ge R *et al* 2017 Exclusive formation of formic acid from CO_2 electroreduction by a tunable Pd–Sn alloy *Angew. Chem.* **56** 12219–23

[37] Choi S Y, Jeong S K, Kim H J, Baek I H and Park K T 2016 Electrochemical reduction of carbon dioxide to formate on tin–lead alloys *ACS Sustain. Chem. Eng.* **4** 1311–8

[38] Ma S, Sadakiyo M, Heima M, Luo R, Haasch R T and Gold J I *et al* 2016 Electroreduction of carbon dioxide to hydrocarbons using bimetallic Cu–Pd catalysts with different mixing patterns *J. Am. Chem. Soc.* **139** 47–50

[39] Watanabe M 1991 Design of alloy electrocatalysts for CO_2 reduction *J. Electrochem. Soc.* **138** 3382

[40] Rasul S, Anjum D H, Jedidi A, Minenkov Y, Cavallo L and Takanabe K 2014 A highly selective copper–indium bimetallic electrocatalyst for the electrochemical reduction of aqueous CO_2 to CO *Angew. Chem. Int. Ed.* **54** 2146–50

[41] Zhang A, He R, Li H, Chen Y, Kong T and Li K *et al* 2018 Nickel doping in atomically thin tin disulfide nanosheets enables highly efficient CO_2 reduction *Angew. Chem.* **57** 10954–8

[42] Cometto C, Ugolotti A, Grazietti E, Moretto A, Bottaro G and Armelao L *et al* 2021 Copper single-atoms embedded in 2D graphitic carbon nitride for the CO_2 reduction *npj 2D Mater. Appl.* **5** 63

[43] Wang K, Liu D, Liu L, Liu J, Hu X and Li P *et al* 2022 Tuning the local electronic structure of oxygen vacancies over copper-doped zinc oxide for efficient CO_2 electroreduction *eScience* **2** 518–28

[44] Zhong X, Liang S, Yang T, Zeng G, Zhong Z and Deng H *et al* 2022 Sn Dopants with synergistic oxygen vacancies boost CO_2 electroreduction on CuO nanosheets to CO at low overpotential *ACS Nano.* **16** 19210–9

[45] Cai F, Hu X, Gou F, Chen Y, Xu Y and Qi C *et al* 2023 Ultrathin $ZnIn_2S_4$ nanosheet arrays activated by nitrogen-doped carbon for electrocatalytic CO_2 reduction reaction toward ethanol *Appl. Surf. Sci.* **611** 155696

[46] Yuan Y, Qi L, Gao Z, Guo T, Zhai D and He Y *et al* 2023 Performance exploration of Ni-doped MoS_2 in CO_2 hydrogenation to methanol *Molecules* **28** 5796–6

[47] Dong H, Zhang L, Li L, Deng W, Hu C and Zhao Z J *et al* 2019 Abundant Ce^{3+} Ions in Au–CeO_x nanosheets to enhance CO_2 electroreduction performance *Small* **15** 1900289

[48] Wang M, Ren X, Yuan G, Niu X, Xu Q and Gao W *et al* 2020 Selective electroreduction of CO_2 to CO over co-electrodeposited dendritic core–shell indium-doped Cu@Cu_2O catalyst *J. CO2 Utiliz.* **37** 204–12

[49] Cho W S, Hong D M, Dong W J, Lee T H, Yoo C J and Lee D *et al* 2023 Porously reduced 2-dimensional Bi_2O_2 CO_3 petals for strain-mediated electrochemical CO_2 reduction to HCOOH *Energy Environ. Mater.* e12490

[50] Lee J, Liu H, Chen Y and Li W 2022 Bismuth nanosheets derived by *in situ* morphology transformation of bismuth oxides for selective electrochemical co_2 reduction to formate *ACS Appl. Mater. Interfaces* **14** 14210–7

[51] Luo W, Zhang Q, Zhang J, Moioli E, Zhao K and Züttel A 2020 Electrochemical reconstruction of ZnO for selective reduction of CO_2 to CO *Appl. Catal.* B **273** 119060

[52] Wang X, Klingan K, Klingenhof M, Möller T, Ferreira de Araújo J and Martens I *et al* 2021 Morphology and mechanism of highly selective Cu(II) oxide nanosheet catalysts for carbon dioxide electroreduction *Nat. Commun.* **12** 794

[53] Miao C C and Yuan G Q 2018 Morphology-controlled Bi_2O_3 nanoparticles as catalysts for selective electrochemical reduction of CO_2 to formate *Chem. Electro. Chem.* **5** 3741–7

[54] Chen L, Hao Y C, Li J, Hu L, Zuo X and Dai C *et al* 2023 Controllable crystallization of two-dimensional bi nanocrystals with morphology-boosted CO_2 electroreduction in wide pH environments *Small* **19** 2301639

[55] Chi D, Yang H, Du Y, Lv T, Sui G and Wang H *et al* 2014 Morphology-controlled CuO nanoparticles for electroreduction of CO_2 to ethanol *RSC Adv.* **4** 37329–32

[56] Brea C and Hu G 2022 Shifting and breaking scaling relations at transition metal telluride edges for selective electrochemical CO_2 reduction *J. Mater. Chem.* A **10** 10162–70

[57] Hong X, Chan K, Tsai C and Nørskov J K 2016 How doped MoS_2 breaks transition-metal scaling relations for CO_2 electrochemical reduction *ACS Catal.* **6** 4428–37

[58] Mao X, Wang L, Xu Y and Li Y 2020 Modulating the MoS_2 edge structures by doping transition metals for electrocatalytic CO_2 reduction *J. Phys. Chem.* C **124** 10523–9

[59] Tang S, Zhou X, Zhang S, Li X, Yang T and Hu W *et al* 2018 Metal-free boron nitride nanoribbon catalysts for electrochemical CO_2 reduction: combining high activity and selectivity *ACS Appl. Mater. Interfaces* **11** 906–15

[60] Zhang W, Hu Y, Ma L, Zhu G, Zhao P and Xue X *et al* 2018 Liquid-phase exfoliated ultrathin Bi nanosheets: uncovering the origins of enhanced electrocatalytic CO_2 reduction on two-dimensional metal nanostructure *Nano Energy* **53** 808–16

[61] Bi W, Li X, You R, Chen M, Yuan R and Huang W *et al* 2018 Surface immobilization of transition metal ions on nitrogen-doped graphene realizing high-efficient and selective CO_2 reduction *Adv. Mater.* **30** 1706617–7

[62] Li F, Xue M, Li J, Ma X, Chen L and Zhang X *et al* 2017 Unlocking the electrocatalytic activity of antimony for CO_2 reduction by two-dimensional engineering of the bulk material *Angew. Chem.* **129** 14910–4

[63] Abbasi P, Asadi M, Liu C, Sharifi-Asl S, Sayahpour B and Behranginia A *et al* 2016 Tailoring the edge structure of molybdenum disulfide toward electrocatalytic reduction of carbon dioxide *ACS Nano.* **11** 453–60

[64] Qin Q, Sun M, Wu G and Dai L 2022 Emerging of heterostructured materials in CO_2 electroreduction: a perspective *Carbon Capture Sci. Technol* **3** 100043

[65] Wang C, Wu Z, Liu G, Bai S, Guo L and He L *et al* 2022 Highly efficient electrochemical CO_2 reduction over crystalline–amorphous In_2O_3–CeO_x heterostructures *Inorg. Chem. Front.* **9** 5926–31

[66] Wu D, Chen P, Feng D, Song J and Tong Y 2021 Highly efficient electrochemical reduction of carbon dioxide to formate on Sn modified Bi_2O_3 heterostructure *Dalton Trans.* **50** 14120–4

[67] Fu H Q, Liu J, Bedford N, Wang Y, Sun J W and Zou Y *et al* 2022 Synergistic Cr_2O_3@Ag heterostructure enhanced electrocatalytic CO_2 reduction to CO *Adv. Mater.* **34** 2202854
[68] Gu H, Shi G, Zhong L, Liu L, Zhang H and Yang C *et al* 2022 A two-dimensional van der waals heterostructure with isolated electron-deficient cobalt sites toward high-efficiency CO_2 electroreduction *J. Am. Chem. Soc.* **144** 21502–11
[69] Zhang H, Ouyang T, Li J, Mu M and Yin X 2021 Dual 2D CuSe/g-C_3N_4 heterostructure for boosting electrocatalytic reduction of CO_2 *Electrochim. Acta* **390** 138766–6
[70] Wang K, Liu D, Deng P, Liu L, Lu S and Sun Z *et al* 2019 Band alignment in Zn_2SnO_4/SnO_2 heterostructure enabling efficient CO_2 electrochemical reduction *Nano Energy* **64** 103954–4
[71] Ning S, Wang J, Xiang D, Huang S, Chen W and Chen S *et al* 2021 Electrochemical reduction of SnO_2 to Sn from the bottom: *in situ* formation of SnO_2/Sn heterostructure for highly efficient electrochemical reduction of carbon dioxide to formate *J. Catal.* **399** 67–74
[72] Shen C, Li K, Ma Y, Liu S, Wang X and Xu J *et al* 2023 Electrochemical reduction of CO_2 via a CuO/SnO_2 heterojunction catalyst *Chem. Phys. Lett.* **818** 140438

IOP Publishing

Engineered 2D Materials for Electrocatalysis Applications

Chandra Sekhar Rout

Chapter 7

Engineered 2D materials for N_2 reduction reaction (N_2RR)

Mansi Pathak and Chandra Sekhar Rout

As a vital component of agricultural production, synthetic ammonia (NH_3) contributes significantly to the human and economic development of the world. Comparing NH_3 to H_2, it is simpler to carry and keep and has a higher energy density. Also, NH_3 may act as a medium for storing H_2 and, if needed, break down to provide H_2. The application of nitrogen reduction reaction (N_2RR) under ambient settings for electrochemical NH_3 synthesis has garnered significant attention as a potential replacement for the traditional Haber–Bosch method. The principles and mechanism of electrochemical nitrogen fixing are outlined in this chapter. A thorough summary is provided of the N_2RR performances for each significant electrocatalyst, with an emphasis on two-dimensional (2D) materials. It also includes theoretical insights from current research. Owing to the many techniques that may be employed in constructing various 2D materials. These investigations have led to the development of recommendations for obtaining high N_2RR catalytic activity and good selectivity. Important aspects of the effects of environmental pollution are also covered. the most recent development of the idea of using defects, doping, alloying, strain-stress, crystalline, edge modification, and heterostructures to enhance the performance of electrocatalytic nitrogen reduction reaction (N_2RR) is covered. The purpose of this is to provide an overview of the influence of defects on the performance of the synthesis of NH_3. Finally, the possibility of further routes for using defect engineering to create catalysts for N_2RR is presented in figure 7.1.

7.1 Introduction

Synthetic ammonia (NH_3) plays a significant role in the growth of the overall economy and humanity as a necessary raw ingredient for agricultural output. NH_3 has a greater energy density than hydrogen (H_2) while also being easier to transport

 © IOP Publishing Ltd 2024

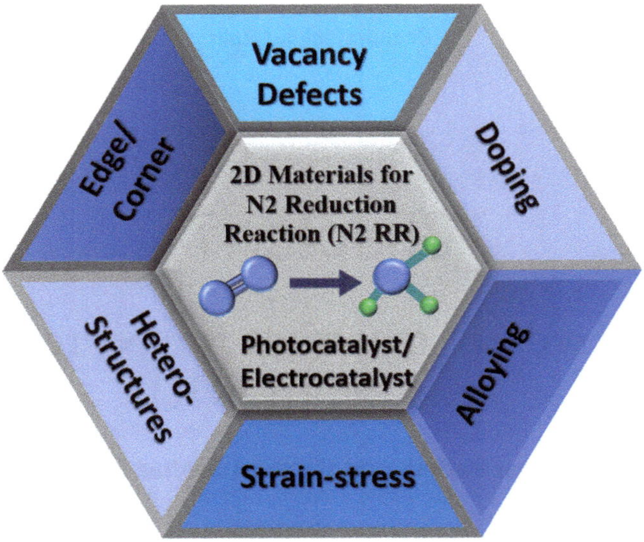

Figure 7.1. Schematic representation for various approaches in engineering 2D Materials for N_2 reduction reaction (N_2RR).

and preserve. However, NH_3 can serve as an H_2 storage medium and, if necessary, disintegrate to release H_2 [1, 2]. Additionally, a substitute for hydrogen (H_2), NH_3 with a 17.6 wt% hydrogen content is also regarded as a significant hydrogen transporter since it is easier to store and carry than gaseous H_2. The full combustion of NH_3 produces nitrogen (N_2) and water (H_2O), primarily executing as a carbon-free energy storage intermediary and preventing the release of hazardous gases like carbon monoxide (CO) and greenhouse gases like carbon dioxide (CO_2) into the environment. The potential of NH_3 chemistry is therefore quite intriguing. The present NH_3 synthesis method, however, primarily uses the Haber–Bosch reaction, which may be realized under extreme conditions of temperature and pressure, since the nitrogen (N_2) molecules are exceedingly persistent and challenging to activate [3–5]. Around 1%–2% of the total amount of energy consumed worldwide is used in this process. The utilized H_2 is mostly derived through the natural gas restructuring process, which releases millions of tons of CO_2 annually alongside the high energy inputs [6, 7]. Therefore, alternative procedures should be developed to design an environmentally compatible and efficient NH_3 production method to accomplish highly-efficient N_2 reduction for NH_3 synthesis in light of these energy and environmental costs.

The catalyst is crucial for the production of NH_3, and several researchers have dedicated their efforts to designing catalysts that can function at standard temperatures and pressure. Through the nitrogenase enzyme, many plants can transform the N_2 in the air into NH_3 fertilizer, supplying resources for their own accord advancement. However, because nitrogenase reacts slowly during the N_2 fixation process [8, 9] research is focused on creating other, more effective strategies. For instance, more efficient ways to reduce N_2 into NH_3 at ambient temperature and pressure include photocatalytic and electrocatalytic N_2 reduction processes (N_2RR) [10]. Its starting materials are H_2O and N_2, and its final product is $3H_2O + N_2 \rightarrow 2NH_3 + 3/2O_2$. N_2 may be continually converted to NH_3 using just infinite solar energy or clean power [11]. Converting clean energy from light and electricity into chemical energy and achieving effective utilization and transformation of renewable energy is of utmost importance for energy and environmental protection. A lot of attention has been paid to directly synthesizing NH_3 from N_2 and H_2O under ambient settings using newly developed photo- and electrocatalytic reduction methods. Solar power or electrical energy from photovoltaic cells and turbines powered by wind are two sources of sustainable energy that make these intriguing catalytic processes routine. Other benefits include moderate reaction conditions, straightforward infrastructure, and reductions in environmental pollutants [12]. Through a better understanding of the overall N_2RR process, catalyst nanostructure design offers an additional means of reducing N_2RR energy. In order to initiate N_2RR, the adsorption of nitrogen on the catalyst must be encouraged. A feature of 2D catalysts is their comparatively large specific surface area, which enhances the amount of N_2 that is adsorbed and improves the catalyst–N_2 interaction in heterogeneous catalytic processes [13, 14]. Additionally, a significant number of electrons must be available at or close to the catalytic site for the N–N triple bond to be activated, and those electrons must be successfully transferred to the adsorbed N_2 molecule. In this instance, the activation and reduction of N_2 molecules are significantly aided by the electron-rich sites of defects or unsaturated sites exposed to 2D materials [12, 14]. Catalysts have a number of drawbacks, including inadequate conductivity in electrocatalysis and low effectiveness of photogenerated carrier dispersion in photocatalysis, in addition to difficulties in activating N_2 molecules and inadequate certain surface area and active sites. 2D materials often have shorter carrier diffusion paths, a distinct electronic structure, more vacancy-type defects, and exposed edge sites. Furthermore, on an ultrathin 2D sheet, defects such as structure, crystalline twinning, and crumpled sheet easily arise due to the many accessible atoms on the surface. This may have an impact on the inherent qualities and raise the catalytic site activities [15–20]. The inherent characteristics of 2D electrocatalysts and their capacity to employ engineering to methodically enhance the properties are the sources of their beneficial qualities. For instance, by taking into account the edges, flaws, dopants, shift, and size of a material, an extensive number of active sites may be created on a 2D surface. These properties are advantageous for the separation of photogenerated carriers and the enhancement of conductivity [21–23]. In general, N_2RR performance in both photocatalysis and electrocatalysis may be significantly enhanced by building 2D structures to address these limiting issues.

7.2 N_2RR-mechanism and insights

The triple bond on the nitrogen molecule cannot be broken without a higher energy of 941 kJ mol^{-1}, which makes the molecule inert. Its broad energy gap of 10.08 eV, strong proton affinity (493.8 kJ mol^{-1}), negative electron affinity of ~1.9 eV, high ionization potential of 15.58 eV, and other characteristics make the electron transfer procedure challenging [24]. It is an endothermic reaction ($\Delta H° = +37.6$ kJ mol^{-1}) for the H atom to join the N_2 molecule, and the initial bond cleavage is kinetically difficult. The initial bond must be broken with a high energy of 410 kJ mol^{-1}, or about half of the overall dissociation energy. Although it is possible thermodynamically for the N_2 molecule to hydrogenate to NH_3, the process cannot occur independently. Protons and electrons are needed by the N_2RR in an electrolytic system in order to produce ammonia. HER, on the other hand, competes with NRR since it needs just two electrons, whereas N_2RR needs six, therefore lowering the Faradaic efficiency (FE). Diazene (N_2H_2) and hydrazine (N_2H_4) are the two potential intermediates that might occur in the protonation processes during N_2RR after the first protonation step. The following equations provide the equilibrium potentials needed for the various N_2RR and HER [6, 25, 26].

$$2H^+ + 2e^- \leftrightarrow H_2\ (\Delta E° = 0\ \text{V versus SHE at pH} = 0) \tag{7.1}$$

$$N_2 + 6H^+ + 6e^- \leftrightarrow 2NH_3\ (E° = -0.057\ \text{V versus SHE at pH} = 0) \tag{7.2}$$

$$2H_2O + 2e^- \leftrightarrow H_2 + 2OH^-\ (E° = -0.828\ \text{V versus SHE at pH} = 14) \tag{7.3}$$

$$N_2 + 6H_2O + 6e^- \leftrightarrow 2NH_3 + 6OH^-\ (E° = -0.736\ \text{V versus SHE at pH} = 14) \tag{7.4}$$

$$N_2 + e^- \leftrightarrow N_2^-\ (E° = -3.37\ \text{V versus RHE at pH} = 14) \tag{7.5}$$

$$N_2 + e^- \leftrightarrow N_2H\ (E° = -3.2\ \text{V versus RHE at pH} = 0) \tag{7.6}$$

$$N_2 + 2H^+ + 2e^- \leftrightarrow N_2H_2\ (E° = -1.1\ \text{V versus RHE at pH} = 0) \tag{7.7}$$

$$N_2 + 4H^+ + 4e^- \leftrightarrow N_2H_4\ (E° = -3.36\ \text{V versus RHE at pH} = 0) \tag{7.8}$$

Many electrocatalysts kinetically prefer hydrogen evolution reaction (HER) over the NRR because the NRR equilibrium potentials are near to the HER (from equations (7.1) versus (7.2) and (7.3) versus (7.4)) [27]. The partially hydrogenated intermediate molecules, N_2H_2 (ΔH_f°= 212.9 kJ mol^{-1}) and N_2H_4 (ΔH_f° = 95.35 kJ mol^{-1}), have far higher formation enthalpies than the completely hydrogenated of the degree (ΔH_f° = −92.22 kJ mol^{-1}) intermediate molecule. Proton affinities of N_2H_2 (803 kJ mol^{-1}) and N_2H_4 (853.2 kJ mol^{-1}) are nearly twice as high as those of N_2 (493.8 kJ mol^{-1}), suggesting that the protonation of these molecules is simpler than that of N_2. Additionally, the energy gap of N_2H_4 is 7.44 eV, which is less compared to the N_2 molecule. Hence, compared to the subsequent hydrogenation stages, the initial hydrogenation process (equation (7.6)) is more kinetically challenging [28].

In general, there are three main steps involved in nitrogen reduction on the heterogeneous catalyst surface: (1) chemisorption of hydrogen and nitrogen atoms on the surface of the catalyst; (2) dissociation or association of nitrogen and reductive addition of H atoms; and (3) desorption of the product from the surface.

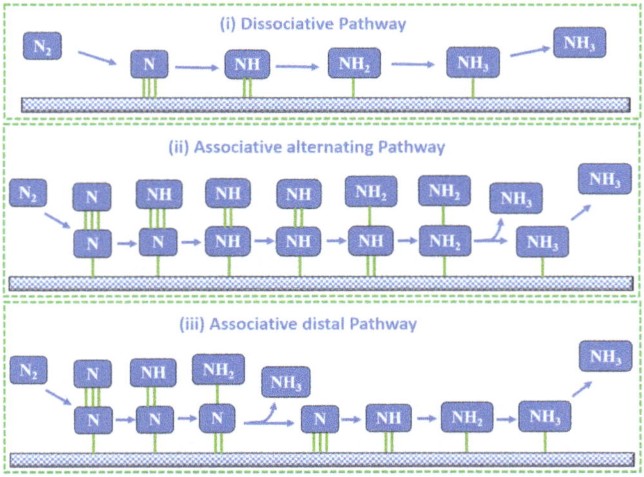

Figure 7.2. Schematic representation of N_2RR mechanism (i) dissociative pathway; (ii) associative alternating pathway and (iii) associative distal pathway.

Depending on the shape of the N_2 molecule, there are three possible ways that the NRR on the catalytic surface might proceed: dissociative, associative, and enzymatic routes [29]. The dissociative mechanism is shown in figure 7.2, where N_2 adsorbs on the surface and is followed by the breakdown of the triple bond between nitrogen atoms prior to protonation. Each individual N atom was then individually transformed into ammonia by the incorporation of hydrogen atoms, and the molecules of ammonia desorb over the surface after that. The dissociative pathway is followed by the Haber–Bosch process, and the nitrogen triple bond (941 kJ mol^{-1}) has a strong thermodynamic bond [30–32].

Briefly, the N_2 adsorption, $N_{2(ads)}$ activation and breakage of N–N bonds, the hydrogenation process of the deposited N_2 molecules ($N_{2(ads)}$) or N atom ($N_{(ads)}$), and NH_3 desorption are the six electron–proton processes that comprise N_2RR. There have been several suggested processes for the reduction of N_2 to NH_3, with the dissociative and associative pathways being the most often mentioned. As distinct from the associative process, which involves the N_2 molecule combining with the catalyst surface and direct hydrogenation of the N_2 molecule, the dissociative pathway involves the full breaking of the N–N triple bond of N_2, followed by hydrogenation on the N atom. A last N–N bond break is required for the formation of NH_3 molecules. Different hydrogenation sequences lead to the division of the associative route into distal and alternating pathways. The distal route involves hydrogenation preferably on the N atom most distant from the catalyst surface, releasing one NH_3 molecule at a time before the hydrogenation maintains to produce another NH_3 molecule. In the alternating route, two N atoms on the surface of the catalyst are hydrogenated alternately, and an additional NH_3 molecule is produced after the first NH_3 molecule is released. The triple bond's dissociation energy (941 kJ mol^{-1}) must often be overcome with a significant energy input as the dissociative route includes breaking the N–N triple bond. The majority of photocatalytic and electrocatalytic N_2RR at normal temperature and pressure is generally thought to follow the associative pathway, in which N_2 chemisorption and electron transfer from the catalyst to N_2 lead to the lower energy for N–N bond breaking. It should be noted that the Haber–Bosch reaction at high temperature and high pressure pertains to the dissociative pathway [11, 12, 38].

7.3 Different engineered approaches to enhance N_2 reduction reaction performance

Defects in oxides are gradually recognized as a positive parameter for enhancing the N_2RR performance, in contrast to traditional studies that concentrate on conventional materials for N_2RR. Various types of oxide defects are also proposed, which advances the discipline of defect engineering for N_2RR. Though N_2 reduction to NH_3 has a low potential (N_2 + 6H+ + 6e$^-$ → 2NH_3, 0.148 V versus reversible hydrogen electrode (RHE)) [13], slower kinetics are caused by repeated electron–proton transfer and subsequent intermediates. Furthermore, the competitive HER stances are a major obstacle to the N_2 reduction reaction (N_2RR) since the N_2 reduction potential is similar to that of HER. The majority of conventional, unaltered catalysts are unable to activate N_2. 2D materials possess certain inherent characteristics that set them apart from traditional structural electrocatalysts. These characteristics include a sizable specific surface area, tunability, evenly exposed lattice planes, and distinct electronic states. These materials enable the acquisition of a comparatively simple form of the active site with almost the same coordination number, while the ultrahigh specific surface area and atomic thickness make the total number of atoms and the number of active sites equal [13, 39]. The following are essential for achieving a good photocatalytic high NH_3 yield that features optimal light capture, appropriate valence bands and conduction band locations, effective charge separation, and, the maximum number of active sites for the adsorption, activation, and reduction of N_2. These issues must be carefully taken into account and resolved when designing and optimizing photocatalysts. In particular, 2D photocatalyst offers the following benefits viz; (i) The shortened carrier diffusion pathways derived from ultrathin nature will greatly improve the transmission and separation efficiency of photogenerated carriers, and due to the quantum size effect, the band position may be changed by varying the number of layers. (ii) Defects on 2D nanostructures assist in reducing the bandgap and extend the absorption spectrum of photocatalysts. In light of this, 2D photocatalysts often exhibit superior photocatalytic activity when compared to their bulk counterparts. (iii) Specific surface area is greatly improved, allowing the majority of anion vacancies or metal cation vacancies and active sites to be exposed on the surface and participate in the photocatalytic reaction. These advantages greatly aid the photocatalytic N_2RR by promoting the adsorption and activation of N_2 molecules as well as the involvement of additional photogenerated carriers in this process [40–44].

Similarly, for the electrocatalytic N_2RR activity there are factors like surface shape, electronic structure, and effective active site. It is possible to enhance their N_2RR performance by expanding 2D nanostructures. The chemical features of 2D electrocatalysts are generally thought to be responsible for their remarkable properties. These properties involve: (i) the extensively large surface area, which encourages the adsorption of N_2 molecules; (ii) an abundance of defect sites, including edge, topological, and vacancy defects, which can be utilized as N_2RR active sites; (iii) 2D structure of atomic layer thickness the which is useful to study the interaction between active sites and catalytic processes that provides precise regulation of their electronic structures through heteroatom doping, defect engineering, and surface engineering to enhance conductivity and generate new active sites [45–49]. These characteristics offer exceptional potential for N_2 molecule adsorption, induction, and electroreduction as well as atomic-level control active sites for N_2RR progression. The local coordinating conditions for metal cores in semiconductor photocatalysts may be controlled and modified through the use of defect engineering. It is quite possible to adjust the band edges and bandgap widths in semiconductor photocatalysts, improve photoexcited electron transport, and encourage N_2 adsorption and activation on semiconductor surfaces using defect engineering [50, 51]. Since nanosized photocatalysts have extremely high specific areas, defects are often highly common, meaning that a sizable portion of the material's total atoms are low-coordinated surface atoms. Evidently, surface atom energies differ significantly from bulk atom energies.

7.3.1 Vacancy defects

It has long been thought that N_2 fixation is a very difficult process, despite the fact that N_2 is exothermic when converted to NH_3 at ambient temperature and atmospheric pressure. The rate-determining phase in N_2RR, the activation, and dissociation of N_2 molecules, is made extremely challenging by the strong N–N triple bond [32–34]. Two N atoms are joined to form molecule N_2 by an effective N–N triple bond. An N atom possesses five valence electrons that reside in its 2s and 2p orbits, three of these are unpaired. Four bonding orbitals (two σ orbitals and two π orbitals) and four anti-bonding orbitals (two σ^* orbitals and two π^* orbitals) are formed from the hybrid atomic orbitals after bonding. N_2 molecules are activated by giving electrons to bonding orbitals and receiving electrons from anti-bonding π^* orbitals. This lengthens the bond and lowers the bond energy, allowing for more N_2 reduction. lately, researchers used N_2 chemisorption at the unsaturated and defective sites on the catalyst interface to activate the strong N–N bond, which changed the active N_2 molecule into NH_3 [3, 35, 36].[1] For instance, Li, Hao *et al* [37] have shown that the N_2 molecules can combine a pair of Bi atoms in the BiOBr (001) surface that has an end-on bonded structure to be chemisorbed on oxygen vacancies. Via a charge back-donation, the local electrons on oxygen vacancies with the noticeable

[1] Lately researchers used N_2 chemisorption at the unsaturated and defective sites on the catalyst interface to activate the strong N–N bond, which changed the active N_2 molecule into NH_3.

electron-rich centre may initiate the N–N triple bond (figures 7.3(a)–(d)), The potential N≡N bond activation is shown by the charges that back transport from oxygen vacancies to the adsorbed N_2, which may be accomplished by extending the bond length to 1.133 Å from the 1.078 Å for a free N_2 molecule. The enhanced indirect radiative interaction with photons in oxygen vacancies-abundant BiOBr demonstrated the oxygen vacancies as electron-trapping sites, as shown in figure 7.3(g), the photon energy (2.06 eV), far less than the pure BiOBr (2.81 eV). Their studies reveal that the vicinity around oxygen vacancies is usually electron-rich, and minimal or unusually valence metal cations with a distinct local electronic structure from the bulk lattice are frequently the consequence. The electron-rich

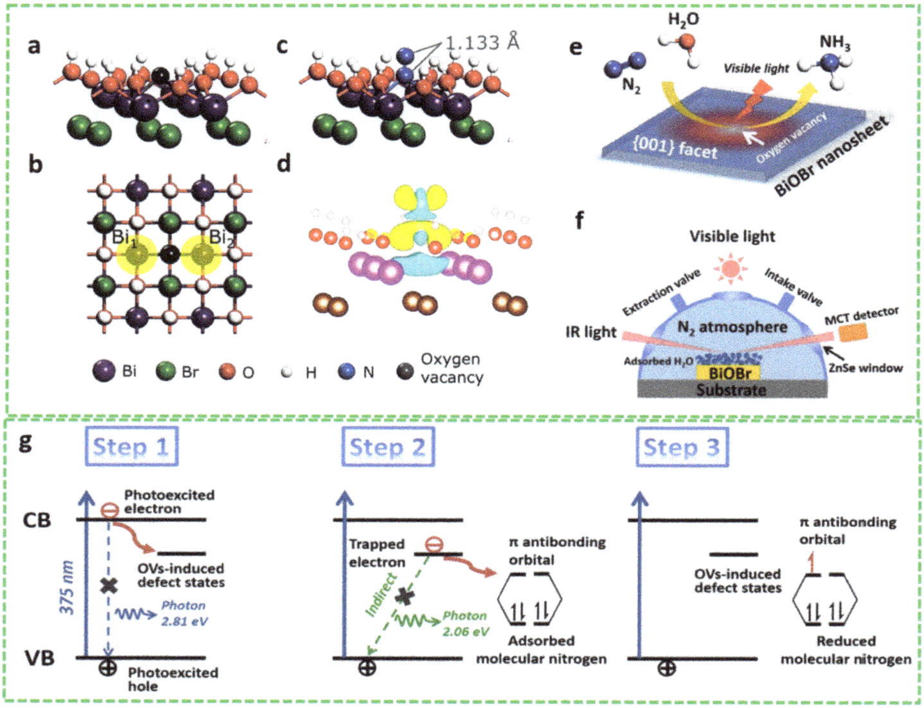

Figure 7.3. (a) N_2 activation on the OV of BiOBr (001) surface is predicted theoretically. (a) Top view and (b) side view of BiOBr (001) surface with an OV. The N_2 adsorption shape on the BiOBr (001) surface OV is shown in (c). (d) The N_2-adsorbed (001) surface charge density differential. Charge depletion and buildup in the space are represented by the blue and yellow isosurfaces, respectively. (e) A schematic representation of the photocatalytic N_2 fixation model, whereby water acts as both a proton source and a solvent. (f) A designed reaction cell to record *in situ* infrared signals. (g) Enhanced interfacial electron transport mechanisms caused by oxygen vacancies are depicted schematically. Steps 1 and 2—the directly excited electrons from the BiOBr CB are initially dynamically trapped by oxygen vacancies-induced defect states, which prevents charge carrier recombination. Step 3—the trapped electrons may then be effectively transported to fill the vacant anti-bonding orbitals of adsorbed N_2, suppressing the indirect recombination of trapped electrons with photo-excited holes. Reprinted with permission from [37]. Copyright (2015) American Chemical Society.

cations surrounding these oxygen vacancies (OV) sites promote the reduction of the adsorbed N_2 by electron transfer, allowing for the restoration of a normal valence. These oxygen-vacancy sites bind N_2 effectively.

Materials composed of carbon nitride have a unique 2D van der Waals layered crystal framework. In carbon nitride (g-C_3N_4) structure is a hexagonal carbon scaffold with nitrogen-atom-substituted carbons created via the hybridization of sp^2 of carbon and nitrogen atom, analogous to graphene. The electrochemical method is often used to examine the nitrogen vacancies (NVs) of g-C_3N_4 for the N_2RR [52–54]. In order to convert dinitrogen in the presence of NV defects, metal-free polymeric carbon nitride was initially used. Peng *et al* have reported that effective electrocatalytic nitrogen reduction that has high FE was made possible by a nitrogen-defective polymeric carbon nitride nanolayer [55].

Figure 7.4(a) shows the simple synthesis technique for CN/C_{600} fiber. The SEM image in figures 7.4(c)–(d) shows that the CN layer was synthesized on carbon fiber, with a 2D layer thickness. Continuous layered structures are shown in the TEM micrographs of the CN thin layer in figure 7.4(e) that displays the electron diffraction pattern of the targeted area shows the (002) diffraction peak, which suggested that the material was densely packed. CN/C_{500} and CN/C_{600} N_2RR performance was assessed using the indophenol blue technique. A standard plot was used to assess the degree of NH^{4+} in the electrolyte, and the results show that concentration, as well as absorbance, have a very linear relationship. The

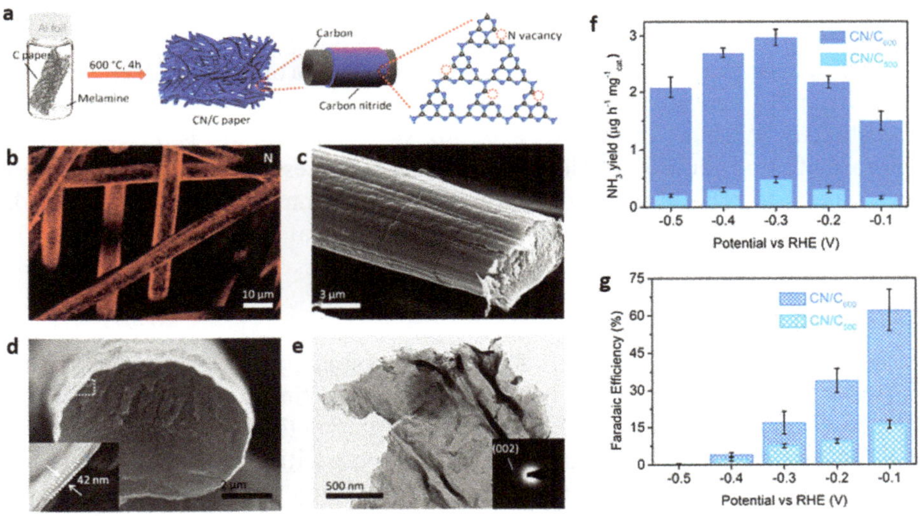

Figure 7.4. (a) Schematic representation of the N vacancies-containing carbon nitride framework and the synthesis of CN/C_{600}. The grey and blue spheres stand in for the atoms of carbon and nitrogen, respectively. (b) Nitrogen element mapping on CN/C_{600}, (c) SEM image of a single CN-coated C fiber. (d) SEM cross-sectional image of a single C fiber coated with CN, (e) TEM image of a CN sheet that has separated from the CN/C_{600} fiber (inset- SAED pattern with the (002) diffraction), (e) NH_3 yield rates at different potentials for CN/C_{500} and CN/C_{600}, and (f) a comparison of the FE of CN/C_{500} and CN/C_{600} at different potentials. Reprinted with permission from [55]. Copyright (2020) American Chemical Society.

accompanying chemical impurity list reveals that the blank, 0.1 m aqueous HCl, had ammonium impurities, which accounts for the non-negligible NH_3 baseline in the blank 0.1 M HCl. The generation of ammonia improved as the applied voltage was raised; at −0.3 V, the highest output was 2.87 μg h^{-1} mg^{-1}. A larger negative potential had no effect on performance, either, since the output was reduced with higher applied potentials. In comparison to the reversible hydrogen electrode, the FE at −0.3 V was 16.8%, whereas at −0.1 and −0.2 V, it was 62.1% and 33.9%, respectively, as shown in figure 7.4(f). Furthermore, the CN/C_{600} layer exhibited N_2RR performance at −0.3 V that was >21 times better than CN_{600} powder. The improved N_2RR findings highlight how important the electrical conductivity of CN is to the N_2RR performance.

Similar to nitrogen and oxygen vacancies, sulfur vacancies (SVs) have been employed in studies on photocatalytic activities [56–59]. Owing to its exceptional catalytic activity towards a variety of electrochemical processes, such as hydrodesulfurization [60], oxygen reduction reaction (ORR) [61], the reduction of carbon dioxide reaction (CO_2RR) [62], and HER [63], transition metal dichalcogenide materials have recently attracted a great deal of interest. Transition metal disulfides with ultrathin layers possess appealing optical, photoelectric, and electrical characteristics [36]. According to several studies, tightly bonded excitons can be formed by pairs of photoexcited electron holes in ultrathin transition metal disulfides. For the first time, Sun *et al* have demonstrated the Trion-induced N_2RR on ultrathin 2D MoS_2 [64]. Ultrathin MoS_2 is an n-type semiconductor with a high number of free electrons. N_2 molecules are possibly trapped at the locations of sulfur vacancies (SVs) in 2D MoS_2, where they are activated by lending electrons through bonded orbitals and accepting electrons to triple anti-bonding orbitals. The NH^{4+} evolution rate and the N_2 photoreduction of the MoS_2 samples as produced under various circumstances. Without the use of a cocatalyst or sacrificial agent, N_2 molecules undergo a Trion-induced six-electron reduction process that activates them and converts them to NH_3 at a rate of 325 μmol gh^{-1}. Commercial bulk MoS_2 has little action, suggesting that ultrathin 2D MoS_2 has a special benefit.

7.3.2 Doping and alloying

Another approach is to add heteroatoms to the majority of semiconductor photocatalysts in order to change the local electronic structure of the photocatalyst for better N_2 fixation performance. Heteroatom doping is thought to be an additional useful method for modifying catalyst chemical composition, surface characteristics, and electronic structures. Non-metal and metal doping are the two main types of heteroatom doping. For example, non-metal doping with B, N, S, Cl, and P towards carbon materials can favour electron transfer while also controlling the adsorption/desorption behaviors of reactants, intermediates, and products on the catalyst surface by generating charge and spin densities on C atoms near dopants [45]. One of the most promising 2D electrocatalysts is graphene-based material, which has achieved remarkable strides in both enhancing catalytic performance and comprehending catalytic actions [46, 65]. However, as a result of its high specific surface area and good electrical conductivity, graphene is a valuable substrate for the N_2RR. Although pure graphene has been used in different ways, there are difficulties and possibilities in using it to convert dinitrogen to ammonia under normal conditions despite the fact that pure graphene has found widespread application. The foremost limitation is that the delocalized π bonding system of graphene shows less catalytic activity, consequential in a high barrier for dinitrogen adsorption. Nevertheless, considering its large specific surface area and superior electrical conductivity, graphene is a promising substrate for the N_2RR [65, 66]. Currently, heteroatom doping and defect engineering can lower a material's spin density, Bader charge, and reaction barrier; as a result, there is rivalry between N_2RR and HER due to synergistic electron transfer exchanges among the dopants and adjoining carbon atoms [67–69]. The N_2RR activity is often significantly connected with its topological flaws, edge defects, and dopant-derived defects, and the impact of each defect is mostly determined by its local electronic structures. Moreover, the electroneutrality of sp^2 carbon material may be broken, strain and stress introduced, and new active sites can be generated as a result of the heteroatom and carbon atom having different inherent electronegativity and sizes. This increases the catalytic activity of the material. The long-term stable, low-cost metal-free nanomaterials have grown in importance as a component of N_2RR catalytic technology [70–74]. According to Yu *et al*, among the most active N_2RR electrocatalysts is 2D metal-free boron-doped graphene (B-doped graphene) [75]. The boron doping in the graphene framework causes a shift of electron density, wherein the boron site that lacks electrons increases the adsorption energy with N_2 molecules. The NH_3 generation rate and Faraday efficiency in the B-doped graphene sample are 5 and 10 times greater than those of the undoped graphene, accomplishing 9.8 $\mu g\ h^{-1}\ cm^{-2}$ and 10.8% at −0.5 V versus RHE, respectively. The NH_3 production rate and Faraday efficiency value remained consistent at 96% of the original activity during a 10-hour continuous electrochemical N_2RR test. The B-doping site's local electron-deficient environment offers Lewis bases a strong binding site, increasing the adsorption energy of N_2 molecules. The reaction paths of N_2RR on BC_3, BC_2O, BCO_2, and C structures, together with the related energy changes are computed using density functional theory (DFT). The lowest energy barrier for the electroreduction of N_2 is provided by the BC_3 structure among the several B-doped graphene configurations. The most often used carbon-based electrocatalyst in experimental work is N-doped graphene. Furthermore, on the periodic table, N is located just to the right near carbon. They are more electronegative and have comparable atomic radii for nitrogen. Nitrogen doping allows for the modification of graphene's electrical structure while maintaining its 2D planar structure. In addition, compared to metal-based electrocatalysts, N-doped graphene (metal-free) has special qualities and is ecologically safe [65–67]. For instance, Liu and co-workers used simple carbonization to create the first known N-atom-doped porous carbon-based electrocatalyst (NPC) which demonstrate notable N_2RR activity which results in a high ammonia generation rate (1.40 mmol $g^{-1}\ h^{-1}$ at −0.9 V versus RHE), the NPC proved successful in fixing N_2 to ammonia. Pyridinic and pyrrolic N were shown to be active sites for ammonia synthesis by experiments and

DFT calculations; their concentrations were essential for stimulating ammonia production on NPC [76]. Yang *et al* found that N-doped porous carbon with a large specific surface area of 1547.1 m^2 g^{-1} possesses an NH_3 production rate of 15.7 μg h^{-1} mg^{-1} and a Faraday efficiency of 1.5%, which is superior to the results reported in the literature [77].

Furthermore, past studies revealed that the interstitial alloys produced by adding C, N, and O to the transition metal lattice structure have strong catalytic activity for the process involved in the manufacture of ammonia [111]. Carbon vacancies can be introduced to mitigate the HER and the buildup of H atoms, as demonstrated by Matanovica *et al* [112]. They eliminated half of the carbon atoms from the cubic MoC structure to produce a sub-stoichiometric $MoC_{0.5}$ composition in order to study the impact of the metal–carbon ratio on the selective and activity properties of molybdenum carbide (MoC). Carbon vacancies were added, causing the structure to become tetragonal. The DFT studies reveal that the adsorption energy of N_2 on the surfaces of MoC [98] and $MoC_{0.5}$ (100) is −0.23 and −0.44 eV, respectively. These results also show that the surface of MoC [108] has the greater catalytic capability for HER and hinders the N_2RR, but they also suggest that elevating the metal–carbon ratio might enhance the free energy transformation of hydrogen desorption as well as decrease the activity of HER to a certain extent. According to certain theories, carbon vacancies can significantly enhance the degree of N_2 molecules and N atoms interaction.

7.3.3 Hetero-structures/hetero-junctions

For photocatalytic-driven nitrogen fixation, the band energy alignment at the interface between the two components is essential to the concept of heterostructure formation. With the right material selection, the resulting heterostructured photocatalysts might increase the separation of the photogenerated electrons and holes, which is crucial to achieving high photocatalytic efficiency, in addition to gaining a larger spectrum range of light absorption. It is often possible to suppose that two or more materials are the 'defect' of one another, hence a comprehensive definition of 'defect' that includes heterostructures might be put forward. Each component has a distinct impact on every other one, which together alters the heterostructure photocatalyst's capacity for photocatalytic nitrogen fixation. To improve photocatalysis performance, there has been a lot of attention paid to the creation of heterostructured photocatalysts during the last few years [50, 78, 79]. In this regard, Feng *et al*, have studied the $LnCO_3OH$/g-C_3N_4 (Ln–CN, Ln=La, Pr) heterojunction for photocatalytic nitrogen fixation. La–N bonds at the CN–$LaCO_3OH$ interface acted as electron transfer channels, transferring photogenerated electrons from the CB of $LaCO_3OH$ to the VB of CN. This decreased the rate at which charges recombined on each semiconductor. In addition, the Z-scheme heterojunction maintained CN's proper CB position. The high reduction energy required to change nitrogen into ammonia was provided by it. In comparison with CN and $LaCO_3OH$ separately, the photocatalytic N_2 fixation on La–CN was improved based on the aforementioned variables.

MoS_2 is anticipated to actively participate in N_2RR given the significance of the Mo and S components in nitrogenase. For the first time, Sun and the group examined the effect of the electrical structure of MoS_2 on N_2RR activity. The Mo edge in MoS_2 may be an N_2RR active site, similar to HER sites [80]. Through electrochemical testing, this catalyst is able to get a high NH_3 production (8.08×10^{-11} mol s^{-1} cm^{-1}) and FE of 1.17% at −0.5 V when compared to a reversible hydrogen electrode in 0.1 m Na_2SO_4. MoS_2 continues to be active for the N_2RR in acidic environments, where a significant HER takes place. The N_2 adsorption and activation sites are Mo atoms on the edge of defects, which lowers the energy barrier of the potential regulating phase for N_2RR. Well-explored MoS_2-reduced graphene oxide (rGO) heterojunction has also been applied to electrocatalysis owing to its large surface area, conductivity, and electrical and chemical interaction impacts [81]. For instance, Li *et al* have presented the first report on the MoS_2-rGO heterojunction exhibiting exceptional N_2RR performance [82]. The MoS_2/RGO hybrid, in which MoS_2 nanosheets are evenly wrapped on RGO sheets, is depicted in figure 7.5(a). Numerous curved nanosheets are seen on the surface in the transmission electron microscopy (TEM) micrograph figure 7.5(b) which is in agreement with the SEM finding. Compared to MoS_2 and other reported N_2RR electrocatalysts, it has a far greater NH_3 production of 24.8 μg h^{-1} mg^{-1} at −0.45 V versus RHE in 0.1 $liClO_4$. MoS_2-rGO/CPE exhibits a minor variation in NH_3 yielding and FEs over the six rounds of cycling experiments, as shown in figure 7.5(d). The associative alternating pathway on the MoS_2-rGO heterojunction is more practicable, according to DFT calculations than the distal pathway. $*NHNH_2 \rightarrow *NH_2NH_2$ is the step that determines the potential of the proton–electron coupling transference process, the steps involved for the free energy profiles are illustrated in figure 7.5(e). Each of these studies demonstrated the enormous potential of 2D MoS_2 for N_2RR.

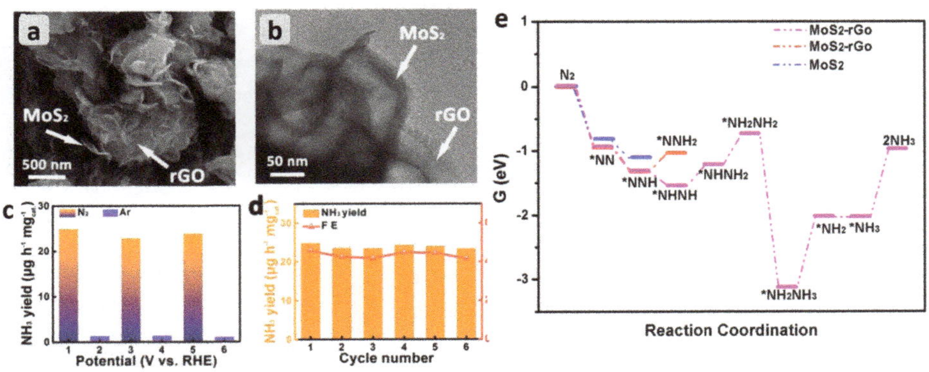

Figure 7.5. (a) and (b) FESEM and TEM images of MoS_2@rGO heterojunction; (a) NH_3 yields for MoS_2-rGO/CPE with N_2- and Ar-saturated electrolytes alternated in 2 h cycles, (c) recycling test of MoS_2-rGO/CPE at −0.45 V and (d) MoS_2-rGO (red/pink dotted line) and MoS_2 (blue dotted line) edge sites with regard to free-energy profiles for N_2RR. To indicate the adsorption location, an (*) mark is used. Reproduced from [82] with permission from the Royal Society of Chemistry.

The effective separation of charge carriers was mostly responsible for the enhanced photocatalytic performance, aside from the increased surface area contribution. The recombination of charge carriers was much inhibited by the loaded carbon layer and the MoS_2 nanoparticles functioning as electron trappers. In this context, a ternary MoS_2/C–ZnO heterostructure composite was effectively constructed by decorating the carbon layer and then photo depositing MoS_2 nanoparticles, as reported by Xing *et al* [83], in order to transmit electrons to the doped species by acting as a photosensitizer. Under conditions similar to those of artificial sunlight, ZnO produced both electrons and holes. The electrons moved easily to the carbon layer and then to the MoS_2 nanoparticles that were decorated. Additionally, the maximum rate of ammonia formation under these conditions was attained because of the capacity of surface coordinately unsaturated Mo atoms to chemisorb and activate N_2. The efficiency of 1% MoS_2/CZ_{300} hybrid photocatalytically fixing N_2 in the presence of various sacrificial agents is shown in figure 7.6(a–b). Due to its low absorption of visible light, pure ZnO exhibits almost negligible activity, as seen by the result in figure 7.6(c). Additionally, MoS_2 exhibits little photocatalytic activity in the fixation of N_2. This result is acceptable given its low activity under simulated sunshine. In contrast to ZnO or MoS_2, the C–ZnO sample exhibits strong

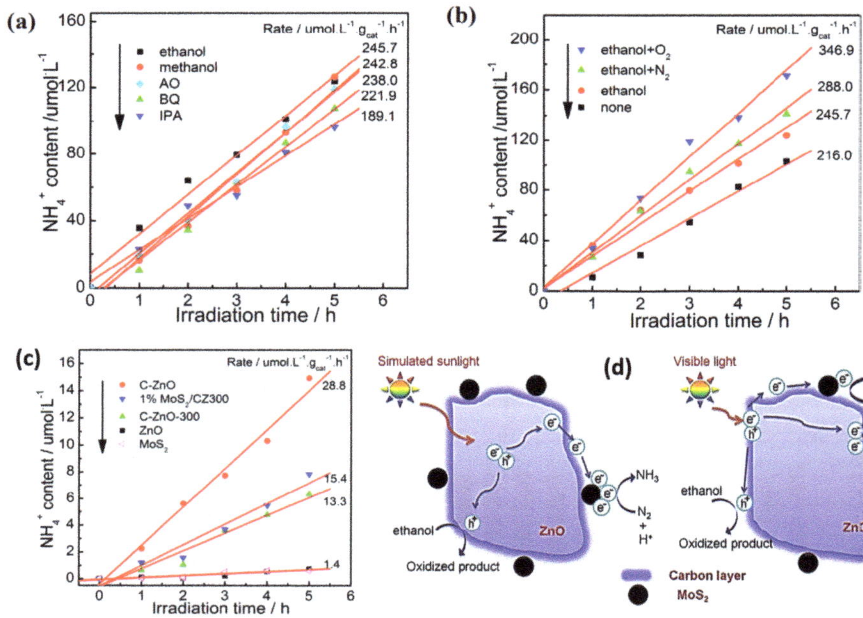

Figure 7.6. Efficacy of 1% MoS_2/CZ_{300} hybrid photocatalytically fixing N_2 in the presence of various sacrificial agents. (a) The impact of OH^-, O^{2-}, and h^+ trappers; (b) N_2 and O_2 effects; (c) the N_2 fixation performance of ZnO, MoS_2, C–ZnO, and MoS_2/CZ_{300} composites by photocatalysis in the presence of visible light. The proposed charge transfer mechanism of the MoS_2/CZ_{300} composite exposed to visible light and simulated sunlight. Reprinted with permission from [83]. Copyright (2018) American Chemical Society.

photocatalytic N_2 fixing activity when exposed to visible light. At 28.8 μmol l^{-1} g^{-1} h^{-1}, the NH_3 production rate is reached, which is approximately half of that under simulated sunlight. As illustrated in figure 7.6(d), the first process was mostly responsible for the observed light under the simulated sunshine, while the carbon photosensitization method appeared to have little effect. Subsequently, the MoS_2/C–ZnO system yielded a lower ammonia production than when exposed to visible light, as the second mechanism dominated the charge transfer process in the carbon layer alone loaded ZnO.

Polymeric C_3N_4 structure has received significant interest as a metal-free visible-light response photocatalyst because of its exceptional chemical stability and distinctive 2D structure [54]. In addition to being affordable and stable and meeting the prerequisites for photocatalysts, C_3N_4 also possesses the ease of manipulation of its chemical properties and energy band structure. As such, it is regarded as one of the most significant materials in the field of photocatalysis that merit more investigation. The strongly delocalized π conjugate system is formed when sp^2 hybridizes with the C and N atoms in the structure [84, 85]. The construction of a hybrid heterojunction including a 2D/2D C_3N_4/rGO composite was carried out by Wu *et al* [86]. The N_2 fixation activity exhibited a significant increase of 8.3 times when the 2D structures were employed, owing to their effective charge separation capabilities, as compared to the utilization of C_3N_4 unaccompanied. Following that, the researchers successfully synthesized honeycomb iron (Fe) doped C_3N_4 material, which exhibited exceptional N_2 photo fixation capability [87]. The findings of the characterization indicate that the Fe^{3+} ion is incorporated into an Interstitial site inside the electron-rich C_3N_4 material, forming a stable Fe–N bond. The Fe^{3+} site has the ability to chemisorb and activate N_2 molecules. Subsequently, it can facilitate the transfer of photogenerated electrons from C_3N_4 to the adsorbed N_2 molecules. This process leads to a significant 13.5-fold enhancement in activity. Several 2D TiO_2 composites have been developed for photocatalytic activity. Yang *et al* [88] demonstrated that N_2 photoreduction could be achieved using the 'working-in-tandem' process when Au nanocrystals were anchored on an ultrathin TiO_2 nanosheet containing OVs defects. The OVs in 2D TiO_2 chemisorbs causes the N_2 molecule to become activated. The active N_2 is subsequently reduced to NH_3 by the photogenerated heated electron resulting from the plasmon resonance of Au nanocrystals. The quantum efficiency at 550 nm is 0.82%, the highest value yet observed. Alongside the Au/TiO_2-OV 2D heterojunction, subsequently, a single-atom Ru composite supported by 2D TiO_2-rich-in OVs was created by Sun and co-workers [89]. Figure 7.7(a) describes the decoration of Ru atoms presented by bright spots on TiO_2 nanosheets. 56.3 μmol gh^{-1} NH_3 generation rate is the consequence of the synergy of Ru and OVs, which improves N_2 molecule chemisorption, weakens the N≡N link, and encourages the separation of photogenerated carriers. The most stable binding states and their corresponding adsorption energies of Ru atomic sites supported on the perfect, O-2v, and O-3v defective surfaces (bottom layer), and the optimized structure of the TiO_2(001) surface (top layer) is given in figure 7.7(b). The ammonia yield rates over TR-0.1, TR-0.5, TR-1.0, TR-2.0, and TR-ND over TiO_2-NS are shown in figure 7.7(c) which suggests that with good repeatability, the

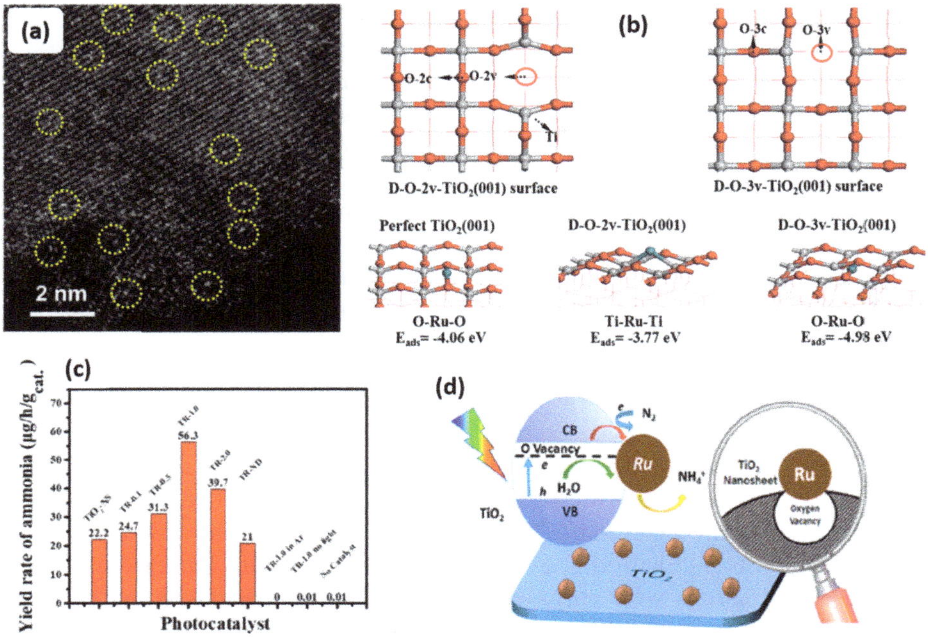

Figure 7.7. (a) HRTEM image of Ru–TiO_2. (b) The top layer of the TiO_2(001) surface has an optimized structure, while the bottom layer displays the most stable binding states and their corresponding adsorption energies of Ru atoms sites that are supported on O-2v, O-3v, and ideal surfaces. (c) The ammonia yield rates over TR-0.1, TR-0.5, TR-1.0, TR-2.0, and TR-ND over TiO_2-NS. (d) Diagrammatic representation of the mechanism of N_2 photoreduction on single Ru site-loaded TiO_2 nanosheets void of oxygen. Reprinted with permission from [89]. Copyright (2019) American Chemical Society.

TR-1.0 catalyst yields the greatest ammonia production rate of 56.3 μg h^{-1} g_{cat}^{-1} and figure 7.7(d) gives a representation of the mechanism of N_2 photoreduction on single Ru site-loaded TiO_2 nanosheets void of oxygen. Decreased hydrogen evolution, increased N_2 chemisorption, improved charge carrier separation, and enhanced N_2 photoreduction to ammonia were all facilitated by isolated Ru atoms that were potentially situated at the oxygen vacancies of TiO_2.

'Mxenes,' are a novel class of 2D materials of TM carbides and nitrides, with several catalytic applications. Mxenes are made up of a limited number of thick layers of TM carbides and nitride atoms [90]. A hybrid electrocatalyst $TiO_2/Ti_3C_2T_x$ with oxygen vacancies achieved a high FE of 16.07% at around −0.45 V versus RHE in a different investigation demonstrated by Fang *et al.* Due to the localized electron deficiency produced by these anion defects, the π back-donation was able to facilitate the N_2 adsorption and activation [91]. Thus, the produced electrocatalyst shows an impressive RHE in 0.1 m HCl and an NH_3 production of 32.17 μg h^{-1} $mg^{-1}{}_{cat}$ at −0.55 V versus RHE. Furthermore, when compared to TiO_2 (101) or $Ti_3C_2T_x$ alone, the DFT calculations validate TiO_2 (101)/$Ti_3C_2T_x$ possesses the lowest N_2RR energy barrier (0.40 eV), positioning it as one of the most promising N_2RR

electrocatalysts. Another work was presented by Kong *et al* [92] wherein, MnO_2-decorated $Ti_3C_2T_x$ MXene was used as an active catalyst to obtain electrochemical N_2RR. The NH_3 production was 34.12 μg h^{-1} mg^{-1} using 0.1 M HCl as the electrolyte, and the FE was 11.39% at –0.55 V versus RHE. With unsaturated Mn atoms acting as active sites, the MnO_2–$Ti_3C_2T_x$ composite catalyst surface effectively absorbed and activated N_2 molecules throughout the electrochemical reaction process. Their study shows that electrons are transported from N–N bonds to Mn atoms, forming Mn–N bonds that weaken N–N triple bonds and activate N_2 molecules. This is demonstrated by the charge density of the MnO_2–$Ti_3C_2T_x$ heterostructure catalyst while absorbing N_2 molecules.

7.3.4 Edge/corners engineering

The surface atoms along the edge and corner of nanocrystals are usually very reactive due to their high surface energy. The activation of inert molecules has been shown to be greatly lowered by such edge and corner defect configurations. In addition to other types of faulty materials, porous materials also have additional functionality that usually appears near their rough edges. Additionally, the 2D structure can facilitate the adsorption and activation of N_2 molecules by exposing a significant number of edge active sites. In 2D materials comprising usual graphene, transition metal carbides, nitrides, or carbonitrides (MXenes), and transition metal dichalcogenides, edge engineering is essential for controlling the growth kinetics and morphological evolution. This allows for a variety of edge structures as well as the provision of distinct electronic structures and functionalities [93–98]. Owing to their very thin thickness, they often have a higher concentration of exposed edge metal sites and vacancy-type defects, which differs from bulk electrical characteristics and improves N_2 molecule chemisorption and activation on the catalyst surface. The edge C atoms of graphene, for instance, were shown to have larger charge densities and to provide more active sites than the basal-plane C, demonstrating effective electrocatalytic activity in ORR. Thus, it is imperative to understand the edge development of 2D materials in order to construct optimal and useful N_2RR catalysts [99, 100].

MXenes are widely employed in catalytic processes, energy storage, and separation [101–103]. According to a recent theoretical study, MXenes with M atoms making up the terminal surface might activate N_2 molecules, showing promising results for NH_3 synthesis [104]. Contritely, the basal plane of 2D MXene showed poor binding capacity to N_2, terminating with oxygen-containing groups (e.g., OH* and O*) that directly bonded with surface M atoms. This led to reduced catalytic effectiveness in NRR when compared with the competing HER [104, 105]. Lou and co-workers have demonstrated an enhancing technique to provide a 2D catalyst ($Ti_3C_2T_x$ MXene) with exceptional N_2RR performance by expanding the active areas for nitrogen absorption and activation and preventing the development of hydrogen [106]. Due to their ability to bind N_2 molecules with the highest adsorption energy, Ti atoms are particularly active sites of N_2 adsorption in the $Ti_3C_2O_2$ structure. A $Ti_3C_2T_x$ MXene with a T–Ti–C–Ti–C–Ti–T structure was shown to be an effective NRR electrocatalyst by purposefully exposing enough active sites at the edge plane (T = O, F). Low energy barriers of 0.64 eV in $*N_2 \rightarrow *NNH$ and 0.52 eV in $*NNH \rightarrow *NNH_2$ for NH_3 synthesis were achieved by the N_2 molecules interacting preferentially with the middle Ti on the edge plane, as demonstrated by DFT calculations. This was to overcome thermodynamic barriers by developing 2D $Ti_3C_2T_x$ MXene with lower size and good dispersion in a vertically aligned metal host (FeOOH nanosheets). Notably, 2D $Ti_3C_2T_x$ MXene hosted on vertically aligned nanosheet materials with more sluggish HER activity after shrinking in size achieved a higher FE under near ambient conditions than those aligned on a parallel host (stainless steel mesh [SSM]) with better HER activity. A high FE of 5.78% was obtained at −0.2 V (versus RHE) for NH_3 synthesis when a vertically oriented FeOOH with slow activity towards HER was applied to host $Ti_3C_2T_x$ nanosheets. In an acidic electrolyte, $Ti_3C_2T_x$ (T = F, OH) nanosheets demonstrated an ammonia production of 20.4 μh^{-1} $mg^{-1}{}_{cat}$. and a higher FE of 9.3% at ~0.4 V versus RHE (figures 7.8(a)–(d)). Rather than the metal atoms positioned on basal

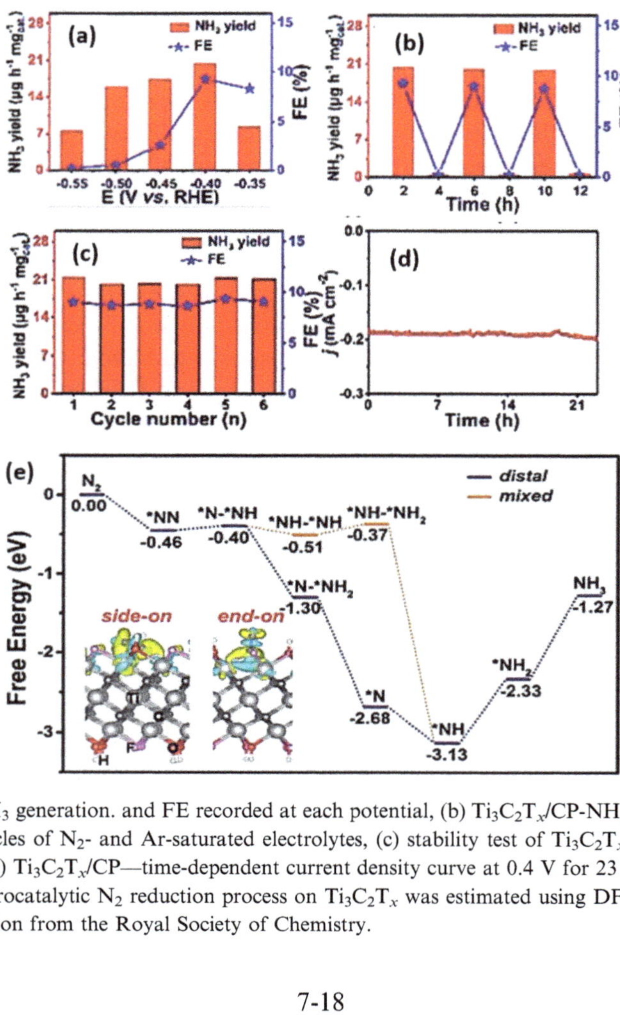

Figure 7.8. (a) NH_3 generation. and FE recorded at each potential, (b) $Ti_3C_2T_x$/CP-NH_3 yields and FEs with alternating 2 h cycles of N_2- and Ar-saturated electrolytes, (c) stability test of $Ti_3C_2T_x$/CP–during repeated N_2RR at 0.4 V, (d) $Ti_3C_2T_x$/CP—time-dependent current density curve at 0.4 V for 23 h, and (e) the energy profile of the electrocatalytic N_2 reduction process on $Ti_3C_2T_x$ was estimated using DFT. Reproduced from [107] with permission from the Royal Society of Chemistry.

planes, the Ti atoms at the defective edges of $Ti_3C_2T_x$ were shown to be responsible for the N_2RR activity as shown in figure 7.8(e). The DFT results showed that the protonation of $*NH_2$ was the rate-limiting step and that N_2RR followed the distal mechanism while N_2 was voluntarily deposited on the edge site Ti atom [107].

Analogous to MXene, the MoS_2 edges were essential for polarizing and activating the N_2 molecules, but the basal plane was neutral for N_2RR. Zhang *et al* have presented that the N_2 reduction reaction (N_2RR) is initially catalyzed by MoS_2 at ambient temperature and atmospheric pressure, in accordance with theoretical expectations [80]. Through electrochemical evaluation, this catalyst is able to get a high NH_3 production (8.08×10^{-11} mol s^{-1} cm^{-1}) and FE (1.17%) at −0.5 V when compared to an RHE in 0.1 M Na_2SO_4. MoS_2 is still effective for the N_2RR in acidic environments where a robust hydrogen evolution process takes place. Designing potential catalysts for high-performance N_2 reduction under ambient settings is made possible by the ease with which the edge engineering method may be applied to other 2D catalysts to control their catalytic characteristics.

7.3.5 Strain, stress, and non-crystallinity

Surface strain is often observed in nanocrystal photocatalysts, leading to notable alterations in bond lengths and bond angles in comparison to the bulk state. Highly effective photocatalysts for N_2 fixation and other photoreactions may be created by taking advantage of strain-induced lattice symmetry distortions at the surfaces of the photocatalyst or at heterojunctions in hybrid photocatalyst systems [108, 109]. The description of strain in nanostructures may be done in four ways viz; directly from HRTEM pictures (or by extrapolating from them), through computer simulations, XRD, and using x-ray absorption measurements. HRTEM images are used to generate strain maps, which show the pattern of distribution of strain on the nanoscale.

Particularly in 2D nanosheets, layered double hydroxides (LDHs) and TiO_2 have been shown experimentally. Cu-doped TiO_2 and CuCr-LDHs nanosheet photocatalysts were produced for N_2 fixing by Zhang and colleagues [22, 32]. Ti^{4+} (0.64 Å) was substituted with bigger Cu^{2+} (0.73 Å), resulting in many OVs. These were caused by Jahn–Teller distortions, concurrent lattice distortion, and compressive strain. The TiO_2-OV strain demonstrated accelerated N_2 adsorption and a favorable hydrogenation step of $N_2^* \rightarrow N_2H^*$, which was attributed to these positive factors. The result was a lower reaction energy (0.365 eV) and higher adsorption energy (−0.37 eV) compared to the TiO_2-OV (adsorption energy: −0.25 eV, reaction energy: 0.893 eV) and TiO_2-Pure (adsorption energy: −0.17 eV, reaction energy: 2.115 eV. This resulted in the designed catalyst displaying excellent photocatalytic activity over a wide range of solar absorption, allowing for NH_3 yield rates of 78.9 μmol h^{-1} g^{-1} under full solar illumination, 1.54 μmol h^{-1} g^{-1} (quantum yield of 0.08%), and 0.72 μmol h^{-1} g^{-1} (quantum yield of 0.05%) under 600 and 700 nm irradiations, respectively. The VOs and Jahn–Teller distortions about the Cu^{2+} centres lead to the production of significant compressive strain in these materials. It was discovered that this strain enhanced N_2 adsorption as well as electron transfer from the photocatalyst to N_2. It also stabilized VOs [110, 111]. As was already established, strain typically occurs at photocatalytic material contacts. The strain in $SrTiO_3$ bi-crystals with [001] symmetric tilt grain borders was investigated by Choi and co-workers [112]. The degree of tilt has a significant impact on the strain situation. It is noteworthy that the generation of VO in transition metal oxides is strongly influenced by both shear and biaxial stresses (compressive and tensile) (figure 7.9). The relative production energy of VO decreases with increasing strain. Thus, strain and VO in transition metal oxides often have mutually causative interactions, and the existence of strain-VO defect structures acts to improve photocatalytic processes in a complementary manner. In order to create a unique NRR electrocatalyst (Zr–TiO_2), Zr^{4+} was first doped into TiO_2 due to its identical d-electron configuration, oxide structure, and appropriate atom size [112]. Stable Zr^{4+} doping produced both compressive and tensile strain, which, in contrast to the well-known low-valance metal dopants used in the production of OVs, facilitated the generation of numerous OVs connected with nearby bi-Ti^{3+} sites. These bi-Ti^{3+} sites on anatase (101) surfaces significantly increased the chemisorption of N_2, as the theoretical simulations demonstrated. The developed Zr–TiO_2 demonstrated a notably increased

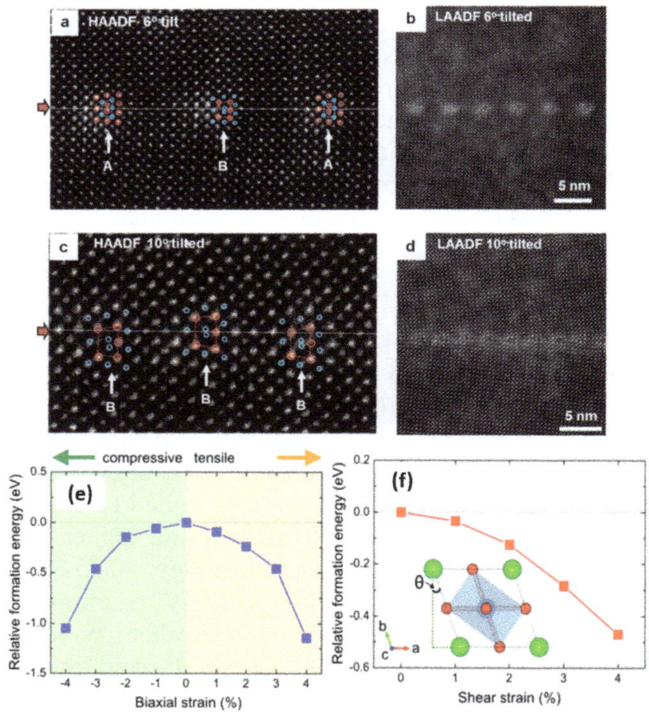

Figure 7.9. (a and c) HAADF STEM images and (b and d) LAADF-STEM images of the 6°-tilt and 10°-tilt borders, respectively, show the grain boundary planes, while the red and blue circles represent the Sr–Sr and Ti–O atomic columns, (e) biaxial strain, and (f) shear strain. Reprinted with permission from [112]. Copyright (2015) American Chemical Society.

performance in N_2RR, with a high FE of 17.3% at −0.45 V (versus RHE) and NH_3 production rate of 8.90 μg h^{-1} cm^{-2}, and a greatly enhanced Ti^{3+} ratio of 29.1% (8.5% of undoped TiO_2). the primary function of surface VO is to provide coordinatively unsaturated metal active sites that facilitate N_2 adsorption and activation. Since they have a high density of VO and unsaturated coordinated metal active sites localized to the surface region, which do not impact photoexcited charge transfer in the crystalline bulk, photocatalysts with amorphous surface structures are effective. In most cases, bulk vacancies act as electron-trapping sites in photocatalysts and are generally detrimental to photocatalysis by preventing photoexcited electrons from reaching surface adsorbates. It has been demonstrated that amorphous layers on semiconductor surfaces greatly improve photocatalytic N_2 fixation [108–112].

7.4 Summary and outlook

The development of suitable catalysts for N_2RR that greatly enhance catalytic performance is made possible through defect engineering. In contrast to noble metals and nanosized catalysts, the development of defect-containing catalysts for N_2RR is still in its infancy, despite some great results obtained by using the flaws in the materials to increase the N_2RR performance. Nitrogen- and oxygen-vacancy faults are the main topics of the present defect engineering research for N_2RR. The oxygen-vacancy defects are crucial for electrocatalytic N_2RR because they not only alter the catalyst's electrical charge and electron density distribution but also serve as reactive sites for reactant adsorption, lowering the activation energy barrier.

In terms of heterostructure photocatalysts, in addition to producing photogenerated electron transfer and lifting separation efficiency of conventional photogenerated carriers, the tactful superimposition of the advantages of each material results in more energetically appropriate reaction pathways and mechanisms that improve the photocatalytic nitrogen fixation ability. Defect sites, surface structures, and the chemical reaction mechanism on the catalyst from various scales must all be investigated in order to better understand the connection between the 2D structures and N_2RR activity. The conductivity, N_2 activation sites, and charge separation efficiency of the catalyst may all be further optimized by creating certain flaws in 2D structures; however, this is still mostly unexplored territory. We have tremendous optimism for lowering our reliance on fossil fuels and lessening the effects of climate change through the heterogeneous catalytic synthesis of NH_3 under moderate circumstances, notwithstanding the immense obstacles involved. Researchers are working tirelessly, and we are confident that in the near future, more stable, effective, and selective N_2RR catalysts will be built, leading to the eventual industrial production of NH_3.

Herein, we have provided a first overview of the latest advancements and innovations concerning 2D catalysts for both photocatalytic and electrocatalytic N_2RR. Effective N_2RR catalyst design was made possible by a thorough investigation of the effects of 2D structures on the excitation, migration, and separation of photogenerated carriers as well as the adsorption and activation of N_2 at the active sites. There are still a number of obstacles to be overcome in this developing subject, despite its promise. Economic and practical concerns show that the existing situation is far from adequate. In addition to offering a cost-effective means of storing renewable energy, this technique has the potential to drastically cut the energy use and greenhouse gas emissions of the ammonia sector by displacing the Haber–Bosch process. The significant advancements in N_2RR have shown that catalytic N_2 reduction on heterogeneous catalyst surfaces is feasible in ambient environments. Versatile solutions for rational catalyst design have been widely proposed to increase N_2RR performance. The goal is to inhibit hydrogen evolution, the principal side reaction competing for protons and electrons, and significantly improve energy utilization efficiency.

Acknowledgments

The authors gratefully acknowledge financial assistance from the SERB Core Research Grant (Grant No. CRG/2022/000897), Department of Science and Technology (DST/NM/NT/2019/205(G)), and Minor Research Project Grant, Jain University (JU/MRP/CNMS/29/2023).

References

[1] Erisman J W, Sutton M A, Galloway J, Klimont Z and Winiwarter W 2008 How a century of ammonia synthesis changed the world *Nat. Geosci.* **1** 636–9

[2] Foster S L, Bakovic S I P, Duda R D, Maheshwari S, Milton R D, Minteer S D, Janik M J, Renner J N and Greenlee L F 2018 Catalysts for nitrogen reduction to ammonia *Nat. Catal.* **1** 490–500

[3] Hoffman B M, Lukoyanov D, Yang Z-Y, Dean D R and Seefeldt L C 2014 Mechanism of nitrogen fixation by nitrogenase: the next stage *Chem. Rev.* **114** 4041–62

[4] Siddharth K, Hong Y, Qin X, Lee H J, Chan Y T, Zhu S, Chen G, Choi S-I and Shao M 2020 Surface engineering in improving activity of Pt nanocubes for ammonia electro-oxidation reaction *Appl. Catal.* B **269** 118821

[5] Lan R, Irvine J T S and Tao S 2012 Ammonia and related chemicals as potential indirect hydrogen storage materials *Int. J. Hydrogen Energy* **37** 1482–94

[6] Cui X, Tang C and Zhang Q 2018 A review of electrocatalytic reduction of dinitrogen to ammonia under ambient conditions *Adv. Energy Mater.* **8** 1800369

[7] Rafiqul I, Weber C, Lehmann B and Voss A 2005 Energy efficiency improvements in ammonia production—perspectives and uncertainties *Energy* **30** 2487–504

[8] Li C *et al* 2019 Dendritic Cu: a high-efficiency electrocatalyst for N_2 fixation to NH_3 under ambient conditions *Chem. Commun.* **55** 14474–7

[9] Liu J, Ma K, Ciais P and Polasky S 2016 Reducing human nitrogen use for food production *Sci. Rep.* **6** 30104

[10] Suryanto B, Du H-L, Wang D, Chen J, Simonov A and MacFarlane D 2019 Challenges and prospects in the catalysis of electroreduction of nitrogen to ammonia *Nat. Catal.* **2** 290–6

[11] Zhang G, Li Y, He C, Ren X, Zhang P and Mi H 2021 Recent progress in 2D catalysts for photocatalytic and electrocatalytic artificial nitrogen reduction to ammonia *Adv. Energy Mater.* **11** 2003294

[12] Zhang G, Sewell C D, Zhang P, Mi H and Lin Z 2020 Nanostructured photocatalysts for nitrogen fixation *Nano Energy* **71** 104645

[13] Guo X, Du H, Qu F and Li J 2019 Recent progress in electrocatalytic nitrogen reduction *J. Mater. Chem.* A **7** 3531–43

[14] Xu Y, Liu X, Cao N, Xu X and Bi L 2021 Defect engineering for electrocatalytic nitrogen reduction reaction at ambient conditions *Sustain. Mater. Technol.* **27** e00229

[15] Tan C *et al* 2017 Recent advances in ultrathin two-dimensional nanomaterials *Chem. Rev.* **117** 6225–331

[16] Li X-F, Li Q-K, Cheng J, Liu L, Yan Q, Wu Y, Zhang X-H, Wang Z-Y, Qiu Q and Luo Y 2016 Conversion of dinitrogen to ammonia by FeN_3-embedded graphene *J. Am. Chem. Soc.* **138** 8706–9

[17] Fan M *et al* 2021 The modulating effect of N coordination on single-atom catalysts researched by Pt-Nx-C model through both experimental study and DFT simulation *J. Mater. Sci. Technol.* **91** 160–7

[18] Bat-Erdene M, Bati A S R, Qin J, Zhao H, Zhong Y L, Shapter J G and Batmunkh M 2022 Elemental 2D materials: solution-processed synthesis and applications in electrochemical ammonia production *Adv. Funct. Mater.* **32** 2107280

[19] Sun Y, Gao S, Lei F and Xie Y 2015 Atomically-thin two-dimensional sheets for understanding active sites in catalysis *Chem. Soc. Rev.* **44** 623–36

[20] Yin H, Dou Y, Chen S, Zhu Z, Liu P and Zhao H 2020 2D electrocatalysts for converting earth-abundant simple molecules into value-added commodity chemicals: recent progress and perspectives *Adv. Mater.* **32** 1904870

[21] Ahmed M I, Liu C, Zhao Y, Ren W, Chen X, Chen S and Zhao C 2020 Metal–sulfur linkages achieved by organic tethering of ruthenium nanocrystals for enhanced electrochemical nitrogen reduction *Angew. Chem. Int. Ed.* **59** 21465–9

[22] Zhao Y, Zhao Y, Shi R, Wang B, Waterhouse G I N, Wu L-Z, Tung C-H and Zhang T 2019 Tuning oxygen vacancies in ultrathin TiO_2 nanosheets to boost photocatalytic nitrogen fixation up to 700 nm *Adv. Mater.* **31** 1806482

[23] Wu X, Dai J, Zhao Y, Zhuo Z, Yang J and Zeng X C 2012 Two-dimensional boron monolayer sheets *ACS Nano.* **6** 7443–53

[24] Shilov A E 2003 Catalytic reduction of molecular nitrogen in solutions *Russ. Chem. Bull.* **52** 2555–62

[25] Arashiba K, Kinoshita E, Kuriyama S, Eizawa A, Nakajima K, Tanaka H, Yoshizawa K and Nishibayashi Y 2015 Catalytic reduction of dinitrogen to ammonia by use of molybdenum–nitride complexes bearing a tridentate triphosphine as catalysts *J. Am. Chem. Soc.* **137** 5666–9

[26] Bazhenova T A and Shilov A E 1995 Nitrogen fixation in solution *Coord. Chem. Rev.* **144** 69–145

[27] Eberle U, Felderhoff M and Schüth F 2009 Chemical and physical solutions for hydrogen storage *Angew. Chem. Int. Ed. Engl.* **48** 6608–30

[28] Jia H-P and Quadrelli E A 2014 Mechanistic aspects of dinitrogen cleavage and hydrogenation to produce ammonia in catalysis and organometallic chemistry: relevance of metal hydride bonds and dihydrogen *Chem. Soc. Rev.* **43** 547–64

[29] Skúlason E, Bligaard T, Gudmundsdóttir S, Studt F, Rossmeisl J, Abild-Pedersen F, Vegge T, Jónsson H and Nørskov J K 2012 A theoretical evaluation of possible transition metal electrocatalysts for N_2 reduction *Phys. Chem. Chem. Phys.* **14** 1235–45

[30] Ertl G 2008 Reactions at surfaces: from atoms to complexity (nobel lecture) *Angew. Chem. Int. Ed.* **47** 3524–35
[31] Somorjai G A and Li Y 2011 Impact of surface chemistry *Proc. Natl Acad. Sci.* **108** 917–24
[32] Niu X, Shi A, Sun D, Xiao S, Zhang T, Zhou Z, Li X and Wang J 2021 Photocatalytic ammonia synthesis: mechanistic insights into N_2 activation at oxygen vacancies under visible light excitation *ACS Catal.* **11** 14058–66
[33] Bian X, Zhao Y, Zhang S, Li D, Shi R, Zhou C, Wu L-Z and Zhang T 2021 Enhancing the supply of activated hydrogen to promote photocatalytic nitrogen fixation *ACS Mater. Lett.* **3** 1521–7
[34] Chen X, Li N, Kong Z, Ong W-J and Zhao X 2018 Photocatalytic fixation of nitrogen to ammonia: state-of-the-art advancements and future prospects *Mater. Horizons* **5** 9–27
[35] John J, Lee D-K and Sim U 2019 Photocatalytic and electrocatalytic approaches towards atmospheric nitrogen reduction to ammonia under ambient conditions *Nano Converg.* **6** 15
[36] Hao Q, Liu C, Jia G, Wang Y, Arandiyan H, Wei W and Ni B-J 2020 Catalytic reduction of nitrogen to produce ammonia by bismuth-based catalysts: state of the art and future prospects *Mater. Horiz.* **7** 1014–29
[37] Li H, Shang J, Ai Z and Zhang L 2015 Efficient visible light nitrogen fixation with BiOBr nanosheets of oxygen vacancies on the exposed {001} facets *J. Am. Chem. Soc.* **137** 6393–9
[38] Shipman M A and Symes M D 2017 Recent progress towards the electrosynthesis of ammonia from sustainable resources *Catal. Today* **286** 57–68
[39] Zhao R *et al* 2019 Recent progress in the electrochemical ammonia synthesis under ambient conditions *Energy Chem.* **1** 100011
[40] Gupta U and Rao C N R 2017 Hydrogen generation by water splitting using MoS_2 and other transition metal dichalcogenides *Nano Energy* **41** 49–65
[41] Murali G, Reddy Modigunta J K, Park Y H, Lee J-H, Rawal J, Lee S-Y, In I and Park S-J 2022 A review on MXene synthesis, stability, and photocatalytic applications *ACS Nano.* **16** 13370–429
[42] Su T, Shao Q, Qin Z, Guo Z and Wu Z 2018 Role of interfaces in two-dimensional photocatalyst for water splitting *ACS Catal.* **8** 2253–76
[43] Ganguly P, Harb M, Cao Z, Cavallo L, Breen A, Dervin S, Dionysiou D D and Pillai S C 2019 2D nanomaterials for photocatalytic hydrogen production *ACS Energy Lett.* **4** 1687–709
[44] Ida S and Ishihara T 2014 Recent progress in two-dimensional oxide photocatalysts for water splitting *J. Phys. Chem. Lett.* **5** 2533–42
[45] Chia X and Pumera M 2018 Characteristics and performance of two-dimensional materials for electrocatalysis *Nat. Catal.* **1** 909–21
[46] Wang Q, Lei Y, Wang Y, Liu Y, Song C, Zeng J, Song Y, Duan X, Wang D and Li Y 2020 Atomic-scale engineering of chemical-vapor-deposition-grown 2D transition metal dichalcogenides for electrocatalysis *Energy Environ. Sci.* **13** 1593–616
[47] Anantharaj S, Karthick K and Kundu S 2017 Evolution of layered double hydroxides (LDH) as high-performance water oxidation electrocatalysts: a review with insights on structure, activity and mechanism *Mater. Today Energy* **6** 1–26
[48] Sun T, Zhang G, Xu D, Lian X, Li H, Chen W and Su C 2019 Defect chemistry in 2D materials for electrocatalysis *Mater. Today Energy* **12** 215–38
[49] Singh A, Price C C and Shenoy V B 2022 Magnetic order, electrical doping, and charge-state coupling at amphoteric defect sites in Mn-doped 2D semiconductors *ACS Nano.* **16** 9452–60

[50] Cheng M, Xiao C and Xie Y 2019 Photocatalytic nitrogen fixation: the role of defects in photocatalysts *J. Mater. Chem.* A **7** 19616–33
[51] Mao C, Wang J, Zou Y, Li H, Zhan G, Li J, Zhao J and Zhang L 2019 Anion (O, N, C, and S) vacancies promoted photocatalytic nitrogen fixation *Green Chem.* **21** 2852–67
[52] Dong G, Ho W and Wang C 2015 Selective photocatalytic N_2 fixation dependent on g-C_3N_4 induced by nitrogen vacancies *J. Mater. Chem.* A **3** 23435–41
[53] Lin W *et al* 2022 Creating frustrated lewis pairs in defective boron carbon nitride for electrocatalytic nitrogen reduction to ammonia *Angew. Chem. Int. Ed.* **61** e202207807
[54] Ong W-J, Tan L-L, Ng Y H, Yong S-T and Chai S-P 2016 Graphitic carbon nitride (g-C_3N_4)-based photocatalysts for artificial photosynthesis and environmental remediation: are we a step closer to achieving sustainability? *Chem. Rev.* **116** 7159–329
[55] Peng G, Wu J, Wang M, Niklas J, Zhou H and Liu C 2020 Nitrogen-defective polymeric carbon nitride nanolayer enabled efficient electrocatalytic nitrogen reduction with high faradaic efficiency *Nano Lett.* **20** 2879–85
[56] Wei F, Ni L and Cui P 2008 Preparation and characterization of N–S-codoped TiO_2 photocatalyst and its photocatalytic activity *J. Hazard. Mater.* **156** 135–40
[57] Izumi Y, Itoi T, Peng S, Oka K and Shibata Y 2009 Site structure and photocatalytic role of sulfur or nitrogen-doped titanium oxide with uniform mesopores under visible light *J. Phys. Chem.* C **113** 6706–18
[58] Du C, Zhang Q, Lin Z, Yan B, Xia C and Yang G 2019 Half-unit-cell $ZnIn_2S_4$ monolayer with sulfur vacancies for photocatalytic hydrogen evolution *Appl. Catal.* B **248** 193–201
[59] Liu Y, Wang H, Yuan X, Wu Y, Wang H, Tan Y Z and Chew J W 2021 Roles of sulfur-edge sites, metal-edge sites, terrace sites, and defects in metal sulfides for photocatalysis *Chem Catal* **1** 44–68
[60] Varakin A N, Mozhaev A V, Pimerzin A A and Nikulshin P A 2018 Comparable investigation of unsupported MoS_2 hydrodesulfurization catalysts prepared by different techniques: advantages of support leaching method *Appl. Catal.* B **238** 498–508
[61] Zhao K, Gu W, Zhao L, Zhang C, Peng W and Xian Y 2015 MoS_2/nitrogen-doped graphene as efficient electrocatalyst for oxygen reduction reaction *Electrochim. Acta* **169** 142–9
[62] Asadi M *et al* 2019 Highly efficient solar-driven carbon dioxide reduction on molybdenum disulfide catalyst using choline chloride-based electrolyte *Adv. Energy Mater.* **9** 1803536
[63] Ding Q, Song B, Xu P and Jin S 2016 Efficient electrocatalytic and photoelectrochemical hydrogen generation using MoS_2 and related compounds *Chem.* **1** 699–726
[64] Sun S, Li X, Wang W, Zhang L and Sun X 2017 Photocatalytic robust solar energy reduction of dinitrogen to ammonia on ultrathin MoS_2 *Appl. Catal.* B **200** 323–9
[65] Jin H, Guo C, Liu X, Liu J, Vasileff A, Jiao Y, Zheng Y and Qiao S-Z 2018 Emerging two-dimensional nanomaterials for electrocatalysis *Chem. Rev.* **118** 6337–408
[66] Dai L, Xue Y, Qu L, Choi H-J and Baek J-B 2015 Metal-free catalysts for oxygen reduction reaction *Chem. Rev.* **115** 4823–92
[67] Zhao S, Lu X, Wang L, Gale J and Amal R 2019 Carbon-based metal-free catalysts for electrocatalytic reduction of nitrogen for synthesis of ammonia at ambient conditions *Adv. Mater.* **31** 1805367
[68] Li X and Zhi L 2018 Graphene hybridization for energy storage applications *Chem. Soc. Rev.* **47** 3189–216

[69] Wang H, Li X-B, Gao L, Wu H-L, Yang J, Cai L, Ma T-B, Tung C-H, Wu L-Z and Yu G 2018 Three-dimensional graphene networks with abundant sharp edge sites for efficient electrocatalytic hydrogen evolution *Angew. Chem. Int. Ed.* **57** 192–7

[70] Qiu B, Xing M and Zhang J 2018 Recent advances in three-dimensional graphene based materials for catalysis applications *Chem. Soc. Rev.* **47** 2165–216

[71] Xu Y, Wang M, Ren K, Ren T, Liu M, Wang Z, Li X, Wang L and Wang H 2021 Atomic defects in pothole-rich two-dimensional copper nanoplates triggering enhanced electrocatalytic selective nitrate-to-ammonia transformation *J. Mater. Chem.* A **9** 16411–7

[72] Wang W *et al* 2023 Filling the gap between heteroatom doping and edge enrichment of 2D electrocatalysts for enhanced hydrogen evolution *ACS Nano.* **17** 1287–97

[73] Hu C and Dai L 2016 Carbon-based metal-free catalysts for electrocatalysis beyond the ORR *Angew. Chem. Int. Ed.* **55** 11736–58

[74] Gong K, Du F, Xia Z, Durstock M and Dai L 2009 Nitrogen-doped carbon nanotube arrays with high electrocatalytic activity for oxygen reduction *Sci.* **323** 760–4

[75] Yu X, Han P, Wei Z, Huang L, Gu Z, Peng S, Ma J and Zheng G 2018 Boron-doped graphene for electrocatalytic N_2 reduction *Joule* **2** 1610–22

[76] Liu Y, Su Y, Quan X, Fan X, Chen S, Yu H, Zhao H, Zhang Y and Zhao J 2018 Facile ammonia synthesis from electrocatalytic N_2 reduction under ambient conditions on N-doped porous carbon *ACS Catal.* **8** 1186–91

[77] Yang X *et al* 2018 Nitrogen-doped porous carbon: highly efficient trifunctional electrocatalyst for oxygen reversible catalysis and nitrogen reduction reaction *J. Mater. Chem.* A **6** 7762–9

[78] Yuan Y-P, Ruan L-W, Barber J, Joachim Loo S C and Xue C 2014 Hetero-nanostructured suspended photocatalysts for solar-to-fuel conversion *Energy Environ. Sci.* **7** 3934–51

[79] Wang H, Zhang L, Chen Z, Hu J, Li S, Wang Z, Liu J and Wang X 2014 Semiconductor heterojunction photocatalysts: design, construction, and photocatalytic performances *Chem. Soc. Rev.* **43** 5234–44

[80] Zhang L, Ji X, Ren X, Ma Y, Shi X, Tian Z, Asiri A M, Chen L, Tang B and Sun X 2018 Electrochemical ammonia synthesis via nitrogen reduction reaction on a MoS_2 catalyst: theoretical and experimental studies *Adv. Mater.* **30** 1800191

[81] Li Y, Wang H, Xie L, Liang Y, Hong G and Dai H 2011 MoS_2 nanoparticles grown on graphene: an advanced catalyst for the hydrogen evolution reaction *J. Am. Chem. Soc.* **133** 7296–9

[82] Li X, Ren X, Liu X, Zhao J, Sun X, Zhang Y, Kuang X, Yan T, Wei Q and Wu D 2019 A MoS_2 nanosheet–reduced graphene oxide hybrid: an efficient electrocatalyst for electrocatalytic N_2 reduction to NH_3 under ambient conditions *J. Mater. Chem.* A **7** 2524–8

[83] Xing P, Chen P, Chen Z, Hu X, Lin H, Wu Y, Zhao L and He Y 2018 Novel ternary MoS_2/C-ZnO composite with efficient performance in photocatalytic NH_3 synthesis under simulated sunlight *ACS Sustain. Chem. Eng.* **6** 14866–79

[84] Zhang H, Cao J, Kang P, Tang Q, Sun Q and Ma M 2018 Ag nanocrystals decorated g-C_3N_4/Nafion hybrid membranes: one-step synthesis and photocatalytic performance *Mater. Lett.* **213** 218–21

[85] Wang X, Maeda K, Thomas A, Takanabe K, Xin G, Carlsson J M, Domen K and Antonietti M 2009 A metal-free polymeric photocatalyst for hydrogen production from water under visible light *Nat. Mater.* **8** 76–80

[86] Hu S, Zhang W, Bai J, Lu G, Zhang L and Wu G 2016 Construction of a 2D/2D g-C_3N_4/rGO hybrid heterojunction catalyst with outstanding charge separation ability and nitrogen photofixation performance via a surface protonation process *RSC Adv.* **6** 25695–702
[87] Hu S, Chen X, Li Q, Li F, Fan Z, Wang H, Wang Y, Zheng B and Wu G 2017 Fe^{3+} doping promoted N_2 photofixation ability of honeycombed graphitic carbon nitride: the experimental and density functional theory simulation analysis *Appl. Catal.* B **201** 58–69
[88] Yang J, Guo Y, Jiang R, Qin F, Zhang H, Lu W, Wang J and Yu J C 2018 High-efficiency 'Working-in-Tandem' nitrogen photofixation achieved by assembling plasmonic gold nanocrystals on ultrathin titania nanosheets *J. Am. Chem. Soc.* **140** 8497–508
[89] Liu S *et al* 2019 Photocatalytic fixation of nitrogen to ammonia by single Ru atom decorated TiO_2 nanosheets *ACS Sustain. Chem. Eng.* **7** 6813–20
[90] Pathak M and Rout C S 2021 Miniaturized energy storage: microsupercapacitor based on two-dimensional materials *Fundamentals and Supercapacitor Applications of 2D Materials* ed C S Rout and D J B T Late (Amsterdam: Elsevier) pp 311–58 ch 11
[91] Fang Y, Liu Z, Han J, Jin Z, Han Y, Wang F, Niu Y, Wu Y and Xu Y 2019 High-performance electrocatalytic conversion of N_2 to NH_3 using oxygen-vacancy-rich TiO_2 *in situ* grown on $Ti_3C_2T_x$ MXene *Adv. Energy Mater.* **9** 1803406
[92] Kong W, Gong F F, Zhou Q, Yu G, Ji L, Sun X, Asiri A M, Wang T, Luo Y and Xu Y 2019 An MnO_2–$Ti_3C_2T_x$ MXene nanohybrid: an efficient and durable electrocatalyst toward artificial N_2 fixation to NH_3 under ambient conditions *J. Mater. Chem.* A **7** 18823–7
[93] Voiry D, Mohite A and Chhowalla M 2015 Phase engineering of transition metal dichalcogenides *Chem. Soc. Rev.* **44** 2702–12
[94] Chhowalla M, Shin H S, Eda G, Li L-J, Loh K P and Zhang H 2013 The chemistry of two-dimensional layered transition metal dichalcogenide nanosheets *Nat. Chem.* **5** 263–75
[95] Geim A K 2009 Graphene: status and prospects *Sci.* **324** 1530–4
[96] Wang Q H, Kalantar-Zadeh K, Kis A, Coleman J N and Strano M S 2012 Electronics and optoelectronics of two-dimensional transition metal dichalcogenides *Nat. Nanotechnol.* **7** 699–712
[97] Kibsgaard J, Chen Z, Reinecke B N and Jaramillo T F 2012 Engineering the surface structure of MoS_2 to preferentially expose active edge sites for electrocatalysis *Nat. Mater.* **11** 963–9
[98] Zhao X, Sun W, Geng D, Fu W, Dan J, Xie Y, Kent P R C, Zhou W, Pennycook S J and Loh K P 2019 Edge segregated polymorphism in 2D molybdenum carbide *Adv. Mater.* **31** 1808343
[99] Jaramillo T F, Jørgensen K P, Bonde J, Nielsen J H, Horch S and Chorkendorff I 2007 Identification of active edge sites for electrochemical H2 evolution from MoS_2 nanocatalysts *Sci.* **317** 100–2
[100] Shen A, Zou Y, Wang Q, Dryfe R A W, Huang X, Dou S, Dai L and Wang S 2014 Oxygen reduction reaction in a droplet on graphite: direct evidence that the edge is more active than the basal plane *Angew. Chem. Int. Ed.* **53** 10804–8
[101] Anasori B, Lukatskaya M R and Gogotsi Y 2017 2D metal carbides and nitrides (MXenes) for energy storage *Nat. Rev. Mater.* **2** 16098
[102] Kurtoglu M, Naguib M, Gogotsi Y and Barsoum M W 2012 First principles study of two-dimensional early transition metal carbides *MRS Commun.* **2** 133–7

[103] Peng J, Chen X, Ong W-J, Zhao X and Li N 2019 Surface and heterointerface engineering of 2D MXenes and their nanocomposites: insights into electro- and photocatalysis *Chem.* **5** 18–50

[104] Li X, Zeng C and Fan G 2015 Ultrafast hydrogen generation from the hydrolysis of ammonia borane catalyzed by highly efficient bimetallic RuNi nanoparticles stabilized on $Ti_3C_2X_2$ (X = OH and/or F) *Int. J. Hydrogen Energy* **40** 3883–91

[105] Azofra L M, Li N, MacFarlane D R and Sun C 2016 Promising prospects for 2D d2–d4 M3C2 transition metal carbides (MXenes) in N_2 capture and conversion into ammonia *Energy Environ. Sci.* **9** 2545–9

[106] Luo Y, Chen G-F, Ding L, Chen X, Ding L-X and Wang H 2019 Efficient electrocatalytic N_2 fixation with MXene under ambient conditions *Joule* **3** 279–89

[107] Zhao J, Zhang L, Xie X-Y, Li X, Ma Y, Liu Q, Fang W-H, Shi X, Cui G and Sun X 2018 $Ti_3C_2T_x$ (T = F, OH) MXene nanosheets: conductive 2D catalysts for ambient electrohydrogenation of N_2 to NH_3 *J. Mater. Chem.* A **6** 24031–5

[108] Shi R, Zhao Y, Waterhouse G I N, Zhang S and Zhang T 2019 Defect engineering in photocatalytic nitrogen fixation *ACS Catal.* **9** 9739–50

[109] Wang H, Li Y and Xiao X 2023 Facile synthesis of Ni-doped WO_{3-x} nanosheets with enhanced visible-light-responsive photocatalytic performance for lignin depolymerization into value-added biochemicals *Catalysts* **13** 1205

[110] Jin H, Kim S S, Venkateshalu S, Lee J, Lee K and Jin K 2023 Electrochemical nitrogen fixation for green ammonia: recent progress and challenges *Adv. Sci.* **10** 2300951

[111] Cao N, Chen Z, Zang K, Xu J, Zhong J, Luo J, Xu X and Zheng G 2019 Doping strain induced bi-Ti^{3+} pairs for efficient N_2 activation and electrocatalytic fixation *Nat. Commun.* **10** 2877

[112] Choi S-Y *et al* 2015 Assessment of strain-generated oxygen vacancies using $SrTiO_3$ bicrystals *Nano Lett.* **15** 4129–34

IOP Publishing

Engineered 2D Materials for Electrocatalysis Applications

Chandra Sekhar Rout

Chapter 8

Engineered 2D materials for methanol oxidation reaction (MOR)

Sithara Radhakrishnan†, K A Sree Raj† and Chandra Sekhar Rout

Two-dimensional (2D) materials piqued great interest in the field of electrocatalysts due to their diverse and tunable electronic, chemical and optical properties. 2D materials like graphene, MXene and transition metal dichalcogens (TMDs) supported catalysts have been recognized as a stable and reliable substrate for supporting a variety of Pt, Pd and other metal nanoparticles designed specifically for cathodic oxygen reduction process. These materials have been shown to be particularly effective in enhancing oxygen reduction kinetics, prompting researchers to look into their potential as catalyst supports for methanol oxidation processes. These 2D material-supported electrocatalysts exhibit improved kinetics in methanol oxidation, indicating substantial promise as long-term catalyst supports for alcohol fuel cells. Despite this, researchers have discovered some limitations and challenges that must be overcome before 2D materials can be used as catalytic supports in these applications to their full potential. In light of these advancements, we are discussing here the fundamentals and then the catalyst design methods used in 2D materials to improve catalytic activity in depth.

8.1 Introduction

The direct alcohol fuel cell (DAFC) is considered as an effective and alternative energy system in green energy transitions. Alcohol fuel cell technologies have multidimensional benefits in the form of cost effectiveness, environmental friendliness, portability and high energy density [1–3]. In the domain of DAFCs, direct methanol fuel cells (DMFCs) are widely explored due to the higher energy density of methanol (4820 Wh l^{-1}) over hydrogen [4]. Apart from the higher theoretical energy density, DMFC has other advantages like safe operation, quick refuelling, long life, facile

† SR AND KA were equally contributed.

doi:10.1088/978-0-7503-5719-7ch8 © IOP Publishing Ltd 2024

storage of the fuel, low working temperature and higher efficiency in energy conversion [5, 6]. Even though considerable developments were accomplished, DMFCs still struggle with hindering factors such as the methanol crossover from anode to cathode through the proton exchange membrane and the slow anode reaction kinetics which affects the cathode reaction as well as the fuel efficiency [7]. To overcome these challenges of DMFCs and improve the methanol oxidation reaction (MOR) at the anode, researchers are trying to implement high performing electrocatalysts into these systems. Methanol is a simple molecule consisting of one carbon atom, therefore, in contrast with other alcohols methanol exhibits higher CO_2 selectivity [8]. Platinum is widely considered as the best electrocatalyst in existence for DMFCs. Studies have shown that in micro-DMFC the methanol conversion during MOR is enhanced by increasing the loading of catalysts [9]. This process can trigger the methanol crossover and shoot up the cost of the DMFC systems. To overcome these hurdles, scholars across the globe are in search for a suitable, functional, cost effective and green catalyst for MOR to advance and commercialize DMFCs in the near future [10].

Platinum as an electrocatalyst can boost electrocatalytic efficiency, as it has high corrosion resistance and effective electric properties. Apart from the cost and unavailability, Pt is also deactivated by the CO formation during the methanol oxidation in a fuel cell. The carbon monoxide intermediate adsorbed reacts with the Pt catalyst and reduces its life span [11–13]. Considering all these factors, moving beyond Pt as catalyst is a challenging task for the research community yet it is inevitable for the development of fuel cell technologies. In addition to the electrocatalysts the catalyst support also contributes to the electrochemical methanol oxidation in a fuel cell [14, 15]. Numerous electrocatalyst support materials have been reported in the last few decades which include both carbonaceous and non-carbonaceous materials. Materials such as activated carbons, graphene, carbon nanotubes, porous carbons, carbon black, carbides, nitrides, borides, metal oxides, polymers, mesoporous silica etc, are some of the electrocatalytic support materials reported over the years [16, 17]. To attain the optimum MOR performance the support materials should possess certain capabilities such as good electrical conductivity, high surface area, strong interaction with the catalyst, corrosion resistance and catalyst recovery. Based on these criteria electrocatalytic support is selected [18, 19]. Recent research focuses heavily on the implementation of nanostructured materials for the use of both electrocatalyst and catalyst support. An advanced electrocatalyst with well-controlled dimensions in nanoscale can produce dramatic enhancement in the MORs. In these nanostructured materials, 2D materials gathered great attention in the area of MOR due to their unique properties compared to their bulk counterparts [20, 21].

2D nanosheets are widely regarded as a cost-effective candidate with large active sites for excellent interfacial charge transfer for electrocatalytic activities. Recently, a large number of 2D materials like graphene, TMDs, layered double hydroxides (LDH), MXenes etc, have been explored for alcohol oxidation reactions (AORs) owing to their shortened charge migration path, large surface area and abundance of diffusion channels for ions/molecule transport [22, 23]. Moreover, 2D nanosheets can overcome the damage of the bubbles generated in the electrocatalyst structure and improve the electrocatalytic reactions and stability. Many of these 2D nanostructures suffer from inherent shortcomings when used for MOR. Some of these shortcomings were addressed in the initial chapters [23, 24]. There are many strategies reported to engineer the structural properties of 2D materials such as morphology control, alloying, doping, stress and strain engineering, edge engineering and heterostructures to promote their electrocatalytic activity. We will provide a detailed discussion on the individual impact of these strategies on the MORs in the coming sections. We will briefly discuss some of the fundamental aspects of MOR first.

8.2 MOR-mechanisms and insights

For a long time, MOR has attracted great attention from scientists due to the interest in its fundamental and application aspects. MOR is a multi-step reaction mechanism with a potential release of reaction intermediates. It is been suggested that apart from adsorbed CO and CO_2, formic acid might be an intermediate during MOR [24, 25]. CO is considered as a catalytic poison, therefore, to overcome this a H_2O molecule will intervene and convert the intermediates into products in the case of many electrocatalysts [20]. The methanol oxidation completes when the C–O bond is established during the multiple phases of dehydrogenation [26]. The catalytic support surface chemistry also plays a crucial role in the MOR activity of the catalysts. The use of nanostructured materials prevents catalyst poisoning from the intermediates which then can improve the MOR process. In the case of DMFC, electrochemical methanol oxidation occurs at the anode. The overall methanol oxidation reaction in the anode involves a six-electron transfer accompanied by surface intermediates which is given as follows [27, 28],

$$CH_3OH + H_2O \rightarrow CO_2 + 6H^+ + 6e^- \tag{8.1}$$

A schematic representation of possible MOR pathways and their corresponding intermediates is shown in figure 8.1 [4]. It can be noted from figure 8.1 that each carbonaceous compound produces a proton/electron pair except the intermediates. Apart from this, water molecules near the anode will also release a proton/electron pair in the form of hydroxyls which then combines with the intermediates. As mentioned earlier the formation of CO after multiple adsorption/desorption depletes the rate of MOR and the stability of the catalyst by blocking the active sites [29, 30]. The formation mechanism of CO during the MOR of a Pt catalyst is provided in figure 8.2. In an indirect reaction, CHO or COH undergoes a direct dehydrogenation which results in the formation of CO. During the formation of CHO or COH due to the hydrogen abstraction of formaldehyde or hydroxymethylene an OH species is formed by extracting a proton/electron pair from the H_2O molecule. The OH group then reacts with the carbonaceous species to make di-oxygenated species or formic acid. CO_2 is produced during the dehydrogenation process when OH is added to either a carboxyl or a formate group [4, 31]. There is also another pathway for CO_2 production which involves the formation of formic acid or dioxymethylene

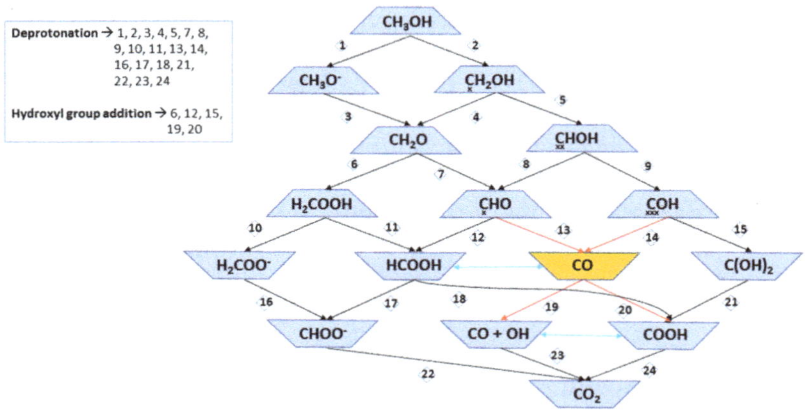

Figure 8.1. Possible methanol oxidation reaction routes (reaction scheme). Reproduced from [4] CC BY 4.0.

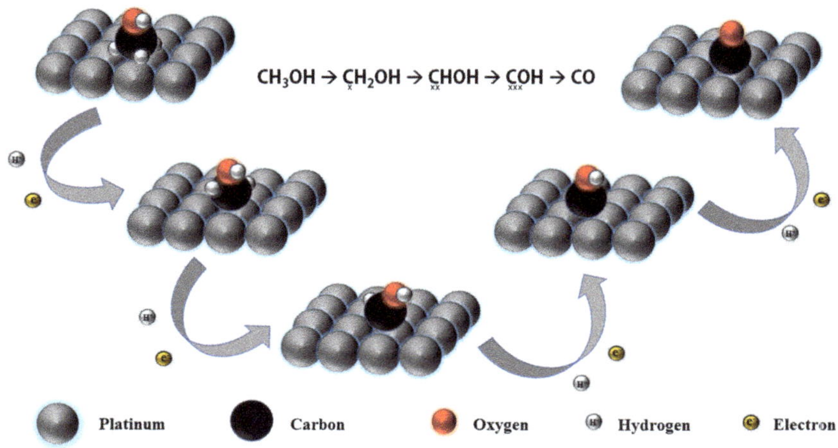

Figure 8.2. Figure depicting the sequential stripping of hydrogen atoms during methanol oxidation in a Pt catalyst. Reproduced from [4] CC BY 4.0.

intermediate when OH addition creates H_2COOH during dehydrogenation. Further dehydrogenation of dioxymethylene will eventually produce CO_2 [32]. It is crucial to understand the MOR kinetics. There are several tools available to evaluate the overall MOR occurring in the electrochemical cell. Therefore, comprehension of these tools is vital in understanding MOR and DMFC.

8.2.1 Polarization curve

The polarization curve is a plot of cell voltage versus current density that offers information about the cell and the anodic and cathodic losses in the cell [32]. Losses like mass transport loss, activation losses and ohmic loss of the cell can be

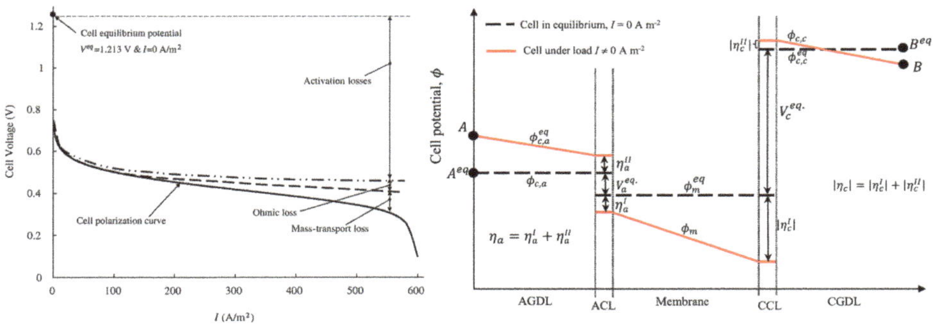

Figure 8.3. (a) A schematic illustration of a polarization curve, and (b) a schematic diagram showing the potential distributions for a DMFC at equilibrium ($I = 0$ Am^{-2}) and at non-equilibrium ($I \neq$ Am^{-2}) conditions, reprinted from [27], Copyright (2023), with permission from Elsevier.

determined using a polarization curve (figure 8.3(a)) [27, 28]. The absence of reactants at the catalyst sites creates mass transport loss which is very noticeable at the high current density regions were meeting the high demand for reactants at catalyst sites is challenging. The voltage loss of the cell is caused by mass transport loss at the limiting current density when either oxygen at the cathode catalyst layer (CCL) or methanol at the anode catalyst layer (ACL) is lowered. In the case of ohmic loss, the potential drop occurs as a result of the resistance in electric charge flow through the medium. In DMFCs, the ohmic loss is usually contributed by the transport of H^+ through the membrane and the catalyst while electronic contribution is often neglected. Activation loss causes the potential loss of the cell at low current density values. It is the required potential of the reaction to move from the equilibrium conditions. The activation loss at the anode is slightly larger than at the cathode due to the slow kinetics of MOR [27].

8.2.2 Kinetics

Kinetics deals with the reaction rates of the non-equilibrium conditions of the electrochemical mechanism. Commonly, there are two types of kinetic modelling approaches exist in DMFC, the Butler–Volmer equation and the Tafel assumption [33]. We will discuss these two fundamental models that are commonly used to explain electrochemical kinetics followed by non-Tafel kinetics.

8.2.2.1 Butler–Volmer model of kinetics

The Butler–Volmer model can explain the anodic and cathodic reaction kinetics of an electrochemical cell. For a generic electrochemical equilibrium reaction of

$$\mathrm{Oe^{-}} \leftrightarrow \mathrm{R} \tag{8.2}$$

The Arrhenius equation may be employed to represent the forward and backward reaction rate constant (k, m^2 s^{-1}),

$$k = \text{A. exp}\left(\frac{-E_\text{A}}{R_\text{U}T}\right) \tag{8.3}$$

where A is the frequency factor, T is the temperature, R_U is the universal gas constant, and E_A represents the activation energy. When the fundamental concept of activation potential is applied to the Arrhenius equation while neglecting the mass transport limit, then the kinetics behind equation (8.2) can be described using the Butler–Volmer reaction,

$$J = aJ_0\left[\exp\left(-\frac{\alpha_c F}{R_\text{U}T}\eta\right) - \exp\left(-\frac{\alpha_a F}{R_\text{U}T}\eta\right)\right] \tag{8.4}$$

where, J is the rate of reaction, a is the electrochemical area/unit volume of the catalyst, J_0 corresponds to the exchange current density, η denotes the overpotential (in equation (8.4), $\eta = V - V^{\text{eq}}$) and α is the transfer coefficient where a denotes the anodic reaction (backward, $\eta > 0$) and c denotes cathodic reaction (forward, $\eta < 0$) (in equation (8.4), $\alpha_a + \alpha_c = 0$). It is to be noted that, even though the anodic reaction rate is negative, the absolute value of the rate is considered [27].

8.2.2.2 Tafel kinetics

Tafel kinetics defines the scenario in which the reaction in equation (8.2) transitions in such a strange way that one of the terms in the right-hand side of equation (8.4) is significantly smaller than the other. The commonly accepted Tafel kinetics of the oxygen reduction reaction are obtained by including mass transport and assuming the half-cell's negligible CCL reaction (backward), as shown in equation (8.5).

$$CH_3OH + \frac{3}{2}O_2 \leftrightarrow 2H_2O + CO_2 \tag{8.5}$$

So, the obtained Tafel kinetics is,

$$J_{O_2} = \alpha J_{0,\,O_2}\cdot\left(\frac{C_{O_2}}{C_{O_2}^{\text{Ref}}}\right)\cdot\exp\left(-\frac{\alpha_c F}{R_\text{U}T}\eta_c\right) \tag{8.6}$$

where C is the concentration at the reaction sites while superscripted Ref denotes the reference value. Tafel kinetics frequently used in MOR but it is not quite appropriate for sluggish MOR [27, 34–38]. Similar to ORR, Tafel kinetics of MOR can be expresses as,

$$J_{\text{MeOH}} = \alpha J_{0,\,\text{MeOH}}\cdot\left(\frac{C_{\text{MeOH}}}{C_{\text{MeOH}}^{\text{Ref}}}\right)\cdot\exp\left(-\frac{\alpha_a F}{R_\text{U}T}\eta\right) \tag{8.7}$$

Another key factor in the MOR kinetics is the relation between overpotential (η) and the phase potentials. Figure 8.3(b) depicts a conceptual representation of local potential distributions at equilibrium ($I = 0$ Am^{-2}) and non-equilibrium ($I \neq 0$ Am^{-2}) conditions. Parameters like ϕ_c and ϕ_m denote electrode potential and membrane potential, respectively. Here, an assumption of no methanol crossover through the membrane is considered. At equilibrium the cell shows no potential loss, all the potential lines are horizontal (dashed line) in figure 8.3(b). The overall cell equilibrium voltage (V_{cell}^{eq}) is defined as the ϕ_c difference between the two ends (A and B) of the cell.

$$V_{cell}^{eq} = \phi_{c,\,c}|_{B^{eq}} - \phi_{c,\,a}|_{A^{eq}} = V_c^{eq} - V_a^{eq} = 1.213\ \text{V} \tag{8.8}$$

When the cell is under a load ($I \neq 0\ \text{Am}^{-2}$), electrons and protons transport throughout the medium and there will be some potential losses (continuous lines in figure 8.3(b)) [4]. The overpotentials of anode and cathode catalyst layers are given as,

$$\eta_c = \phi_{c,\,c} - \phi_m - V_c^{eq} \tag{8.9a}$$

$$\eta_a = \phi_{c,\,a} - \phi_m - V_a^{eq} \tag{8.9b}$$

8.2.2.3 Non-Tafel kinetics for MOR

The non-Tafel kinetic model is used when the MOR is carried out in a multi-step process with each step following Butler–Volmer model. There are reports of various sets of multi-step MOR kinetics that have been reported over the years (37–40). The well accepted MOR non-Tafel kinetic model is achieved by taking account of the following elementary and intermediate stages,

$$CH_3OH_{Reservoir} \rightarrow CH_3OH_{ACL} \tag{8.10a}$$

$$CH_3OH_{ACL} + Pt \rightarrow CH_3OH_{ad,Pt} \tag{8.10b}$$

$$CH_3OH_{ad,Pt} \rightarrow CO_{ad,Pt} + 4H^+ + 4e^- \tag{8.10c}$$

$$H2O + Ru \leftrightarrow OH_{ad,\,Ru} + H^+ + e^- \tag{8.10d}$$

$$CO_{ad,Pt} + OH_{ad,\,Ru} \rightarrow CO_2 + H^+ + e^- + Pt + Ru \tag{8.10e}$$

Here the subscript 'ad' corresponds to the surface adsorption of the catalyst mentioned. The common non-Tafel kinetics of MOR is obtained by implementing the Butler–Volmer relation to the above reactions [39, 40],

$$J_{MeOH} = \alpha J_{0,MeOH} \cdot \frac{C_{MeOH}}{C_{MeOH} + \Gamma \exp\left(\frac{\alpha_a F}{R_U T}\eta\right)} \cdot \exp\left(\frac{\alpha_a F}{R_U T}\eta\right) \tag{8.11a}$$

$$\text{where the kinetic constant } \Gamma = K \exp\left(\frac{\alpha_a F}{R_U T} V_a\right) \tag{8.11b}$$

where K is calculated from the reaction rate constant of equation (8.10c) and (8.10e). Non-Tafel kinetic provides a considerable concentration dependency of the factor $\frac{c_{MeOH}}{c_{MeOH} + \Gamma \exp\left(\frac{\alpha_a F}{R_U T}\eta\right)}$ to the MOR in comparison to Tafel kinetics (equation (8.7)). Tafel kinetic provides the linear relation between reaction rate and methanol concentration using an experimentally fitted reference value of C_{MeOH}^{Ref}.

8.3 Two-dimensional (2D) materials as an anode catalyst

The discovery of graphene has inspired significant scientific interest in the realm of 2D materials. These are layered materials consisting of sp^2 hybridized carbon atoms with hexagonal structure having stronger σ in-plane bonding than the weaker out-of-plane π bonding. The electronic conductivity increases due to the presence of mobile networks. Graphene and/or graphene nanocomposites as electrocatalysts/supports with noble metal nanoparticles, metal oxides, nanocrystals, and so on have received a great deal of theoretical and experimental attention. Other 2D materials beyond graphene such as MXene, TMDs and so on have demonstrated outstanding durability and catalytic activity as an electrocatalyst support due to their large conductivity, surface area, stability, selectivity and tuneable properties. Thus, 2D structures give the appropriate composites with good dispersion and large electrochemical surface area (ECSA), which are clearly advantageous to the electrocatalytic stability and performance for DMFCs.

8.3.1 Graphene-based anode catalysts

The unique aspects of properties of graphene include high surface area (up to 2675 $m^2\ g^{-1}$), high mechanical strength of approximately 1 TPa Young's modulus, thermal (3080–5150 $W\ mK^{-1}$), high electrical conductivity (0.5–100 $S\ m^{-1}$) and high electron mobility at room temperature making it an attractive option for electrocatalysis and other applications requiring good thermal and electrical conductivity. According to their shape, size, number of layers, and dopant presence, various graphene configurations exhibit various properties. Graphene exhibits superiority over other dimensions carbon nanostructures as a result of remarkable electrochemical characteristics. The properties of graphene vary based on its dimensions, morphology, later number and the presence of defects. Graphene's competitors include fullerenes, carbon nanotubes (CNTs) and carbon quantum dots, but their scalability prevents them from outperforming graphene in electrochemical applications.

As is well known, the current state-of-the-art of electrocatalyst for MOR are made of metal nanoparticles such as platinum (Pt), palladium (Pd) and rhodium (Rh) and so on, dispersed on graphene. When compared to commercial pristine Pt/c, graphene-supported Pt/c outperforms it. Similarly, graphene aerogel-supported Pd exhibits higher activity than commercial Pd/C. Modern electroactive metals Pt and Pd are utilized in the oxidation of fuels, but because they are easily contaminated by intermediates like CO, researchers have been forced to create corrosion-resistant alloys like Pt–Fe, Pt–Co, Pt–Ni, Pd–Fe, Pd–Co and Pd–Ni in order to increase the efficiency and stability of the catalyst. For methanol oxidation fuel cells, Pt- and Pd-loaded graphene-supported electrocatalysts such as PtCo/reduced graphene oxide (rGO), PdCo/rGO, Pd/rGO and PtPd/rGO have also been produced. Additionally, adding dopants such as nitrogen, boron, phosphorus and sulfur to replace the carbon atoms in graphene enhances graphene's conductivity and electrochemical characteristics.

8.3.1.1 Pt/GO and Pt/rGO hybrids for MOR

Several research works published since 2009 that used graphene/Pt nanoparticles as an electrocatalyst for MOR have opened up new avenues for developing electrocatalysts for fuel cells. Yoo *et al* compare the performance of Pt/GO and Pt/carbon black catalyst for the first time. This study shows that carbon atoms containing functional groups such as oxygen or dangling bonds act as a defect site in graphene nanosheets and that the metal salt precursor reacting with these defects results in a stable electrocatalyst. The CV curve shown in figure 8.4(a) demonstrates that compared to Pt/carbon black catalyst, Pt/GNS shows excellent current response. The catalyst tolerance (ratio of forward peak current density (I_f) to the reverse anodic peak current density (I_b)) was greater (1.47) than Pt/carbon black catalyst (1.27). A low I_f/I_b ratio indicates inadequate anodic methanol-to-carbon dioxide oxidation and excessive carbonaceous residue deposition on the catalyst surface. As a result, when compared to Pt/carbon black in the forward scan, a significant fraction of Pt/GNS intermediate carbonaceous species was oxidized to carbon dioxide. These small Pt clusters were clearly visible in TEM micrographs (figure 8.4(b)), indicating a significant interaction between Pt NPs and GNS. Thus, even after a heating treatment at 400 °C, this hybrid aids in the prevention of platinum cluster aggregation [41, 42].

GO films or sheets are typically insulating and generally exhibit sheet resistances of 10^{12} Ω/sq or greater. The lattice defects and functional groups help to modify the graphene's electronic structure and act as a scattering centre, influencing electrical transport. As a result, GO reduction is concerned not only with the atomic-scale lattice defects and the removal of oxygen-containing functional groups attached to graphene, but also with restoring the graphitic lattice's conjugated network. The restoration of graphene's electrical conductivity and other properties is the result of

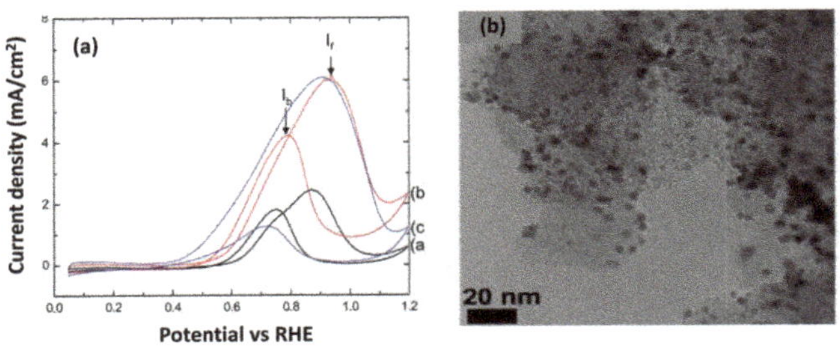

Figure 8.4. (a) *I* versus *V* graph for Pt/GNS (red colour) at room temperature at scan rate 5 mV s^{-1}, Reprinted with permission from [41]. Copyright (2023) American Chemical Society.

these structural changes.[43] Considering these aspects, Seger *et al* synthesized partially reduced GO-Pt catalyst, using a borohydride reduction They followed a two-step reduction method, where the first step involves reduction of GO using $NaBH_4$ along with Pt particles deposition followed by reduction using hydrazine hydrate. They also found out that this partially reduced GO-Pt which is casted on GC with the exposure to hydrazine hydrate shows instability and is sloughed from the GC electrode surface. As a result, they used Torray carbon paper as the electrode substrate, and the FESEM pictures (figure 8.5(a)) demonstrate that graphene sheets that are adhered between the carbon fibres of Torray carbon paper, as well as that Pt nanoparticles are adorned over GO with minimal agglomeration. The electrocatalyst performance of this rGO-Pt in a hydrogen fuel cell emphasises the significance of graphene as an effective support material in the building of an electrocatalyst [42]. Keeping the importance of rGO in the fuel cells, Luo and group synthesized branched Pt nanowires (BPt) using formic acid as reductant (figure 8.5(b)). The TEM images (figure 8.5(c)) clearly demonstrate that (BPt) was grown over rGO under optimal room temperature conditions. This hybrid shows MOR activity around 1.154 mA cm^{-2} at 700 mV which is comparatively higher than the graphene/Pt nanoparticles with activity around 0.120 mA cm^{-2} with CO tolerance around 1.01 [44]. Similarly, Sharma *et al* used a microwave-assisted polyol approach to synthesize Pt/rGO hybrids. They used ethylene-glycol as both a dispersion and reducing agent for both Pt nanoparticles and GO in this one-pot synthesis technique. The ECSA quantifies the catalytically active sites accessible for an electrochemical reaction while also accounting for the conductive channel available for electron flow to and from the electrode surface. The ECSA (figure 8.5(e)) calculated was around 43.1 m^2 g^{-1} for this Pt/rGO hybrid and this

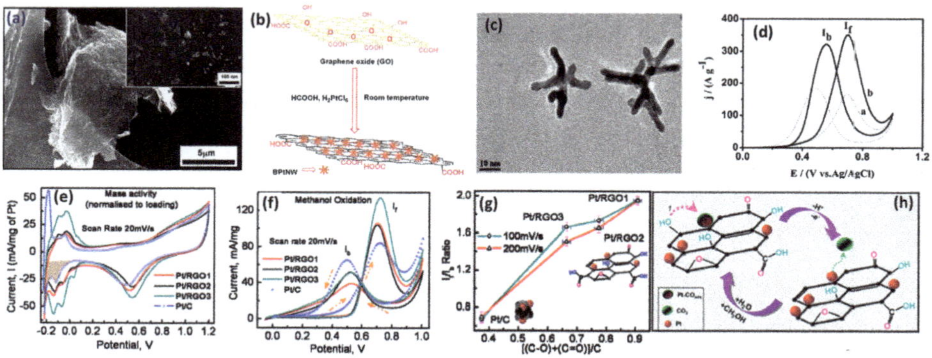

Figure 8.5. (a) FESEM images of GO/Pt deposited on carbon Torray paper. Reprinted with permission from [42]. Copyright (2009) American Chemical Society. (b) Schematic illustration showing the procedure for synthesis of BPtNW/rGO composites. (c) TEM images of BPtNW/rGO composites. (d) CV of BPtNW/rGO composites at scan rate of 50 mV s^{-1} reproduced [44] with permission of the Royal Society of Chemistry. (e) CV curves of Pt/rGO hybrids and Pt/C and (f) in 1 M H_2SO_4 + 4 M CH_3OH (g) dependence of I_f/I_b ratio with respect to the residual oxygen species. (h) Schematic illustration demonstrating the conversion of adsorbed CO_{ads} species to CO_2 using Pt/rGO hybrids. Reprinted with permission from [45]. Copyright (2010) American Chemical Society.

enhancement can be due to the ease of charge transfer at the Pt/rGO interfaces. The forward anodic current here appearing around 0.7 V is due to the methanol oxidation and in the backward scan the oxidation peak appears at 0.5 V, which can be due to the oxidation of CO_{ads} species and when compared to Pt/rGO, Pt/C exhibits lower I_f revealing the higher methanol oxidation activity (figure 8.5(f)). Figure 8.5(g) shows that Pt/rGO has better I_f/I_b ratio, demonstrating excellent ant-poisoning behaviour. Thus, the Pt/rGO remarkable anti-poisoning activity has been linked to the type and surface density of oxygen groups which are covalently bounded to the rGO support. The presence of extra oxygen groups on the surface of graphene can increase the oxidation of CO adsorbed (CO_{ads}), on the Pt sites using the bifunctional mechanism. The proposed mechanism and equations are given in figure 8.5(h) and the following equations (8.12a) and (8.12b):

$$\mathrm{rGO} + \mathrm{H_2O} \rightarrow \mathrm{rGO} - (\mathrm{OH})_{ads} + \mathrm{H}^+ + e^- \text{ Forward scan} \quad (8.12a)$$

$$\mathrm{Pt} - \mathrm{CO}_{ads} + \mathrm{rGO} - (\mathrm{OH})_{ads} \rightarrow \mathrm{CO_2} + \mathrm{H}^+ + \mathrm{Pt} + \mathrm{rGO} + e^- \text{ Reverse scan} \quad (8.12b)$$

As shown in equations (8.12a) and (8.12b), the water molecules adsorbed on the rGO dissociate, generating rGO-(OH) ads near the Pt nanoparticles, which rapidly oxidize CO_{ads} groups on the periphery Pt atoms. The water molecules adsorbed on the rGO undergo dissociation forming rGO-$(OH)_{ads}$ near the Pt nanoparticles, which rapidly oxidize CO_{ads} groups on the peripheral Pt atoms as per equations (8.12a) and (8.12b). rGO's hydrophilic properties enhances water activation and is the primary driving force in this mechanism. Compared to Pt/C, where oxygen groups are scarce, CO_{ad} oxidation is known to proceed through the less effective Langmuir–Hinshelwood (LH) mechanism. The lower efficiency is owing to the high overpotential required for the formation of oxygen-containing species (mostly OH) on the Pt surface via water activation. However, for the Pt/RGO system, it is evident that the process described by equations (8.12a) and (8.12b) would dominate over the LH mechanism, aided by the close proximity of the two systems [45].

Because of the excellent mass transportation ability and reaction kinetics, vertically aligned few-layered graphene (FLG) nanoflakes have gained interest in the field of electrochemical catalysts [46]. Sion *et al* demonstrated that these FLG nanoflakes synthesized using microwave plasma enhanced chemical vapour deposition (MPECVD) technique can be used as an electrocatalyst for MOR. This electrode using FLG nanoflakes demonstrated excellent electron transfer kinetics in the ferrocyanide redox system with an electron transfer rate nearly ΔE_p, of 60 mV. They deposited Pt nanoparticles of size 6 nm diameter over these FLG utilizing magnetron DC sputtering. These electrodes show mass specific peak current density of nearly 62 mA mg cm^{-2} with high I_f/I_b ratio of nearly 2.2 [47]. Similar to this, 3D FLG naoflakes (3D) graphene aerogel (GA) having microporous characteristics has developed as a new generation supporting material capable of greatly reducing graphene restacking while also promoting noble metal nanoparticles deposition. Thus, Li and colleagues used a hydrothermal procedure to construct a 3D architecture composed of graphene aerogel doped with B and N atoms and CNTs. The doping with heteroatoms also helps to improve the noble metal support interaction and at the same time guarantees Pt nanoparticles' uniform dispersion. The FESEM and TEM in figures 8.6(a)–(c) clearly show the interconnected and well-defined 3D architectures with pore size 10 μm. The HRTEM images given in figures 8.3(d) and (e) clearly demonstrate the lattice fringes of both Pt and N, B

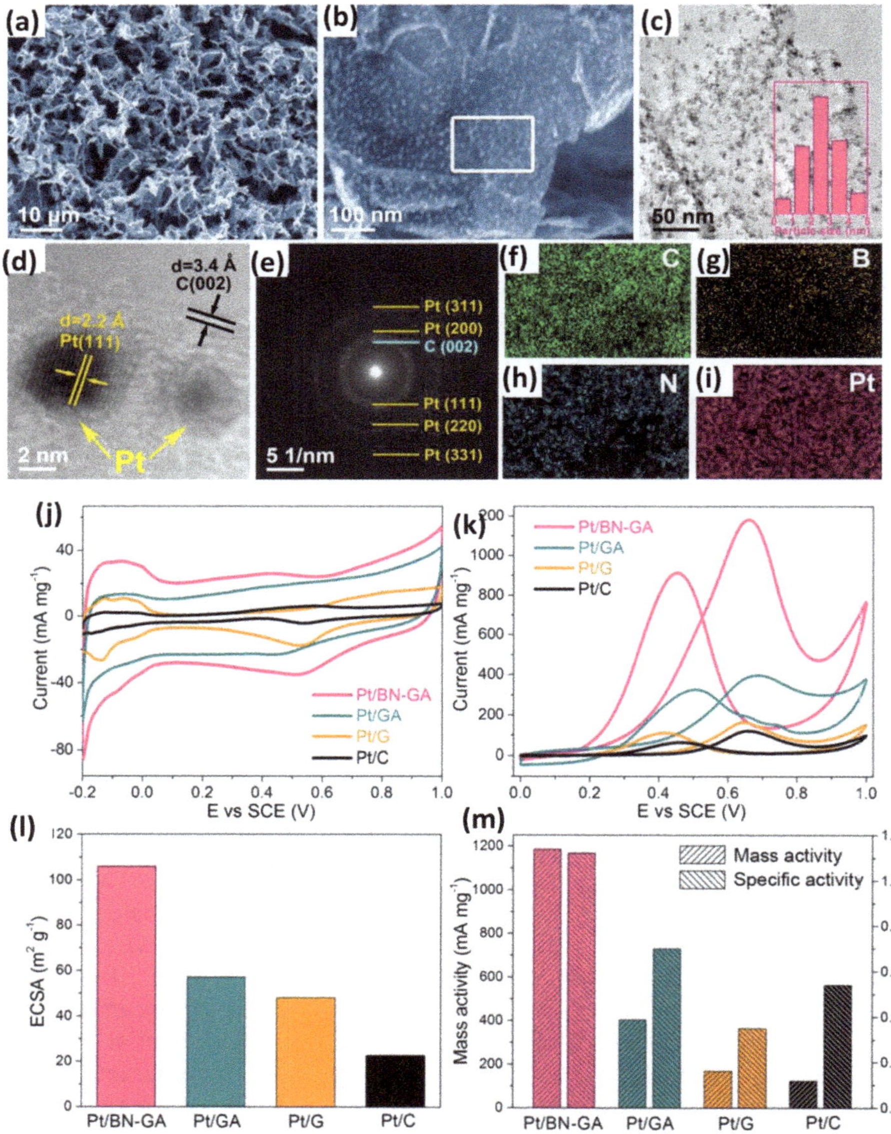

Figure 8.6. (a)–(c) FESEM and HRTEM images of Pt/BN-catalyst, (d and e) HRTEM and SAED pattern of Pt/BN- catalyst with (f–i) elemental mapping, CV curves in (j) 0.5 M H_2SO_4 (k) 0.5 M H_2SO_4 + 1 M methanol (l) ECSA values (m) mass activities. Reprinted with permission from [48]. Copyright (2018) American Chemical Society.

doped graphene and energy-dispersive x-ray (EDX) (figures 8.6(f)–(i)) studies found out that this 3D Pt/BN-GA is made up of B, C, N and Pt. The ECSA calculated from CV (figures 8.6(j)–(l)) in the potential range of −0.2 to 0.1 V show value around 106 $m^2\ g^{-1}$ surpassing commercially available catalysts, such as Pt/C, Pt/G, and Pt/GA. This 3D architecture shows high mass activity around 1184.5 mA mg^{-1} with large ECSA-normalized specific activity around 1.12 mA cm^{-2} (figure 8.6(m)) [48]. Similarly, Yan *et al* synthesized another 3D architecture hybrid containing CNTs and nitrogen doped graphene aerogel using a conventional self-assembly process. This 3D architecture was found to be an excellent electrocatalyst for MOR with mass activity around 871.9 mA mg^{-1} [49]. Zhai *et al* demonstrated that the catalytic activity rGO can be enhanced by its hybrid formation with TMDs such as MoS_2. Zhai and group synthesized Pt supported MoS_2/rGO for MOR. The electrochemical activity of the as-prepared Pt-MoS_2/rGO hybrid was 1.7 and 5.6 times higher than that of commercial Pt-MoS_2 and Pt/C, respectively, indicating its exceptional commercialization potential [50].

8.3.1.2 Pd/GO and Pd/rGO hybrids for MOR

Due to methanol crossover, poor anode kinetics, and CO adsorbate poisoning, the most extensively employed Pt electrocatalysts have lost competence. In fact, Pd is similar to Pt in many ways, and it is regarded as the best substitute for Pt as the MOR electrocatalyst because of its exceptional performance and plentiful availability. Pd-based materials are also less expensive than Pt-based materials and have a high CO tolerance. In alkaline conditions, Pd-based materials exhibit extraordinary electrocatalytic activity while in acidic media, there is essentially no reaction. Furthermore, graphene nanosheets with high electrical conductivity promote charge mobility during the MOR, which is required for improving the electrocatalytic activity of Pd-based catalysts and improving CO resistance. The numerous functional groups on graphene's surface are also advantageous for attaching Pd and modifying the electrocatalysts morphology. It is widely established that Pd-based electrocatalysts are almost inert for MOR under acidic environments. Furthermore, methanol may be virtually electrooxidized at Pd-based materials in alkaline solutions using the following hypothesised mechanism:

$$Pd + CH_3OH \rightarrow Pd - CH_3OH_{ads} \tag{8.13a}$$

$$Pd + OH^- \rightarrow Pd - OH_{ads} + e^- \tag{8.13b}$$

$$Pd - CH_3OH_{ads} + 4OH^- \rightarrow Pd - CO_{ads} + 4H_2O + 4e^- \tag{8.13c}$$

$$Pd - OH_{ads} + Pd - CO_{ads} + OH^- \rightarrow 2Pd + H_2O + CO_2 + e^- \tag{8.13d}$$

Overall reaction:

$$CH_3OH + 6OH^- \rightarrow 5H_2O + CO_2 + 6e^- \tag{8.13e}$$

The increased amounts of OH_{ads} and OH^- on the electrode surface react with CO_{ads}, leading to enhanced electro-performance and tolerance to hazardous intermediates via the Pd metal active surface. The first stage of electrooxidation on the Pd surface is recognised as the creation of Pd–OH_{ads}. The chemisorbed OH^- ions have the ability to oxidize the Pd metal, producing higher valence Pd-oxides with higher potentials.

Huang *et al* for the first time synthesized a hybrid containing Pd and low-defect graphene (LDG) sheets using a soft-chemical method [45]. Liu *et al* synthesized Pd nanoparticles (figure 8.7(a)) having size in the range of 5.5 nm that are well dispersed

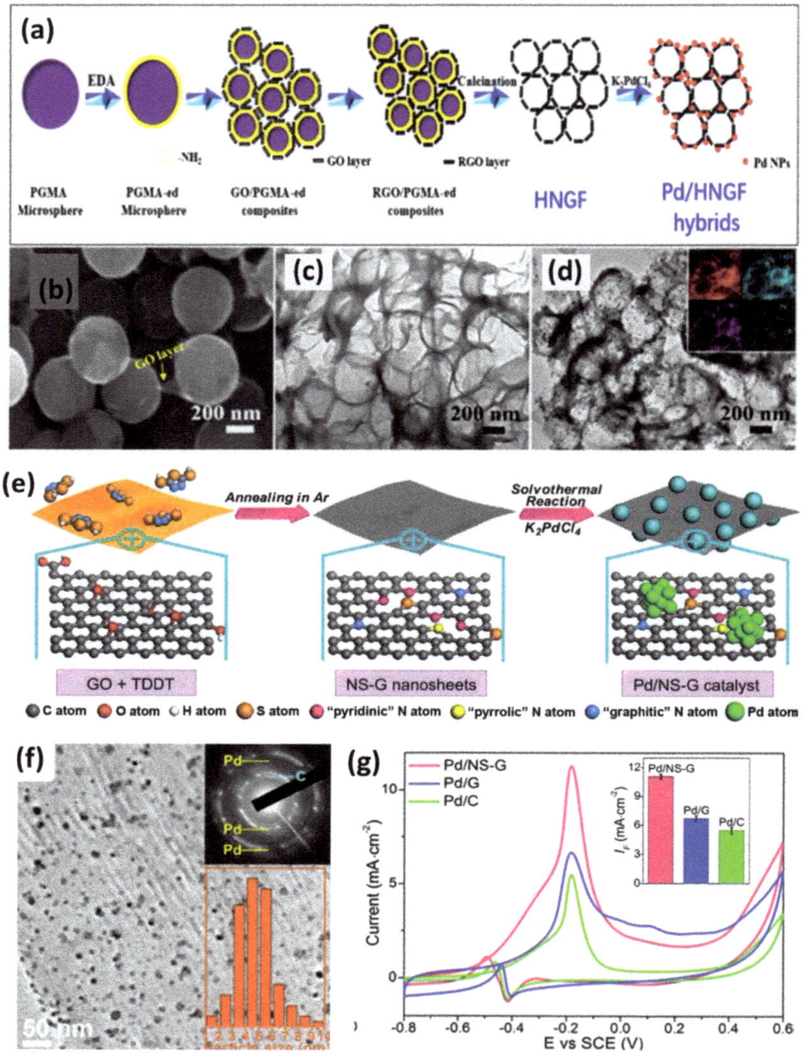

Figure 8.7. (a) Schematic illustration representing the synthesis of Pd/HNGF, (b)–(d) FESEM, TEM and mapping images of Pd/HNGF. Reprinted from [51], Copyright (2016), with permission from Elsevier. (e) Schematic of the synthesis for the Pd/NS-G electrocatalyst. (f) HRTEM images of Pd/NS-G electrocatalyst. (g) CV curve of Pd/NS-G electrocatalyst 0.5 M NaOH + 1 M methanol, reprinted with permission from [54]. Copyright (2016) American Chemical Society.

over N-doped graphene frameworks (HNGF) with the help of poly (glycidyl methacrylate) (PGMA) microspheres as templates. This LDG-supported Pd has more ECSA with more stability. As a result, the Pd/HNGF electrocatalysts enhance MOR activity and have applicability in various chemical processes [51]. Furthermore, Yang *et al* demonstrated the importance of shape-controlled Pd nanoparticles (NPts) placed on graphene surfaces in the design of an electrocatalyst for DMFCs. The Pd/GN exhibits improved electrocatalytic activity because of the electron transport capabilities of graphene and the large surface area and mass ratio of Pd nanoparticles with active crystal facets (110). Pd nanoparticles with an average size of 100 nm are uniformly connected and spread on a wrinkle structure graphene nanosheet. This Pd/GN electrocatalyst outperforms the commercial available catalyst Vulcan XC-72 in terms of stability and tolerance for MOR [52]. In another study, Sawangphruk and group deposited Pd utilizing a direct electrodeposition method in different carbon supports. Here the Pd deposited over carbon supports containing rGO shows enhanced performances. The collaborative impact of 3D ultra porous Pd nanocrystals and fast electron transport due to the large conductivity of rGO nanosheets are thought to play an essential role in improving the catalytic efficiency of Pd/rGO/CFP towards methanol oxidation in alkaline medium [53]. Considering the electrocatalytic of dual-doped graphene nanosheet, Zhang *et al* synthesized a hybrid containing N and S dual-doped graphene and Pd nanoparticles. This dual doping is obtained by utilizing a thermal treatment method followed by solvothermal approach for the growth Pd nanoparticles over dual-doped graphene (figure 8.7(e)). The HRTEM reveals (figure 8.7(f)) atomic lattices of N–S doped graphene and fcc of Pd crystals. This Pd/NS-G (figure 8.7(g)) exhibited highest activity with strong anodic and cathodic peak around −0.2 and −0.4 V versus saturated calomel electrode [54].

8.3.1.3 Rh/GO and Rh/rGO hybrids for MOR

Currently, Pt-based materials are currently the most widely utilized commercial electrocatalysts in the field of fuel cells, despite their high cost, paucity of resources, and limited resistance to poisoning byproducts (e.g. CO) that limit their applications. As a result, major research efforts have been dedicated towards the development of diverse Pt-alternative species that are more affordable. According to recent scientific studies, rhodium (Rh) nanoparticles exhibit competitive catalytic performance and resistance to intermediate CO species [55].

Scattering Rh nanoparticles on appropriate supports, which can offer enough anchoring sites to firmly immobilize Rh nanoparticles and at the same time restrict their overgrowth, is a successful strategy for strengthening the Rh-based catalysts. In this context, noble metal carriers made of carbonaceous materials including CNTs, carbon black and graphene have been used extensively. Kang *et al* for the first time synthesized a MOR electrocatalyst based on Rh/rGO hybrids using a hydrothermal approach. Compared to the commercial Pt/C electrocatalyst, the onset oxidation potential of the MOR at the Rh-NSs/rGO changes negatively by around 120 mV. At 0.61 V potential, the MOR current in the Rh-NSs/rGO heterostructure is 3.6 times more than in the conventional Pt/C electrocatalyst. Furthermore, chronoamperometry experiments show that the Rh-NSs/rGO hybrids have exceptional MOR stability. The electrochemical data demonstrate that the Rh-NSs/rGO hybrids are an exceedingly promising Pt-alternative anode electrocatalyst for the MOR in alkaline medium [56].

8.3.1.4 Non-metal/GO/rGO based catalyst

Development of a cheap, non-noble and highly active electrocatalyst for MOR is hugely desirable. Considering these aspects, Narayanan *et al* synthesized highly electroactive $NiCo_2O_4$/rGO nanorods using a template-free hydrothermal approach followed by calcination. $NiCo_2O_4$/rGO nanorods show a current density of around 38 mA cm^{-2}, which is 12.5 times larger than pristine Co_3O_4/rGO and 2.5 times greater than NiO/rGO. The mechanism is as follows:

$$NiCo_2O_4 + OH^- + H_2O - 3e^- \rightarrow NiOOH + 2CoOOH \quad (8.14a)$$

$$NiOOH + CH_3OH + 1.25O_2 \rightarrow Ni(OH)_2 + CO_2 + 1.5H_2O \quad (8.14b)$$

$$2CoOOH + 2CH_3OH + 2.5O_2 \rightarrow 2Co(OH)_2 + 2CO_2 + 3H_2O \quad (8.14c)$$

This enhancement in the electrochemical performance of this electrocatalyst can be due to

(1) The mixed valence state of $NiCo_2O_4$ results in two active centres made up of redox pairs such as Ni^{2+}/Ni^{3+} and Co^{2+}/Co^{3+}.
(2) Due to short diffusion distance, the $NiCo_2O_4$ nanorod provides an effective transport pathway for electrons and ions.
(3) During the reaction process, the $NiCo_2O_4$ porous structure acts as a pathway for the rapid exchange of materials.
(4) The π–π conjugation present in the sp^2 hybridized carbon atom provides a conduction channel to enhance the electron transfer rate and thereby improving the electrode kinetics and methanol oxidation efficiency.

Recently, ReS_2 has been considered as a competitor to MoS_2. The weak van der Waals force of attraction between its layers results in excellent electrocatalytic performance over other TMDs and allows ReS_2 to be accessed as a single layer. Additionally, the separated ReS_2 layers increase the exposed edge sites and catalyst's specific surface area. Considering the benefits of ReS_2, Askari *et al* synthesized a binary hybrid containing ReS_2 and Rgo that demonstrated its potential application for MOR. They found out that this binary hybrid exhibited current density of around 38 μA cm^{-2} in acidic media and 198 μA cm^{-2} in alkaline media [57].

8.3.2 MXene-based anode catalyst for MOR

One of the most potential applications of MXenes is as a long-lasting catalyst supporting material for anodic/cathodic processes in the fuel cell. MXene-supported catalysts are currently been recognized as a promising and stable support for different Pt and non-Pt metals suitable for cathodic oxygen reduction reactions, and they have been demonstrated as a good choice for improving oxygen reduction kinetics. This has prompted researchers to investigate MXene as a catalyst support for MOR processes [58].

8.3.2.1 Pt supported MXene for MOR

Wang *et al* published the first study on Pt decorated Ti_3C_2 MXene for improved methanol oxidation. The Pt/Ti_3C_2 catalyst demonstrated very large current density compared to commercial Pt/C catalyst. They used a $NaBH_4$ reduction technique to deposit the Pt nanoparticles over MXene. Morphological examination of the Pt/Ti_3C_2 catalyst demonstrated that the layered structure of MXene is capable of immobilizing Pt nanoparticles and is found to be equally dispersed over MXene surfaces. A strong electronic interaction was also observed between Pt and MXene that may be due to the electronic transfer that happens between Pt and MXene, which may help to enhance the chemisorption of Pt. CV curves of Pt/Ti_3C_2 show two peaks of oxidation, where the peak in forward scanning is due to the methanol oxidation and peak in backward scanning may be due to the oxidation of carbonaceous intermediates that are produced during oxidation of methanol. Furthermore, the Pt/Ti_3C_2 catalyst demonstrated exceptional stability under possible cycle conditions. The Pt/Ti_3C_2 catalyst retained 85% of the ECSA after 1000 potential cycles, this definitely indicates that Pt nanoparticles are more stable on Ti_3C_2 MXene than on carbon [59]. The surface functional groups in MXene are determined by the etching procedure employed to synthesize it from the MAX phase. For example, when the LiF/HCl technique is used, the resulting $Ti_3C_2T_x$ has a negatively charged surface due to an abundance of fluorine and hydroxyl groups, making it difficult to bind metal nanoparticles and resulting in poor dispersion. To overcome these dispersion problems, Yang *et al* functionalized MXene ($Ti_3C_2T_x$) using poly (diallyldimethylammoniumchloride) (PDDA) polymer, as shown in figure 8.8(a). The PDDA adsorbed over MXene helps to overcome the problems of restacking/aggregation and also modify the charges in the surface to enhance the

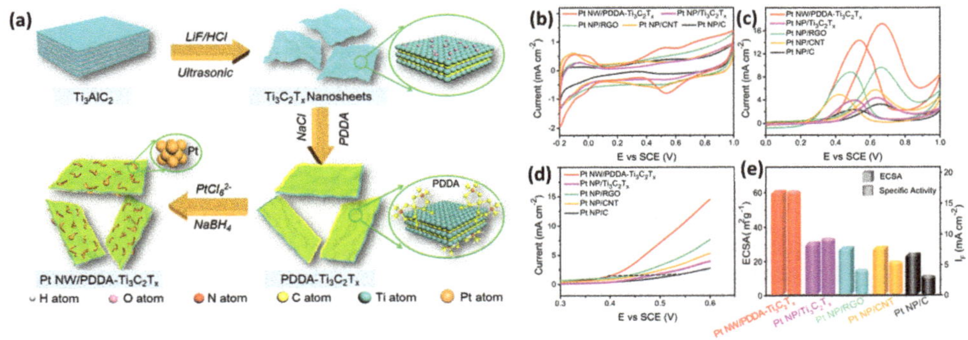

Figure 8.8. (a) Schematic of the synthesis of Pt NW/PDDA-$Ti_3C_2T_x$ catalyst, (b) CV curves of the Pt NW/PDDA-$Ti_3C_2T_x$ and other reference catalysts in 0.5 M H_2SO_4 medium at a scan rate of 50 mV s^{-1}, (c) CV curves, (d) LSV curves, (e) specific ECSA values, reprinted with permission from [60]. Copyright (2020) American Chemical Society.

electrostatic interactions with Pt precursors. This positively charge PDDA also helps in the growth of 1D Pt nanoparticles over MXene that helps to avoid the detachment problems of Pt nanoparticles and results in the stereoassembly of grain boundary-enriched Pt metal nanoparticles. Thus, due to the fascinating structural properties and improved electronic conductivity this Pt/PDDA-$Ti_3C_2T_x$ catalyst shows exceptional electrocatalytic activity with ECSA around 61 m^2 g^{-1}, which is greater than the conventional carbon-supported Pt catalyst. Also, electrocatalyst shows high current density of nearly 17.2 mA cm^{-2} (figures 8.8(b)–(e))[60]. Similarly, to overcome these issues of Pt dispersion Zhu *et al* constructed a 3D crumble $Ti_3C_2T_x$ having abundant Ti vacancies for the confinement of Pt nanoparticles using spray drying process. This Pt/$Ti_3C_2T_x$ electrocatalyst also shows enhanced performances with high mass activity up to 7.32 A $mgPt^{-1}$ [61]. Reducing the concentration of surface –OH and –F and increasing the concentration –O terminal groups in MXene help to enhance the electrochemical activity. Thus, Navjyothi and group utilized a low temperature annealing approach to enrich the –O terminal groups without forming TiO_2. This Ti–O enriched catalyst demonstrated current density of 18.08 mA cm^{-2} with excellent stability [62].

To avoid $Ti_3C_2T_x$ MXene aggregation Zhang *et al* developed a composite containing $Ti_3C_2T_x$ and acid functionalized multi-walled CNTs (MWCNTs). This hybrid may have good electrocatalytic activity because: (i) Pt nanoparticles provide more active sites, which may improve electrochemical activity, and (ii) MWCNTs act as an intercalating agent, resulting in larger interlayer spacing and area, which provides more accessibility to Pt nanoparticles and also acts as a conductive bridge between different layers of MXene. The electrocatalyst containing Pt–$Ti_3C_2T_x$/MWCNT exhibited enhanced ECSA of nearly 175 m^2 g^{-1}. This higher ECSA, results in greater MOR reaction due to the presence of more accessible active sites [63]. In another approach, Yang *et al* constructed a 3D hybrid architecture containing $Ti_3C_2T_x$ and rGO. Due to the higher anti-poisoning ability of Pt/rGO electrocatalyst it was found to be stable for a longer time [64].

8.3.2.2 Pd-supported MXene for MOR

Lang and group synthesized a hybrid of Pd/MXene electrocatalyst for MOR using a formic acid reduction approach. When compared to Pd/C, electrochemical experiments have shown that Pd/MXene catalysts exhibit increased MOR activity by around 60%. This is due to the major contribution of MXene support to favourable MOR kinetics [65]. Using a wet-chemical technique, a 2D/2D heterojunction was made using Pd nanosheets and $Ti_3C_2T_x$. This Pd NSs/MXene heterojunction provides multiple exposed active Pd atoms with an optimized electrical structure, but it also facilitates an intimate Pd/MXene interfacial interaction, resulting in a stable hybrid state. As a result, the Pd NSs/MXene heterojunction formed has good methanol oxidation properties.[66]. Considering the advantages of rGO/MXene hybrid, Zhang *et al* used this catalyst as a support for the growth of Pd using an ethylene-glycol reduction method. This hybrid also shows excellent mass activities for methanol oxidation which is 3.04 times higher than Pd/MXene and Pd/C [67]. PANI-based Pd/MXene hybrid was synthesized by Elancheziyan *et al.* This combination of MXene with electrochemically active PANI/Pd helps to enhance the electrochemical activity. This hybrid demonstrate improved electrocatalytic response to methanol oxidation with a peak current density around 291 mA cm^{-2} and exceptional cyclic stability [68].

8.3.2.3 Non-metal/MXene based catalyst

Despite the fact that Pt and Pd-based catalysts are among the finest electrocatalysts for MOR, the shortage and high price of these precious metals continue to be a barrier to the widespread use of DMFCs. Researchers created a variety of transition metal-based catalysts for different fuel cell processes, including MOR and ORR, to solve this problem. Thus, Yang and group synthesized a hybrid containing $Ti_3C_2T_x$ quantum dots (QDs), MoS_2 QDs and MWCNTs. This ternary hybrid also shows excellent MOR current density [69].

8.3.3 TMDs for MOR

Among the 2D materials, TMDs, which have similar morphology to graphene, have gained a lot of attention in the field of electrochemical conversion reactions due to their high electrocatalytic activity, unique layered structure and anti-corrosion activity and large active surface area.

Askari *et al* for the first-time synthesized a ternary hybrid containing $CoS_2/MoS_2/$ rGO electrocatalyst for MOR which shows maximum current density of around 1.68 mA cm^{-2} [70]. The same group synthesized another porous $Fe_3O@MoS_2$/rGO hybrid as an effective Pt substitute. When compared to Fe_3O/rGO catalyst, the MoS_2 present in the hybrid provided more surface area and active sites, resulting in a greater current density [71]. Beside these studies, Liu *et al* synthesized another hybrid with MoS_2@CoNi-ZIF with zeolitic imidazole framework as a substitute for Pt electrocatalyst [72]. This reported hybrid synthesis was complicated with a complex structure. Thus Gopalakrishnan *et al* reported the utilization of vertically aligned MoS_2 nanostructures grown over a Ni foam using a facie hydrothermal method as an electrocatalyst for MOR (figure 8.9(a)). This hybrid structure is made up of crumpled nanosheet arrays that are adorned with sponge-like nanoparticles. The crumpled shape of nanosheets leads to a large number of electroactive sites, which are useful for electrochemical applications. The enhancement in the current can be observed for CV curves (figure 8.9(b)) of MoS_2/NF-5 with large oxidation and reduction peak. The same catalyst demonstrated 80% of its initial value even after 5000 s (figure 8.9(c)) [73]. We have summarized the MOR electrochemical performance of 2D-based catalysts for the convenience of readers in table 8.1.

8.4 Conclusion and future perspectives

Here we have summarized the 2D material-based electrocatalyst for MOR. Various theoretical and experimental studies proved that when compared to a commercial catalyst available for MOR, 2D material-based electrocatalyst shows improved oxidation current. Beside these, 2D material shows enhanced stability. Most of the 2D materials such as graphene and MXene are suspectable to restacking, which may

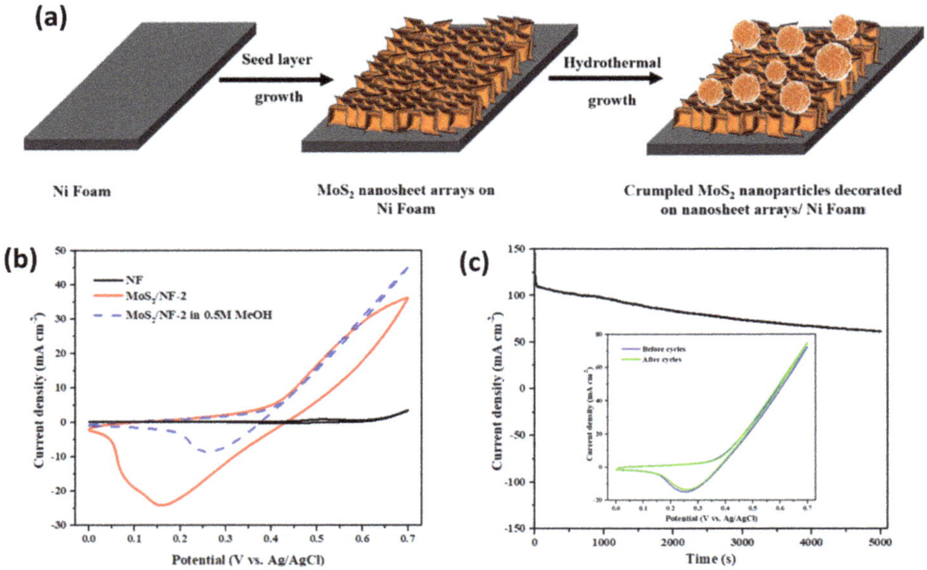

Figure 8.9. (a) Schematic representation of MoS_2/NF over Ni foam, (b) CV responses in 0.5 M MeOH and its reference catalyst (c) stability. Reprinted with permission from [73]. Copyright (2021) American Chemical Society.

reduce its stability. This can be effectively tackled by functionalization with polymers such as PDDA. Beside polymers, MWCNTs are also used as spacers to avoid the restacking. When compared to rGO it was found out that Pt nanoparticles interact more strongly with MXene nanosheets with excellent anti-CO-poisoning properties. However, reports based on MXene so far are much less common compared to those for GO. It is concluded that 2D materials would be a good candidate as a support material for fuel cell anodes; nevertheless, there are numerous issues that must be addressed before 2D materials can compete with typical metal-oxide supports.

8.4.1 Future perspectives

This chapter presents the most significant recent research on the development of 2D materials and their hybrids for MOR applications. 2D materials show excellent electrochemical activity due to their excellent properties such as mechanical robustness, large surface area and ease of mass transfer and charge migrations. Here we summarized the various 2D materials utilized for methanol oxidation reactions and summarized their performances.

(1) A thorough examination of the MOR performance of 2D materials reveals that the hybrid formation with carbon support (MWCNT, rGO) results in enhanced MOR performance and ECSA. Another possible approach may be the formation of 2D/2D heterostructure, which is less explored. While

Table 8.1. Summary of electrochemical performance of 2D materials-supported catalysts for MOR.

Catalyst	Electrolyte	Mass activity (A mg^{-1})	Specific activity (A cm^{-2})	ECSA (m^2 g^{-1})	Stability	References
Commercial Pt/C	Acidic electrolyte 0.1 M $HClO_4$ + 0.5 M CH_3OH	0.19	0.3	62.5	43.65% after 800 cycles	[74]
	Alkaline electrolyte 1 M NaOH + M CH_3OH	0.784	0.022		14.5% after 100 000 s	[75]
Pt/GNS	1 mol dm^{-3} CH_3OH + 0.05 mol dm^{-3} H_2SO_4		0.12 mA cm^{-2}			[41]
BPtNW/rGO	CH_3OH +1 M NaOH	0.299	0.1154	25.9		[44]
Pt/rGO	1 M H_2SO_4 + 4 M CH_3OH	0.157		43.1		[45]
Pt/FLG	1 M H_2SO_4 + 2 M CH_3OH			10.6		[47]
Pt/BN-GA	0.5 M H_2SO_4 + 1 M CH_3OH	1.184	1120	106	39% after 2000 s	[48]
3D Pt/$(LDCNT)_s$-(NG)s	0.5 M H_2SO_4 + 1 M CH_3OH	0.871		132.4	78.1% after 100 cycles	[49]
Pt/MoS_2-rGO	0.5 M H_2SO_4 + 1 M CH_3OH			12.87		[50]
LDG/Pd	0.5 M NaOH + 1 M CH_3OH		0.015	83		[45]
Pd/HNGF	1 M NaOH + 1 M CH_3OH			98.6		[51]
PdNPts/G	0.5 M NaOH + 0.5 M CH_3OH			81.60	83.4%	[52]
PdNPs/G				72.49	94.0%	
PdNPs/V				64.78	95.3% after 100 cycles	
Pd/rGO/CFP	2 M NaOH + 2 M CH_3OH		118.3 mA mg^{-1} cm^{-2}			[53]

(*Continued*)

Table 8.1. (*Continued*)

Catalyst	Electrolyte	Mass activity (A mg^{-1})	Specific activity (A cm^{-2})	ECSA (m^2 g^{-1})	Stability	References
Pd/NS-G,	0.5 M NaOH + 1 M CH_3OH	0.399	0.113	103.6		[54]
ReS_2/rGO	Alkaline- 1 M NaOH + 0.3 M CH_3OH Acidic- 0.5 M H_2SO_4 + 0.3 M CH_3OH			30 cm^2	64% after 2500 s in alkaline media	[59]
Pt/Ti_3C_2	0.5 M H_2SO_4 and 1 M CH_3OH		0.1137	30.2	16.6% after 7200 s	[59]
Pt NW/PDDA-$Ti_3C_2T_x$	0.5 M H_2SO_4 and 1 M CH_3OH		0.175	61	90% after 2000 s	[60]
Pt/$Ti_3C_2T_x$/MWCNT	1 M H_2SO_4 and 2 M CH_3OH	0.922	0.22	175		[63]
Pt/$Ti_3C_2T_x$/rGO	0.5 M H_2SO_4 and 0.5 M CH_3OH			109	21.6 after 3600 s	[64]
Pd/MXene	1 M CH_3OH + 1 M KOH	0.124			100% after 200 s	[65]
Pd/MXene/rGO	1 M CH_3OH + 1 M KOH	0.753	0.2547	97.97		[67]
Pt-on-Pd/$Ti_3C_2T_x$		0.160			37.1% after 200 s	[76]
MoS_2/ $Ti_3C_2T_x$/MWCNTs	1 M CH_3OH + 1 M KOH			160	100% after 1800 s	[69]
CoS_2/MoS_2/rGO	0.3 M CH_3OH + 1 M KOH					[70]
Fe_3O_4/MoS_2/rGO	0.3 M CH_3OH + 0.1 M NaOH					[71]
MoS_2@CoNi-ZIF	0.5 M CH_3OH + 1 M KOH	0.430		95.7	97.5% after 45 h	[72]
MoS_2/NF-5	0.1 M NaOH + 0.5 M MeOH				80% after 5000 s	[73]

there are numerous papers on GO doping, there is less information on the doping of 2D materials such as MXene and TMDs for MOR applications. Other strategies such as heterojunction formation, defect control, single/multi atom loading, confinement effect of 2D materials also need to be studied.

(2) At the moment, the majority of research on catalysts supported by 2D materials for methanol oxidation reactions (MOR) is focused on understanding the critical elements essential for increasing electrocatalytic activity. This understanding is gained by testing catalysts based on 2D materials in perfect liquid electrolyte systems, which are frequently performed in half-cell mode. However, a critical gap exists because none of these catalysts have been tested in a realistic alcoholic fuel cell setting to assess their real-world performance. The examination of deliverable power density under operational conditions in a genuine fuel cell setting, in particular, has yet to be completed. To compete with traditional carbon-supported catalysts, catalysts supported by 2D materials must undergo rigorous testing in actual fuel cell atmospheric conditions.

(3) Long-term stability is a critical criterion for an efficient catalyst in methanol oxidation reactions (MOR). However, investigations dedicated to testing the long-term durability of 2D materials specifically designed for MOR applications are scarce. It is critical to apply these catalysts to lengthy testing durations, particularly in a fuel cell mode, to determine their performance and endurance over longer periods of time.

Acknowledgments

The authors gratefully acknowledge financial assistance from the SERB Core Research Grant (Grant No. CRG/2022/000897), Department of Science and Technology (DST/NM/NT/2019/205(G)), and Minor Research Project Grant, Jain University (JU/MRP/CNMS/29/2023).

Conflict of interest

The authors declare no conflict of interest.

References

[1] Baldauf M and Preidel W 1999 Status of the development of a direct methanol fuel cell *J. Power Sources* **84** 161–6

[2] Verma L K 2000 Studies on methanol fuel cell *J. Power Sources* **86** 464–8

[3] Kamarudin S K, Daud W R W, Ho S L and Hasran U A 2007 Overview on the challenges and developments of micro-direct methanol fuel cells (DMFC) *J. Power Sources* **163** 743–54

[4] Yuda A, Ashok A and Kumar A 2022 A comprehensive and critical review on recent progress in anode catalyst for methanol oxidation reaction *Catal. Rev.* **64** 126–228

[5] Mansor M, Timmiati S N, Lim K L, Wong W Y, Kamarudin S K and Nazirah Kamarudin N H 2019 Recent progress of anode catalysts and their support materials for methanol electrooxidation reaction *Int. J. Hydrogen Energy* **44** 14744–69

[6] Ehsani A, Heidari A A and Asgari R 2019 Electrocatalytic oxidation of ethanol on the surface of graphene based nanocomposites: an introduction and review to it in recent studies *Chem. Rec.* **19** 2341–60
[7] Tong Y, Yan X, Liang J and Dou S X 2021 Metal-based electrocatalysts for methanol electro-oxidation: progress, opportunities, and challenges *Small* **17** 1904126
[8] Hao B, Ye Z, Xu J, Li L, Huang J, Peng X, Li D, Jin Z and Ma G 2021 A high-performance oxygen evolution electrode of nanoporous Ni-based solid solution by simulating natural meteorites *Chem. Eng. J.* **410** 128340
[9] Siwal S S, Thakur S, Zhang Q B and Thakur V K 2019 Electrocatalysts for electrooxidation of direct alcohol fuel cell: chemistry and applications *Mater. Today Chem.* **14** 100182
[10] Ali A and Shen P K 2019 Recent advances in graphene-based platinum and palladium electrocatalysts for the methanol oxidation reaction *J. Mater. Chem.* A **7** 22189–217
[11] Selvarani G, Selvaganesh S V, Krishnamurthy S, Kiruthika G V M, Sridhar P, Pitchumani S and Shukla A K 2009 A methanol-tolerant carbon-supported Pt–Au alloy cathode catalyst for direct methanol fuel cells and its evaluation by DFT *J. Phys. Chem.* C **113** 7461–8
[12] Alegre C, Gálvez M E, Moliner R, Baglio V, Stassi A, Aricò A S and Lázaro M J 2013 Platinum ruthenium catalysts supported on carbon xerogel for methanol electro-oxidation: influence of the catalyst synthesis method *Chem. Cat. Chem.* **5** 3770–80
[13] Bashyam R and Zelenay P 2006 A class of non-precious metal composite catalysts for fuel cells *Nature* **443** 63–6
[14] Zhang J, Lu S, Xiang Y and Jiang S P 2020 Intrinsic effect of carbon supports on the activity and stability of precious metal based catalysts for electrocatalytic alcohol oxidation in fuel cells: a review *Chem. Sus. Chem.* **13** 2484–502
[15] Cheung K-C, Wong W-L, Ma D-L, Lai T-S and Wong K-Y 2007 Transition metal complexes as electrocatalysts—development and applications in electro-oxidation reactions *Coord. Chem. Rev.* **251** 2367–85
[16] Bagkar N C, Chen H M, Parab H and Liu R-S 2009 Nanostructured electrocatalyst synthesis: fundamental and methods *Electrocatalysis of Direct Methanol Fuel Cells* (New York: Wiley) pp 79–114
[17] Ozoemena K I 2016 Nanostructured platinum-free electrocatalysts in alkaline direct alcohol fuel cells: catalyst design, principles and applications *RSC Adv.* **6** 89523–50
[18] Huang T, Mao S, Zhou G, Zhang Z, Wen Z, Huang X, Ci S and Chen J 2015 A high-performance catalyst support for methanol oxidation with graphene and vanadium carbonitride *Nanoscale* **7** 1301–7
[19] Lo A-Y, Hung C-T, Yu N, Kuo C-T and Liu S-B 2012 Syntheses of carbon porous materials with varied pore sizes and their performances as catalyst supports during methanol oxidation reaction *Appl. Energy* **100** 66–74
[20] Holmes S M, Balakrishnan P, Kalangi V S, Zhang X, Lozada-Hidalgo M, Ajayan P M and Nair R R 2017 2D crystals significantly enhance the performance of a working fuel cell *Adv. Energy Mater.* **7** 1601216
[21] Khan K, Khan Tareen A, Aslam M, Zhang Y, Wang R, Ouyang Z, Gou Z and Zhang H 2019 Recent advances in two-dimensional materials and their nanocomposites in sustainable energy conversion applications *Nanoscale* **11** 21622–78
[22] Li Y and Guo S 2019 Noble metal-based 1D and 2D electrocatalytic nanomaterials: recent progress, challenges and perspectives *Nano Today* **28** 100774

[23] Meng T and Cao M 2018 Frontispiece: transition metal carbide complex architectures for energy-related applications *Chem. Eur. J.* **24** 16716–36

[24] Chen Q, Du C, Yang Y, Shen Q, Qin J, Hong M, Zhang X and Chen J 2023 Two-dimensional siloxene as an advanced support of platinum for superior hydrogen evolution and methanol oxidation electrocatalysis *Mater. Today Phys.* **30** 100931

[25] Fadzillah D M, Kamarudin S K, Zainoodin M A and Masdar M S 2019 Critical challenges in the system development of direct alcohol fuel cells as portable power supplies: an overview *Int. J. Hydrogen Energy* **44** 3031–54

[26] Rößner L and Armbrüster M 2019 Electrochemical energy conversion on intermetallic compounds: a review *ACS Catal.* **9** 2018–62

[27] Bahrami H and Faghri A 2013 Review and advances of direct methanol fuel cells: part II: Modeling and numerical simulation *J. Power Sources* **230** 303–20

[28] García-Díaz B L, Patterson J R and Weidner J W 2011 Quantifying individual losses in a direct methanol fuel cell *J. Fuel Cell Sci. Technol.* **9**

[29] Hogarth M P and Hards G A 1996 Direct methanol fuel cells *Platinum Met. Rev.* **40** 150–9

[30] Dinh H N, Ren X, Garzon F H, Piotr Z and Gottesfeld S 2000 Electrocatalysis in direct methanol fuel cells: *in situ* probing of PtRu anode catalyst surfaces *J. Electroanal. Chem.* **491** 222–33

[31] Gokhale A A, Dumesic J A and Mavrikakis M 2008 On the mechanism of low-temperature water gas shift reaction on copper *J. Am. Chem. Soc.* **130** 1402–14

[32] Huang H, Yang S, Vajtai R, Wang X and Ajayan P M 2014 Pt-decorated 3D architectures built from graphene and graphitic carbon nitride nanosheets as efficient methanol oxidation catalysts *Adv. Mater.* **26** 5160–5

[33] Bard A J, Faulkner L R and White H S 2022 *Electrochemical Methods: Fundamentals and Applications* (New York: Wiley)

[34] Wang Z H and Wang C Y 2003 Mathematical modeling of liquid-feed direct methanol fuel cells *J. Electrochem. Soc.* **150** A508

[35] Xu C, Zhao T S and Yang W W 2008 Modeling of water transport through the membrane electrode assembly for direct methanol fuel cells *J. Power Sources* **178** 291–308

[36] Yang W W, Zhao T S and He Y L 2008 Modelling of coupled electron and mass transport in anisotropic proton-exchange membrane fuel cell electrodes *J. Power Sources* **185** 765–75

[37] Yang W W and Zhao T S 2007 A two-dimensional, two-phase mass transport model for liquid-feed DMFCs *Electrochim. Acta* **52** 6125–40

[38] Yang W W and Zhao T S 2007 Two-phase, mass-transport model for direct methanol fuel cells with effect of non-equilibrium evaporation and condensation *J. Power Sources* **174** 136–47

[39] Kareemulla D and Jayanti S 2009 Comprehensive one-dimensional, semi-analytical, mathematical model for liquid-feed polymer electrolyte membrane direct methanol fuel cells *J. Power Sources* **188** 367–78

[40] Meyers J P and Newman J 2002 Simulation of the direct methanol fuel cell: II. Modeling and data analysis of transport and kinetic phenomena *J. Electrochem. Soc.* **149** A718

[41] Yoo E, Okata T, Akita T, Kohyama M, Nakamura J and Honma I 2009 Enhanced electrocatalytic activity of pt subnanoclusters on graphene nanosheet surface *Nano Lett.* **9** 2255–9

[42] Seger B and Kamat P V 2009 Electrocatalytically active graphene-platinum nanocomposites. role of 2-D carbon support in PEM fuel cells *J. Phys. Chem.* C **113** 7990–5

[43] Pei S and Cheng H-M 2012 The reduction of graphene oxide *Carbon* **50** 3210–28
[44] Luo Z, Yuwen L, Bao B, Tian J, Zhu X, Weng L and Wang L 2012 One-pot, low-temperature synthesis of branched platinum nanowires/reduced graphene oxide (BPtNW/RGO) hybrids for fuel cells *J. Mater. Chem.* **22** 7791–6
[45] Sharma S, Ganguly A, Papakonstantinou P, Miao X, Li M, Hutchison J L, Delichatsios M and Ukleja S 2010 Rapid microwave synthesis of CO tolerant reduced graphene oxide-supported platinum electrocatalysts for oxidation of methanol *J. Phys. Chem.* C **114** 19459–66
[46] Zhang Z, Lee C-S and Zhang W 2017 Vertically aligned graphene nanosheet arrays: synthesis, properties and applications in electrochemical energy conversion and storage *Adv. Energy Mater.* **7** 1700678
[47] Soin N, Roy S S, Lim T H and McLaughlin J A D 2011 Microstructural and electrochemical properties of vertically aligned few layered graphene (FLG) nanoflakes and their application in methanol oxidation *Mater. Chem. Phys.* **129** 1051–7
[48] Li M, Jiang Q, Yan M, Wei Y, Zong J, Zhang J, Wu Y and Huang H 2018 Three-dimensional boron- and nitrogen-codoped graphene aerogel-supported pt nanoparticles as highly active electrocatalysts for methanol oxidation reaction *ACS Sustain. Chem. Eng.* **6** 6644–53
[49] Yan M, Jiang Q, Zhang T, Wang J, Yang L, Lu Z, He H, Fu Y, Wang X and Huang H 2018 Three-dimensional low-defect carbon nanotube/nitrogen-doped graphene hybrid aerogel-supported Pt nanoparticles as efficient electrocatalysts toward the methanol oxidation reaction *J. Mater. Chem.* A **6** 18165–72
[50] Zhai C, Zhu M, Bin D, Ren F, Wang C, Yang P and Du Y 2015 Two dimensional MoS_2/graphene composites as promising supports for Pt electrocatalysts towards methanol oxidation *J. Power Sources* **275** 483–8
[51] Liu Q, Lin Y, Fan J, Lv D, Min Y, Wu T and Xu Q 2016 Well-dispersed palladium nanoparticles on three-dimensional hollow N-doped graphene frameworks for enhancement of methanol electro-oxidation *Electrochem. Commun.* **73** 75–9
[52] Yang H, Geng L, Zhang Y, Chang G, Zhang Z, Liu X, Lei M and He Y 2019 Graphene-templated synthesis of palladium nanoplates as novel electrocatalyst for direct methanol fuel cell *Appl. Surf. Sci.* **466** 385–92
[53] Sawangphruk M, Krittayavathananon A, Chinwipas N, Srimuk P, Vatanatham T, Limtrakul S and Foord J S 2013 Ultraporous palladium supported on graphene-coated carbon fiber paper as a highly active catalyst electrode for the oxidation of methanol *Fuel Cells* **13** 881–8
[54] Zhang X, Zhu J, Tiwary C S, Ma Z, Huang H, Zhang J, Lu Z, Huang W and Wu Y 2016 Palladium nanoparticles supported on nitrogen and sulfur dual-doped graphene as highly active electrocatalysts for formic acid and methanol oxidation *ACS Appl. Mater. Interfaces* **8** 10858–65
[55] Qin J, Huang H, Xie Y, Pan S, Chen Y, Yang L, Jiang Q and He H 2022 MXene supported rhodium nanocrystals for efficient electrocatalysts towards methanol oxidation *Ceram. Int.* **48** 15327–33
[56] Kang Y, Xue Q, Jin P, Jiang J, Zeng J and Chen Y 2017 Rhodium nanosheets–reduced graphene oxide hybrids: a highly active platinum-alternative electrocatalyst for the methanol oxidation reaction in alkaline media *ACS Sustain. Chem. Eng.* **5** 10156–62

[57] Askari M B and Salarizadeh P 2019 Ultra-small ReS2 nanoparticles hybridized with rGO as cathode and anode catalysts towards hydrogen evolution reaction and methanol electro-oxidation for DMFC in acidic and alkaline media *Synth. Met.* **256** 116131

[58] Peera S G, Liu C, Shim J, Sahu A K, Lee T G, Selvaraj M and Koutavarapu R 2021 MXene ($Ti_3C_2T_x$) supported electrocatalysts for methanol and ethanol electrooxidation: a review *Ceram. Int.* **47** 28106–21

[59] Wang Y, Wang J, Han G, Du C, Deng Q, Gao Y, Yin G and Song Y 2019 Pt decorated Ti_3C_2 MXene for enhanced methanol oxidation reaction *Ceram. Int.* **45** 2411–7

[60] Yang C, Jiang Q, Huang H, He H, Yang L and Li W 2020 Polyelectrolyte-induced stereoassembly of grain boundary-enriched platinum nanoworms on $Ti_3C_2T_x$ MXene nanosheets for efficient methanol oxidation *ACS Appl. Mater. Interfaces* **12** 23822–30

[61] Zhu J *et al* 2022 Ultrahigh stable methanol oxidation enabled by a high hydroxyl concentration on Pt clusters/MXene interfaces *J. Am. Chem. Soc.* **144** 15529–38

[62] Navjyoti, Sharma V, Bhullar V, Saxena V, Debnath A K and Mahajan A 2023 Modulation of surface Ti–O species in 2D-$Ti_3C_2T_x$ MXene for developing a highly efficient electrocatalyst for hydrogen evolution and methanol oxidation reactions *Langmuir* **39** 2995–3005

[63] Zhang X, Zhang J, Cao H and Li Y 2019 Preparation of Pt/$(Ti_3C_2T_x)_y$-$(MWCNTs)_{1-y}$ electrocatalysts via a facile and scalable solvothermal strategy for high-efficiency methanol oxidation *Appl. Catal.*, A **585** 117181

[64] Yang C, Jiang Q, Li W, He H, Yang L, Lu Z and Huang H 2019 Ultrafine Pt nanoparticle-decorated 3D hybrid architectures built from reduced graphene oxide and MXene nanosheets for methanol oxidation *Chem. Mater.* **31** 9277–87

[65] Lang Z, Zhuang Z, Li S, Xia L, Zhao Y, Zhao Y, Han C and Zhou L 2020 MXene surface terminations enable strong metal–support interactions for efficient methanol oxidation on palladium *ACS Appl. Mater. Interfaces* **12** 2400–6

[66] Huang H, Xiao D, Zhu Z, Zhang C, Yang L, He H, You J, Jiang Q, Xu X and Yamauchi Y 2023 A 2D/2D heterojunction of ultrathin Pd nanosheet/MXene towards highly efficient methanol oxidation reaction: the significance of 2D material nanoarchitectonics *Chem. Sci.* **14** 9854–62

[67] Zhang P, Fan C, Wang R, Xu C, Cheng J, Wang L, Lu Y and Luo P 2019 Pd/MXene ($Ti_3C_2T_x$)/reduced graphene oxide hybrid catalyst for methanol electrooxidation *Nanotechnology* **31** 09LT01

[68] Elancheziyan M, Eswaran M, Shuck C E, Senthilkumar S, Elumalai S, Dhanusuraman R and Ponnusamy V K 2021 Facile synthesis of polyaniline/titanium carbide (MXene) nanosheets/palladium nanocomposite for efficient electrocatalytic oxidation of methanol for fuel cell application *Fuel* **303** 121329

[69] Yang X, Jia Q, Duan F, Hu B, Wang M, He L, Song Y and Zhang Z 2019 Multiwall carbon nanotubes loaded with MoS_2 quantum dots and MXene quantum dots: non-Pt bifunctional catalyst for the methanol oxidation and oxygen reduction reactions in alkaline solution *Appl. Surf. Sci.* **464** 78–87

[70] Askari M B, Salarizadeh P, Seifi M and Rozati S M 2019 Electrocatalytic properties of CoS_2/MoS_2/rGO as a non-noble dual metal electrocatalyst: the investigation of hydrogen evolution and methanol oxidation *J. Phys. Chem. Solids* **135** 109103

[71] Askari M B, Beheshti-Marnani A, Seifi M, Rozati S M and Salarizadeh P 2019 Fe_3O_4@MoS_2/RGO as an effective nano-electrocatalyst toward electrochemical hydrogen

evolution reaction and methanol oxidation in two settings for fuel cell application *J. Colloid Interface Sci.* **537** 186–96

[72] Liu Y, Hu B, Wu S, Wang M, Zhang Z, Cui B, He L and Du M 2019 Hierarchical nanocomposite electrocatalyst of bimetallic zeolitic imidazolate framework and MoS_2 sheets for non-Pt methanol oxidation and water splitting *Appl. Catal.* B **258** 117970

[73] Gopalakrishnan A, Durai L, Ma J, Kong C Y and Badhulika S 2021 Vertically aligned few-layer crumpled MoS_2 hybrid nanostructure on porous Ni foam toward promising binder-free methanol electro-oxidation application *Energy Fuels* **35** 10169–80

[74] Huang L, Zhang X, Wang Q, Han Y, Fang Y and Dong S 2018 Shape-control of Pt–Ru nanocrystals: tuning surface structure for enhanced electrocatalytic methanol oxidation *J. Am. Chem. Soc.* **140** 1142–7

[75] Yuan X, Jiang X, Cao M, Chen L, Nie K, Zhang Y, Xu Y, Sun X, Li Y and Zhang Q 2019 Intermetallic PtBi core/ultrathin Pt shell nanoplates for efficient and stable methanol and ethanol electro-oxidization *Nano Res.* **12** 429–36

[76] Yang C, Jiang Q, Liu H, Yang L, He H, Huang H and Li W 2021 Pt-on-Pd bimetallic nanodendrites stereoassembled on MXene nanosheets for use as high-efficiency electrocatalysts toward the methanol oxidation reaction *J. Mater. Chem.* A **9** 15432–40

Chapter 9

Summary

Pratik V Shinde and Chandra Sekhar Rout

9.1 Summary and future scope

In many electrocatalysis reactions, high-performance electrocatalysts are key when attempting to unlock the barriers that cause the reactions to be sluggish. Consequently, the primary objective in the disciplines of catalytic research and technology is always the development of catalysts with high activity and stability. In light of this motivation, noble metals and their derivatives are always employed to power a majority of catalytic processes because they are capable of producing impressive outcomes. As noble metal-based materials are thought to be the greatest catalysts for this process, the catalytic process itself also poses a difficulty in terms of sustainability. Noble metal materials are highly expensive and they are not found in sufficient sources in Earth's crust, so using them on larger scales is not practical. Therefore, the hunt for alternative non-noble-metal catalysts has become a major topic in the catalytic world. The catalysts for various processes should be effective, abundant on Earth, affordable, simple to use, and cause less pollution overall. As a result, in their pursuit of superior electrocatalysts, researchers have decided to concentrate more on two-dimensional (2D) materials.

The discovery of graphene is widely regarded as an essential turning point in the emergence of 2D nanomaterials. As a material reduces to the nanoscale, its electrical, chemical, mechanical, and optical properties dramatically change. The possibilities for 2D materials have been expanded by the discovery of a variety of 2D materials with diverse compositions, unique properties, and beneficial functionalities. Until now, a portfolio of 2D materials includes graphene, graphitic carbon nitride (g-C_3N_4), hexagonal boron nitride (h-BN), black phosphorus (BP), transition metal dichalcogenides (TMDs), MXenes, perovskites, metal–organic frameworks (MOFs), layered double hydroxides (LDHs), metal oxides, Xenes, and new ones are constantly being added to this. 2D materials offer a unique platform of properties such as large surface area, flexibility, excellent electric conductivity, tunable bandgap, high charge carrier mobility, high thermal stability, and availability of

doi:10.1088/978-0-7503-5719-7ch9 © IOP Publishing Ltd 2024

catalytically active sites/edges. The superior properties of 2D materials open up additional opportunities for utilization in a variety of fields, such as electrical, optical, spintronic, photonics, catalysis, and sensors.

From the perspective of electrocatalysis, non-precious, earth-abundant catalysts with a 2D layered structure have drawn a lot of attention from researchers due to their ability to streamline the catalytic process and reduce expenses. The performance of electrochemical catalysis reactions can be enhanced by using 2D ultrathin structures as a catalyst. The advantage of planar structures is that they can increase a material's specific surface area; as a result, surface reaction rates are higher on these structures than on bulk solid structures. Moreover, the 2D geometry exposes more surface area, easing interaction between the catalyst surface and the adsorbate. It's not always possible to get the desired electrocatalytic performance from pristine 2D materials due to their lack of behaviours. It is particularly desirable in modern electrocatalyst design to be able to not only use 2D materials but also tailor their properties for excellent performance. Therefore, modification or fine-tuning of the properties of 2D materials by using engineering techniques such as doping, alloying, defects, strain, morphology, edge engineering, and heterostructure formation is also important.

In this book, we provided an in-depth examination of recent significant advancements in engineered 2D materials for electrocatalysis applications, along with key tactics to improve their electrocatalytic activity (figure 9.1). Here, we've covered the chapters that address current advances in electrochemical catalytic processes like hydrogen evolution reaction (HER), oxygen evolution reaction (OER), oxygen reduction reaction (ORR), CO_2 reduction reaction (CO_2 RR), N_2 reduction reaction (N_2 RR), and methanol oxidation reaction (MOR). The underlying mechanisms of each reaction are also covered in the chapters, as well as the advantageous aspects of using engineered 2D materials in these reactions. Based on the status of current research, some engineering strategies to improve the functionality and catalytic activity of 2D materials are proposed. The relationship between engineered 2D materials and their impact on catalytic processes was briefly covered in each chapter. The performance of tailored 2D materials for catalytic reactions, as well as the unresolved issue and future directions, are discussed at the conclusion of each chapter. Excellent performance in activity, stability, and selectivity for catalyzing a particular reaction can be achieved with the appropriate engineering of 2D catalysts.

Increased intensive research will assist in addressing the critical challenges and thus promote the development of advanced 2D materials with high electrocatalytic activity and stability (figure 9.2). A revolution in the field of electrocatalysis may be brought about by advancements in currently used synthesis techniques and the application of new technologies to create 2D materials with tailored properties. There is a vast potential for engineering 2D materials using multiple approaches such as introducing defects, doping, alloying, generation of active edges, morphology designing, strain tunable effects, and heterostructure formation. However, there is a need for more in-depth research on the precise control of defects, doping, alloying, and heterostructure formation effects in an optimistic way. In the catalytic process, active sites play a key role in the improvement of performance.

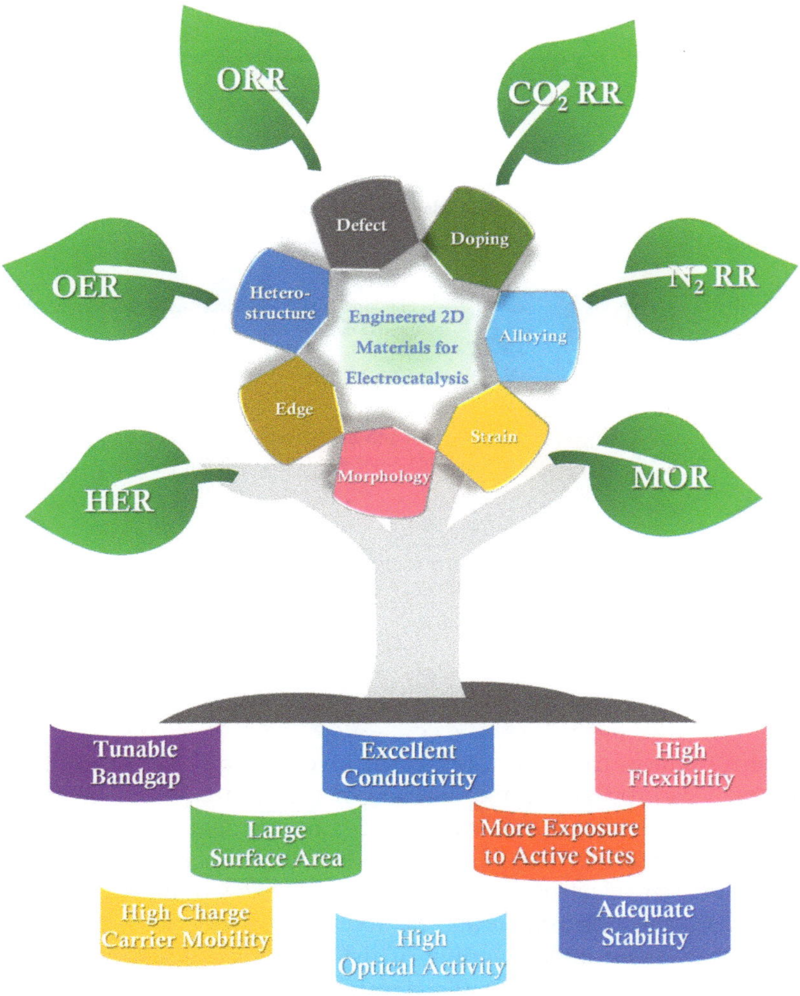

Figure 9.1. A schematic to illustrate the fruitful properties of 2D materials for electrocatalysis application and different engineered approaches utilized to tune them.

In this regard, rational catalyst engineering and preparation can help maximize the number of active sites, resulting in improved catalytic activity. 2D materials have a larger electroactive surface, which increases the contact area between the catalyst and the electrolyte. Additionally, this minimizes the length of the active species' diffusion pathways and boosts the activity of the catalyst. Engineered 2D materials are highly fascinating because of the benefits brought about by the synergy of the constituent parts and distinctive properties at the interface. It is still challenging to precisely regulate and modulate the structural and electrical properties of engineered 2D materials. For better comprehension of the reaction process and enhanced

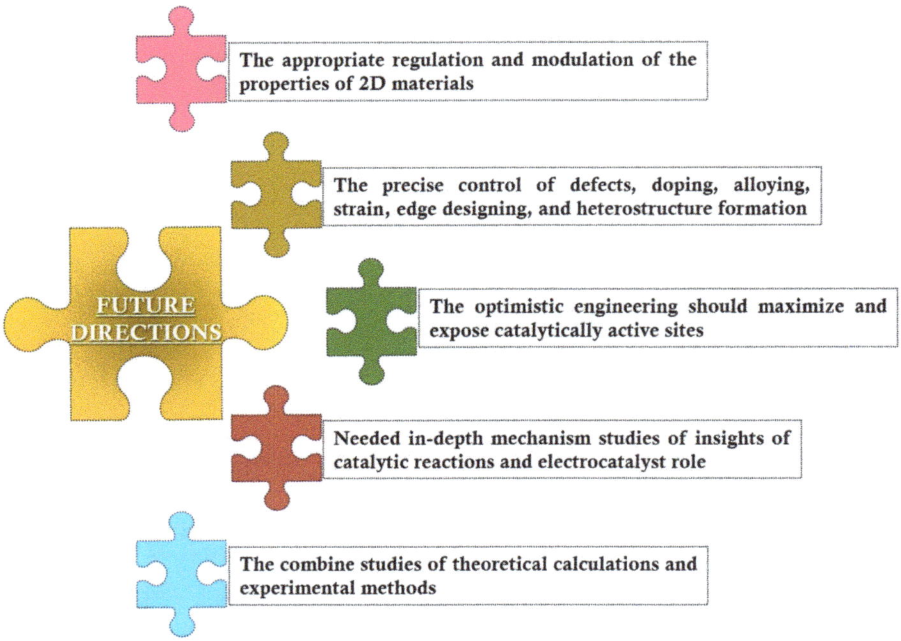

Figure 9.2. A schematic to represent the future direction beneficial for enhancement of performance of 2D materials in electrocatalysis applications

catalytic performance, catalytic mechanism investigations are crucial. To understand the catalysis process and the function of the electrocatalyst in reaction, it is necessary to establish or use existing advanced *in situ* characterization techniques. The bridge between theoretical calculations and experimental results will help to close the gap between electrocatalyst design and its performance.

In conclusion, the potential usage of accessible resources for engineering 2D materials, as well as an understanding of the mechanisms underlying the various catalysis reactions, could lead to the development of high-performance electrocatalysts. The next trend is expected to involve the engineering of 2D materials with controllable designed structures for fruitful electrocatalysis outcomes.

www.ingramcontent.com/pod-product-compliance
Ingram Content Group UK Ltd.
Pitfield, Milton Keynes, MK11 3LW, UK
UKHW051854100926
8556IPUK00005B/628

9 780750 357